Malaysia's Maritime Jurisdictional Limits

Vivian Louis Forbes

Malaysia's Maritime Jurisdictional Limits

An Appraisal

 Springer

Vivian Louis Forbes
University of Western Australia
Crawley, WA, Australia

ISBN 978-3-031-78782-9   ISBN 978-3-031-78783-6   (eBook)
https://doi.org/10.1007/978-3-031-78783-6

This Springer imprint is published by the registered company Springer Nature Switzerland AG
The registered company address is: Gewerbestrasse 11, 6330 Cham, Switzerland

If disposing of this product, please recycle the paper.

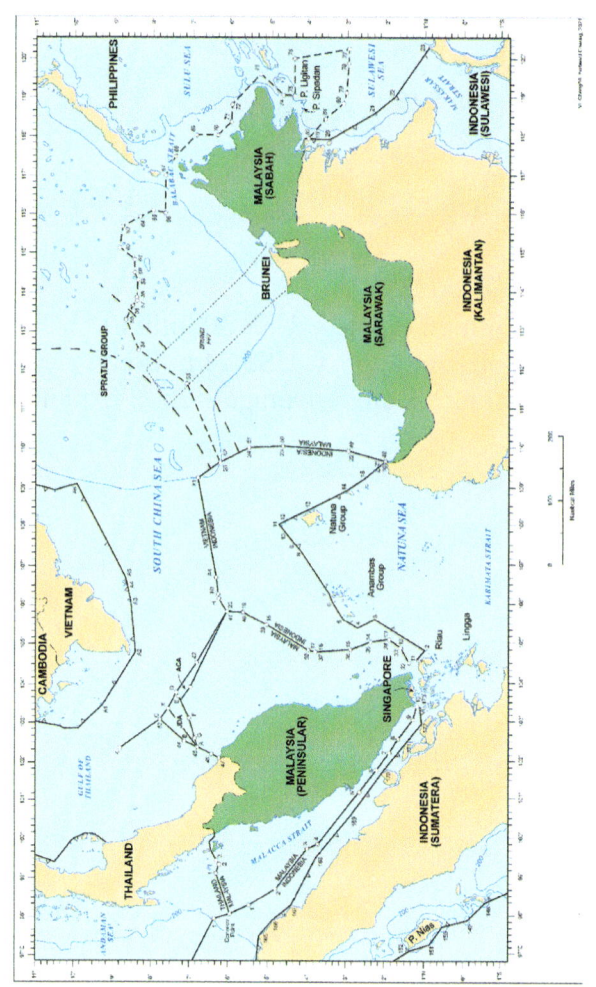

Map A

Fact File: Malaysia

Land surface area: 328,550 km^2

Marine surface area: 614,170 km^2

Inland waters area: 97,310 km^2

Territorial Sea area: 63,670 km^2

Exclusive Economic Zone: 453,190 km^2

Natural Continental Shelf: 476.762 km^2

Population UN Est July'23: 34,308,525

Population growth (in 2022): 1.1%

GDP (2023 US$): 406.31 billion

GDP per capita (2023 US$): 11,972

GDP growth (annual): 8.7%

FDI net inflows (% of GDP): 3.7 (2022)

Exports (in 2022): RM 1290.2 billion (approx. US$ 30 billion)

Imports (in 2022): RM 1084.6 billion (approx. US$ 24 billion)

Ships transiting Malacca Strait: Daily average—250; Annual average—91,000

Map B. *Source* This *Factbook* map is in the public domain. Accordingly, it may be copied freely without written approval from the Central Intelligence Agency (CIA). *The World Factbook 2021.* Washington, DC: Central Intelligence Agency, 2021

Undersea Features of the Southern Sector of the South China Sea in the Vicinity of Malaysia's Maritime Jurisdiction

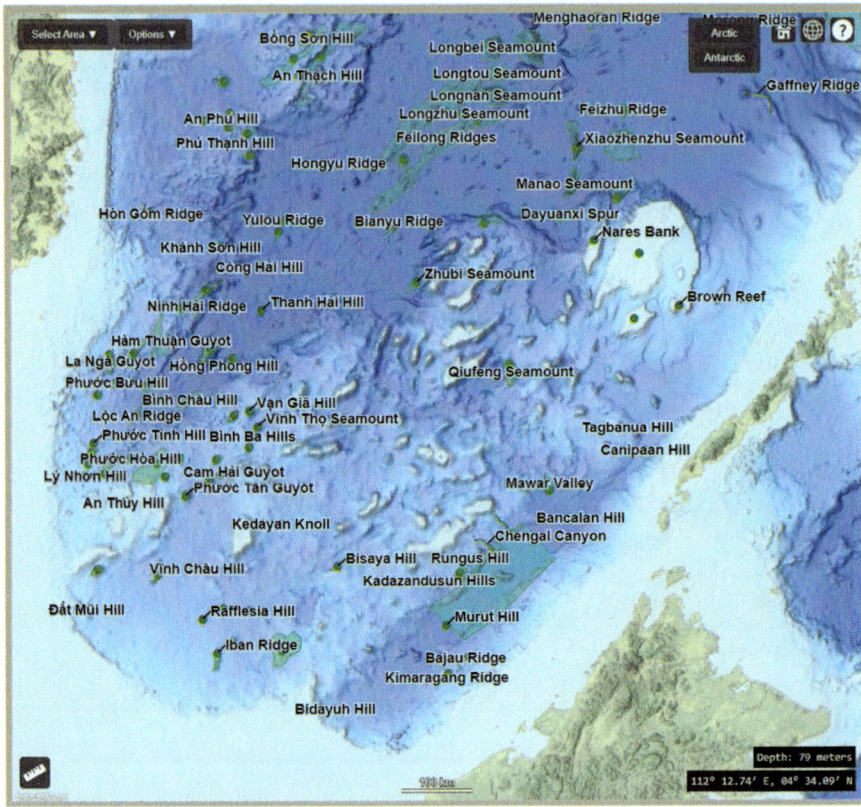

Map C. *Source* NOAA National Centers for Environmental Information. Dataset accessed on 7 May 2024 from Undersea Features Gazetteer (noaa.gov). This image was of a select area screen-capture for the purpose of this presentation. Permission given by NOAA on Wednesday 8 May 2024, 8:56 pm WAST. (https://www.ngds.noaa.gov//gazetter)

This volume is dedicated to 'The People of Malaysia'.

My friends in Malaysia have always extended a warm welcome and were unstinting in their generosity and offers of assistance.

Foreword by Emeritus Professor Dennis Rumley

I have been very fortunate to be closely associated with the academic life of Viv Forbes for more than 40 years—first, as a colleague in the Department of Geography at the University of Western Australia [UWA]; second, as a co-researcher on Indian Ocean issues; third as a principal supervisor for his Ph.D. thesis; and, then as a co-founder of the Indian Ocean Research Group (IORG) in 2002. The latter led to numerous Indian Ocean regional collaborative Conferences and research programmes as well as the creation of an International Journal—*Journal of the Indian Ocean Region*—published by Routledge. Viv Forbes has always been an extremely active cooperative participant in both research and writing in IORG Conferences and books as well as in the Journal, and on the IORG Board.

Viv Forbes' longstanding and ongoing detailed research on regional maritime issues which has resulted in his prolific bibliography and international research profile clearly indicates that he is undoubtedly the Australian doyen of Indian Ocean and Asia Pacific regional maritime boundaries and associated policy appraisal. Viv's remarkable ability to create the ultimate synthesis on key maritime questions is unique. His singular academic contribution and status have been internationally recognized for some years, especially with his very longstanding and ongoing relationship with the Maritime Institute of Malaysia (MIMA) and more recently with Wuhan University and the National Institute for South China Sea Studies in Haikou. The recent UWA award of a D.Litt. is the ultimate recognition of a very fine scholar and researcher.

Dr. Forbes latest book—*Malaysia's Maritime Jurisdictional Limits: An Appraisal*—brings together some of the enormous breadth and depth of his research and policy expertise to focus on a key global issue—the development of the blue economy—and its implications for maritime jurisdiction in Malaysia. In a typically comprehensive treatment, Dr. Forbes discusses the importance of the blue economy concept which was adopted by Malaysia in May 2023. An understanding of the significance and the complexity of delimiting maritime boundaries and jurisdictional zones, the nature of the ocean basin environment and marine safety is shown to be essential to the development and designation of a sustainable blue economy. This is particularly the case about protecting the marine environment and the sustainable use of biotic and mineral resources.

This broad-ranging analysis is very amply illustrated with a plethora of illustrated maps and tables and enhanced with an extremely useful Annexures of primary documents. In the context of Malaysia's 'zone-locked' status and its various sovereignty issues, these are all potentially significant legal instruments in enabling the maximisation of Malaysia's blue economy development.

I strongly commend this book to all who have an interest in Malaysia, in maritime jurisdiction and in blue economy policy formulation.

September 2023   Dennis Rumley
                                                           Emeritus Professor
                                                            Curtin University
                                                             Perth, Australia

Prologue by Emeritus Professor Carlyle A. Thayer

Vivian Forbes' *Malaysia's Maritime Jurisdictional Limits: An Appraisal* is an extraordinary academic study of Malaysia as a maritime state and its quest to develop a Blue Economy. This study is both multi-disciplinary and multi-dimensional because of Forbes' skill in cartography and photographic illustration.

Forbes expertly draws on the literature from the disciplines of history, geography, demography, Earth science, economics, politics, and law. He then takes the reader to a new cognitive level of understanding by combining his textual analysis with 161 specialist maps and selected photographs and 31 tables of vital data.

Malaysia is ordinarily portrayed on maps as two distinct geographical parts, Peninsula Malaysia and East Malaysia along the coast of Borneo. East and West Malaysia are then sub-divided into states. Forbes expands our view by including Malaysia's territorial sea and Exclusive Economic Zone, two maritime areas determined by international law, particularly the United Nations Convention on the Law of the Sea.

Malaysia's Maritime Jurisdictional Limits: An Appraisal is a case study of how one country is attempting to manage the myriad impact of economic development and declining ocean health by pursuing a Blue Economy strategy. Forbes writes that "The concept of a Blue Economy is of a sustainable ocean economy that emerges when the economic activity is in balance with the long-term capacity of ocean ecosystems to remain resilient and healthy."

Forbes identifies two crucial management issues in developing Malaysia's Blue Economy. First, policymakers must understand the range and inter-relationship of coastal and oceanic sustainability issues such as fisheries, pollution, and ecosystem health. Second, policy-managers need to collaborate with a range of stakeholders across the public and private enterprise sectors on a higher scale that previously.

To assist policymakers and stakeholders, Forbes argues that "Maps are of fundamental value to the basis for theories, measurements, and analysis and are useful for those who in engage in planning (in the marine and terrestrial contexts), and applications of marine cadastral, natural hazard assessment, resource development and environmental protection."

Malaysia's Maritime Jurisdictional Limits: An Appraisal employs maps extensively to assist in the analysis of planning and management in seven main areas:

resources and the ecosystem of the Blue Economy; legal and geographical concepts; maritime jurisdiction; ocean basin environment; traffic monitoring and maritime safety; marine parks, protected zones, and marine biotic and mineral resources; and major commercial ports, sea lanes, and piracy.

Vivian Forbes' *Malaysia's Maritime Jurisdictional Limits: An Appraisal* is a pioneering work that solidifies his reputation as the Renaissance Man of Southeast Asian maritime studies. This book will appeal to politicians, government policymakers, academics, and students, civil society advocacy groups, and leading stakeholders in coastal communities concerned about sustainable development on a national scale. It will also serve as a model for similar stakeholders throughout Southeast Asia and beyond.

April 2024

Carlyle A. Thayer
Emeritus Professor
University of New South Wales
Australian Defence Force Academy
Canberra, Australia

Acknowledgements

The compilation and publication of this study would not have been possible without the assistance of Mr. Jimmy Cheung, a spatial scientist, who assisted in the updating of many maps and compiling a few of the new ones. I acknowledge the voluntary works of Ms. Vui Lin Choong and Mr. Felix Lim, both of Perth, Western Australia, who assisted with the compilation of the original maps and thank them for their unstinting assistance with my research over many years. Special thanks to Aimmee and Leeanna, my Granddaughters, who assisted with some of the sketches in this volume.

In Australia, my family, friends, and colleagues were always supportive of my research activities and dedication to writing on topics that were academic in context and of personal interest. I would like to place on record my sincere thanks to Adjunct Professor Patrick H. Armstrong, a true colleague, trusted friend, and excellent mentor over many decades. My thanks also to Emeritus Professor Dennis Rumley, and Emeritus Professor Carlyle A. Thayer for their continued encouragement, advice, and support for my publication record over many decades. I had the pleasure of co-authoring academic articles and books with Dennis and Patrick and Emeritus Professor Donald L. Sparks from which experiences I gained vast knowledge of academic research and techniques. Sincere appreciation to Prof. Stefaan Missinne for his valued comments. Thank you.

In Malaysia, friendly advice, both formal and informal, assistance in the collection of data, and general guidance on matters relating to this study were offered willingly by many friends, associates, and officials are acknowledged with sincere thanks. There are too many to name individually on this page, but I am confident they will understand my predicament.

For this project, I thank the institutions and organisations in China, France, Malaysia, Taiwan, and the UK that welcomed me onto their premises to conduct research within their archives and use their library resources. Minor problems elsewhere were overshadowed by the generosity and support offered by these institutions and friends. I have been able to make use of an abundance of resources to compile an appraisal of Malaysia's Maritime Jurisdictional Limits. I acknowledge the kindness of Google, CIA, NOAA and *VesselFinder* for granting permission to include their graphics as illustrations in this book.

A special word of thanks is extended to my son Andrew, my daughter Jennifer, and her family, all in Perth, who allowed me the space and unhindered time to pursue my research interests. To Datin Sharina Shaukat and Datuk Raja Malik of Kuala Lumpur, thank you for your unstinting generosity and splendid hospitality and for extending warm welcomes whenever I visited Malaysia. A special word of thanks to Commander (Retd.) Ramli Bin Johari of Kuala Linggi International Port for an interesting meeting and conversation on maritime matters in the region, to Christine Lau and Vee Chek of Singapore, and to Sue Suri Hati and Mohd Muzakir of Kuala Lumpur for their collective friendship and permission to use their graphics.

To Asha Jones, Ivan Krus, and Yi Sun (Shirley), all from UWA, my thanks for their assistance, no matter how demanding the requests were on their valued time, whenever they were called upon. To Albert Yuen and the late, Vernon (Vic) Porritt, thank you for stimulating discussions on international matters during our weekly mid-morning informal meetings over many years.

If there are shortcomings and errors in this study, a humble apology is offered. I have endeavoured to produce an up-to-date appraisal of the topics within a self-imposed deadline. The themes of this study are dynamic and are in a flux of change; furthermore, they are much influenced by the geopolitics of the region and elsewhere. Indeed, within each country, internal conflicts exist whenever political ideology and government involvement in decision-making and policy planning are discussed.

Perth, Australia   Vivian Louis Forbes
September 2024

Statement Relating to the Illustrations in This Volume

Regarding Source Acknowledgement of Images

Screenshots of *GOOGLE* (Google Earth Map) images are reproduced here with gratitude. The extract of images (screenshots) is to be found in Figs. 1.1, 1.7, 7.1, 7.7, 7.17 and 7.22.

The following note infers approval of use of images with stated provisions:

"You may use unaltered screenshots of our products for instructional or illustrative purposes in textbooks, instructional books and videos, online articles, blog posts, and other such materials. They may be lightly annotated, but don't distort or modify them in any way, such as changing the appearance of the user interface".

https://about.google/brand-resource-center/guidance/media/ <accessed 18th July 2023>

Sources for the maps at the Frontispiece are Map A the author, Map B from CIA, the USA, and Map C is reproduced courtesy of NOAA National Centers for Environmental Information. To all the agencies and authorities, I offer sincere thanks for use of the extract that are included here purely for educational and illustrative purposes.

Map A depicts, amongst mapped features, the delimited maritime boundaries of Malaysia; Map B portrays Malaysia with its near neighbours and adjacent seas; and Map C delineates the names of undersea features of the southern sector of the South China Sea basin. Readers may be interested to learn that during the early-1980s there were a mere six seamounts within the basin that were named. By 2023, the regional and international cartographic, hydrographic, and bathymetric organisations active in proposing names of 'newly discovered' seamounts within the basin.

Illustrations of shipping density adjacent to Malaysia are from *VesselFinder* (a FREE AIS tracking web site) that were captured at the instant of viewing during the height of the COVID-19 Pandemic, for discussions on the subject matter, are contained in Figs. 5.11–5.18 and Figs. 8.1–8.3, which were captured at the concluding stages of compilation.

A message from VESSELFINDER received at 6.45 pm (WAST) 4th April 2024 granted permission stating: "You are free to use the photos published on our website and any other information…" Source: VesselFinder.

My immediate response was: "Thank you. Truly appreciated".

Most of the maps and many photographs are the property of the present author. The maps and sketches were compiled by the author and his friends who are spatial scientists/cartographers. For all the maps in this volume, cardinal North is at the top of the map, and hence, the arrowhead with an accompanying 'N' is generally NOT depicted. The map scale is not shown on some sketch maps as the graphics are for illustrative purposes.

Photographs used here were sourced from my friends each of whom has granted me verbal permission to include their works in this publication. Due acknowledgement is given above.

Introduction

Marine awareness and ocean governance are concepts that must be given priority if administrators, agencies, and politicians, of a coastal or island state, are to meet the nation's obligations and utilize its sovereign rights under international customary and conventional law. Whilst sovereign rights encompass access to explore and exploit mineral resources and harvest marine biotic resources, such rights also obligate the state to protect the marine environment and to develop the resources sustainably and responsibly within its maritime jurisdictional limits and in the marine area beyond national jurisdiction. However, it is not only the management of marine resources that is of concern but also the orderly administrative processes of maritime space and functionality of marine cadastre within the nation's maritime jurisdictional limits.

Marine boundary delimitation is a subset of international law. One of the many problems in the delimitation of international maritime boundaries is the complex issue of the ownership of the marine biotic and mineral resources of the world's ocean. Determination and delineation of maritime territorial claims and zones of national jurisdiction must be acceptable to the negotiating states and to the international community of nations. Maritime boundary agreements are politically sensitive issues and, ideally, should be accorded the status of international recognition.

This study portrays Malaysia's actual and potential maritime boundaries; port infrastructure; marine resources; maritime themes cartographically and in pictorial format; and marine meteorology and environmental concerns. Perhaps, its greatest significance is the inclusion of a complete suite of primary documents that relates to the maritime boundary agreements that Malaysia negotiated with its maritime neighbours. In addition, it includes details of the Vessel Traffic Information System for the Straits of Malacca and Singapore; the specified corridors that Malaysian-flagged vessels may utilise when plying the Indonesian archipelagic waters in the vicinity of the Anambas and Natuna Archipelagoes.

A suite of eight narrative chapters liberally illustrated with maps, charts and diagrams, tabulations together with a glossary of Malay terms and extensive Annexures provides relevant information to understand the strengths and weaknesses of Malaysia's blue economy advancement. Most of the maps, drawn by the author, illustrate the maritime boundary agreements undertaken by Malaysia, as a nation, since 1969.

This study provides an opportunity to visualize the fragility of Malaysia's adjacent seas and the political and social issues that occur within the legal framework in the context of economic, geophysical, geopolitical, and social issues. It is supplemented with valuable primary documents relating to Malaysia's terrestrial and maritime international political boundaries.

This study adopts an integrated, multidisciplinary approach to the marine political geography of Malaysia. It analyses select provisions of the 1982 UN Law of the Sea Convention and evaluates the national legislation of Malaysia and the maritime boundary agreements between the State and its maritime neighbours.

Vivian Louis Forbes M.Phil., Ph.D., D.Litt.

Professor Vivian Forbes is one of the world's most distinguished authorities on maritime boundaries.

He has a master's degree, a doctorate, and a higher doctorate (D.Litt.) from Australian universities, and his advice is frequently sought by governments and institutions in Asia and Australia. He has lectured on the subject in many counties, displaying both his great learning and his diplomatic sensitivity. While based in Perth Western Australia, he has held important Distinguished Visiting positions in Universities in China, Malaysia and elsewhere.

His erudition is based on a rare combination of practical experience at sea as a merchant navy officer, a very detailed knowledge of the legal instruments of many countries appertaining to the issue and the relevant international agreements, together with an outstanding competence in the fields of navigation, survey, and cartography. These skills are supplemented by competence in administration. He managed one of the largest map libraries in the southern hemisphere with great efficiency for 36 years.

Professor Forbes has a highly distinguished research record: he has published a number of books and several hundred papers in well-known and important journals. He is particularly well-known for his atlases of maritime affairs.

Professor Forbes particular expertise is in the maritime regimes of the Asia-Pacific arena, but he has also published extensively on the boundary problems of the Indian Ocean basin, the Polar regions, and remote islands in many parts of the world.

Patrick Armstrong
Adjunct Professor
University of Western Australia
March 2024

   This study contains expressive maps, and other graphics provide relevant information to understand the strengths and weaknesses of Malaysia's economy. It highlights
the importance of the maritime transportation and the facilities offered by the ports
of Malaysia.

   It provides valuable documentation on Malaysia's maritime jurisdictional limits
in the context of the concept of the Blue Economy and sustainability of marine
biodiversity.

   This book provides an avenue to visualise the fragility of seas and oceans and the
political and economic and social tensions that occur in Malaysia's terrestrial and
maritime frameworks.

   It also reveals the delicate balance of its coasts and islands and the intensity of
economic and population flows in its seas. Importantly, it alludes to the challenges
that Malaysia faces in ensuring that its maritime space is secure and sovereignty
over its adjacent seas is acknowledged by the international community and regional
neighbours.

Contents

1 Mapping to Enhance the Nation's Blue Economy 1
  1.1 Introduction 1
  1.2 Structure of This Study 3
  1.3 Malaysia: A Maritime State 5
  1.4 Malaysia Adopts the Blue Economy Concept 7
      1.4.1 Blue Economy Concept 8
  1.5 Regional Partnership 9
      1.5.1 The 2018 Industry 4.0 Policy 10
  1.6 Monitoring Activities Within Malaysia's Maritime Limits 11
      1.6.1 'Grey Zone' 12
  1.7 Caveat 13
  1.8 Reclamation Projects off the Coast of Malaysia 14
  1.9 Malaysia's Territorial Sea Basepoints: An Analysis 17
  1.10 Joint Statement of 8th June 2023 on Boundary Delimitation 19
  1.11 Summary 24

2 Geographical Setting and Ocean Basin Environment 29
  2.1 Introduction 29
  2.2 Sunda Shelf and Limits of Adjacent Seas 35
      2.2.1 South China Sea (Code 6.1) 54
      2.2.2 Malacca Strait (Code 6.5) 55
      2.2.3 Singapore Strait (Code 6.6) 55
      2.2.4 Natuna Sea (Code 6.4) 55
      2.2.5 Gulf of Thailand (Code 6.3) 55
      2.2.6 Sulawesi Sea (Celebes Sea) (Code 6.22) 56
      2.2.7 Sulu Sea (6.23) 56
      2.2.8 Nature and Quality of the Seabed 56
  2.3 Climate and Marine Meteorology Over the Regional Seas 57
  2.4 Coastal Morphology and Tidal Regime of Malaysia 60

2.4.1   The West Coast of Peninsula Malaysia   60
2.4.2   The East Coast of Peninsula Malaysia   61
2.4.3   Off-Lying Islands and Dangers   61
2.4.4   The Coast of Sarawak   62
2.4.5   The Coast of Sabah   62
2.4.6   Highest and Lowest Tidal Levels at Standard Ports   63
2.5   Summary   64

3   Overview of Literature: Geographical and Legal Concepts   65
3.1   Introduction   65
3.2   Charts and Maps   66
3.2.1   Cartography and GIS (Geographic Information
         SYST)   66
3.3   General Purpose or Reference Map   67
3.3.1   Special Purpose or Thematic Maps   69
3.4   Mapping as an Aid to National Development   70
3.4.1   Scale of the Map   71
3.4.2   Map Projections   73
3.4.3   Marine or Nautical Chart and Topographical Map   75
3.4.4   Geographical Information System   76
3.5   Natural Continental Shelf   80
3.5.1   Hedberg's Proposal   83
3.5.2   Juridical Shelf Edge: Outer Limits of the Continental
         Margin   84
3.6   The Definition of the Continental Shelf and Criteria
        for the Establishment of Its Outer Limit   85
3.7   Summary   85

4   Malaysia's Maritime Jurisdictional Limits   87
4.1   Introduction   87
4.2   Defining, Delimiting and Delineating Maritime Limits   90
4.3   Maritime Zones in Adjacent Seas   91
4.4   Territorial Sea Basepoint and Straight Baseline   91
4.5   Delineation of Baselines; Charts and Publicity   93
4.6   Determination and Delimitation of a Maritime Boundary   95
4.7   Malaysia's Actual and Potential Maritime Limits   97
4.8   Malaysia's Maritime Limits Within the Malacca Strait   99
4.9   Malaysia's Maritime Limits Within Singapore Strait   104
4.10   Malaysia's Maritime Limits: Southern Sector South China
         Sea   112
4.11   Malaysia's Maritime Limits off Sabah   119
4.12   An Agreed Common Area and a Joint Development Area   122
4.13   Malaysia's Outer (Extended) Continental Shelf Claim   123
4.14   The Exclusive Economic Zone in the Celebes Sea   128
4.15   Summary   131

5 Traffic Monitoring and Maritime Safety 133
  5.1 Introduction 133
  5.2 Air Traffic Systems in Malaysian Airspace and Adjacent Seas ... 135
  5.3 Defined Corridors for Malaysian-Flagged Ships Within
      Indonesia's Archipelagic Waters 138
  5.4 Marine Electronic Highway and VTIS 146
  5.5 Maritime Districts of the MMEA 158
  5.6 Maritime Areas of Common Concern 158
  5.7 Submarine Operating Areas 161
  5.8 Landfall Lighthouses of Malaysia 163
  5.9 Summary 165

6 Marine Parks, Protected Zones, and Marine Resources 169
  6.1 Introduction 169
  6.2 Marine Parks off Peninsula Malaysia 171
  6.3 Marine Parks off the Coast of Sabah 177
  6.4 Areas of Concern: Threat of Pollution to Coastal Zone 182
  6.5 Vulnerable Marine Ecosystems: Seagrass Beds 187
  6.6 Commercial Fishing Ports and Fishery Infrastructure 193
  6.7 Oil and Gas Tenement Blocks in Offshore Malaysia 197
  6.8 Tectonic Structure of Seabed: Offshore Sabah and Sarawak,
      South China Sea 200
  6.9 Summary 208

7 Major Commercial Ports of Malaysia, Sea Lanes and Security 211
  7.1 Introduction 211
  7.2 Ports of Malaysia 215
  7.3 Malaysian Ports Within Malacca Strait 215
  7.4 Malaysian Ports Within the Straits of Johor and Singapore 222
  7.5 Malaysian Ports Along the Southern Littoral of S.C.S. 229
  7.6 Prescribed Sea Lanes to Ports of Sabah 241
  7.7 Indonesia's Archipelagic Baseline System (Revised in 2002)
      and Archipelagic Sea Lanes 245
  7.8 Reported Acts of Armed Robbery and Piracy 246
  7.9 Summary 257

8 Conclusion 259
  8.1 Mapping to Enhance the Blue Economy 260
  8.2 Geographical Setting and Ocean Basin Environment 261
  8.3 Geographical and Legal Concepts 261
  8.4 Malaysia's Maritime Jurisdictional Limits 262
  8.5 Traffic Monitoring and Maritime Safety 262
  8.6 Marine Parks, Protected Zones and Marine Biotic
      and Mineral Resources 267
  8.7 Commercial Ports of Malaysia, Sea Lanes and Piracy 267

Annex I: Articles of the 1982 Un Law of the Sea Convention   269

Annex II: Primary Documents Relating to Maritime Boundary
              Delimitation   279

Annex III: Maritime Heritage and Cession of Territory   325

Annex IV: Malaysia as Party to Conventions and Treaties   361

Annex V: Vessel Traffic Information System in the Straits
              of Malacca and Singapore   363

Annex VI: Oil Spills: Contingency Plans and Events
              off the Malaysian Coast   383

Annex VII: Glossary of Malay Words in the Maritime Context   389

Annex VIII: Laws of Malaysia Pertaining to Maritime
              Jurisdiction-Related Issues   393

Annex IX: Landfall Lighthouses of Malaysia   395

Annex X: Sempadan Pelantar Benua Malaysia Ditakrifkan
              Dengan Kordinat-Kordinat Geografis Yang
              Berpandukan Daripada Carta-Carta Admiralty NO.
              771, 793, 1353, 1358, 2414, 2660A DAN 2660B   399

Annex XI: Malaysia's Territorial Sea Base Points List
              of Geographical Coordinates of Points   403

Bibliography   413

About the Author

Professor Vivian Louis Forbes is Adjunct Research Fellow of the University of Western Australia (2023–2026) and is attached to the School of Social Sciences. On 22 July 2022, he was awarded by UWA a higher doctorate degree—a Doctor of Letters—for his extensive research and publications. The Award was based on a submission titled 'Marine Awareness and Ocean Governance: An Interdisciplinary Perspective', an analysis and select collection of his published research material.

Viv Forbes is presently Distinguished Fellow at the Maritime Institute of Malaysia in an affiliation that began with the establishment of the Institute in 1993. He is Adjunct Research Professor at the National Institute for South China Sea Studies, Hainan, PRC (2014–2024), and was Distinguished Visiting Professor at the China Institute for Boundary and Ocean Studies at Wuhan University, China (2013–2016). He was Associate at the Centre for Defence and International Security Studies, National Defence University, Malaysia (2020–2022), and is Foundation Member of the Indian Ocean Research Group.

He was Map Curator and is Professional, Practicing Cartographer, Marine Political Geographer, Lecturer in spatial sciences and Maritime Affairs and Former Merchant Naval Officer. Viv gained a Master of Philosophy Degree from Curtin University of Technology, a Doctor of Philosophy from the University of Western Australia and a Doctor of Letters from UWA.

He has developed expertise in international law particularly as it relates to maritime and terrestrial political boundary determination, law of the sea and associated issues; practical experience in terrestrial and hydrographic surveying; lecturing in Law of the Seas issues, spatial sciences, and political geography; and has many years of practical experience in cartography. He compiled two atlases of Malaysia that MIMA published in 1998 and 2008.

He has an excellent understanding of the problems of maritime boundary determination especially in the regions of Southeast and East Asia and those ocean areas surrounding the Australian continent. He has authored several books, atlases, chapters in books and academic research papers. His prolific bibliography and international research profile clearly indicate that he is according to Emeritus Professor Dennis

Rumley (October 2021) *"the Australian doyen of Indian Ocean maritime boundaries and associated policy questions. Viv's singular academic contribution and status have been internationally recognized for some years, especially with his very longstanding and ongoing relationship with MIMA and more recently with Wuhan University and the National Institute for South China Sea Studies in Haikou"*.

In his assessment of the submission for Viv's higher doctoral degree—Doctor of Letters, Emeritus Professor Carlyle A. Thayer, Canberra, Australia (November 2021), noted, and it was cited in the Award, that:

> *Professor Forbes has produced a stunningly broad and deep body of interdisciplinary research spanning more than 30 years. It has made a substantial and seminal contribution to our knowledge related to maritime studies in the 21st Century. His specialisation and expertise rests on a broad and deep inter-disciplinary base; comprising cartography, geography, history, politics, international law, and international relations. Dr. Forbes' work expertly addresses the most pertinent and vexed questions regarding cooperation and conflict among states over delimiting maritime boundaries and cooperation for good ocean governance, freedom of navigation, maritime safety, and environmental sustainability. As a geographer Dr. Forbes' work demonstrates his excellence in five major areas: cartography and archival research, maritime boundary delimitation, marine geo-politics, ocean governance and marine environment and maritime security. In sum, as one of his examiners succinctly put it, Dr. Forbes can be considered the 'Renaissance man' in maritime studies in the 21st century.*

Other Books of the Author Published by Springer

Australia's Arc of Instability, The Political and Cultural Dynamics of Regional Security
    Dennis Rumley, Vivian L. Forbes, and Christopher Griffin (Editors), 2006
Indonesia's Delimited Maritime Boundaries
    Vivian L. Forbes, 2014

Acronyms

AI	Artificial Intelligence
AIS	Automatic Identification System
AtN	Aids to Navigation
CLCS	Commission on the Limits of the Continental Shelf (UN)
COVID-19	A Pandemic of 2019 (Corona Virus detected in late 2019)
CP	Common Point (with reference to Maritime and Terrestrial Boundary)
CTI	Coral Triangle Initiative
D-WF	Distant-Water Fishing
EEZ	Exclusive Economic Zone (a width of 200 Nautical Miles)
FDI	Foreign Direct Investment
FIR	Flight Information Reporting
GDP	Gross Domestic Product
GEBCO	General Bathymetric Chart of the Oceans
GIS	Geographical Information System
GPS	Global Positioning System
HAT	Highest Astronomical Tide
HWM	High Water Mark
IALA	International Association of Lighthouse Authorities
ICJ	International Court of Justice
IEA	International Energy Agency
IHO	International Hydrographic Organisation (formerly IH Bureau)
IMB	International Maritime Bureau (linked to ICC International Crime Commission)
IMF	International Monetary Fund
IMO	International Maritime Organisation
IoT	Internet of Things
IPCC	Intergovernmental Panel on Climate Change
IUU	Illegal, Unregulated and Unreported fishing activities
JDA	Joint Development Authority (or Malaysia-Thailand Joint Authority)
KFR	Kidnap for Ransom

LAT	Lowest Astronomical Tide
LRIT	Long-Range Identification and Tracking
LWM	Low Water Mark
M&E	Manufacturing and Equipment
MIDA	Malaysian Investment Development Authority
MIMA	Maritime Institute of Malaysia (Malaysian Institute for Maritime Affairs)
MMEA	Malaysia's Maritime Enforcement Agency
MPA	Marine Protected Area
NHC	National Hydrographic Centre of Malaysia
NtM	Notice to Mariners
OBOR	One Belt One Road or (BRI—Belt and Road Initiative—Silk Road)
OCS	Outer Continental Shelf (also referred to as Extended Legal Shelf)
PEMSEA	Partnerships in Environmental Management for Seas of East Asia
PRC	People's Republic of China
PTP	Port of Tanjung Pelapas
ReCAAP	Regional Cooperating Agreement on Combatting Piracy and Armed-Robbery Against Ships in Asia (ISC—Information Sharing Centre or IFC—Information Fusion Centre—linked to ReCAAP)
ROC	Republic of China (also known as Taiwan and Chinese Taipei)
SART	Search and Rescue Transmission
SCS	South China Sea
SLOC	Sea Lane (Line) of Communication (relating to maritime trade)
SOLAS	Safety of Life at Sea Convention
TEU	Twenty-feet Equivalent Units (relating to size/capacity of shipping Container)
TSS	Traffic Separation Scheme (relating to maritime traffic)
UAVs	Unmanned Air Vehicles
ULCC	Ultra Large Crude Carrier
UNCLOS	United Nations Convention for the Law of the Sea, 1982 (Third Convention) (also referred to as 'the 1982 Convention' in the context of this study)
UNCTAD	United Nations Commission on Trade and Development
VLCC	Very Large Crude Carrier
VTIS	Vessel Traffic Information Service
VTS	Vessel Traffic Service (System)
WGS	World Geodetic System

List of Figures

Fig. 1.1   Reclamation work in 2023, off Tanjong Piai, West Johor
           Strait, Malaysia. *Source* Google Earth Map. Scale bar
           depicted at lower right-hand corner of graphic 15

Fig. 1.2   A sea village in Sabah, Malaysia. Image courtesy of
           Mr. Vee Chek, June 2023 16

Fig. 1.3   Lifestyle at a sea village sea in Sabah. Image courtesy
           of Mr. Vee Chek, June 2023 17

Fig. 1.4   Territorial sea straight baselines connecting P. Langkawi,
           P. Perak, and P. Jarak. *Source* Author; North is at the top
           of the map as indeed with all the maps in this volume 18

Fig. 1.5   A working document depicting potential alignment
           of a potential maritime boundary. *Source* Author 21

Fig. 1.6   Proposed territorial sea basepoints on Pulau Sebatik 23

Fig. 1.7   An extract of Google Earth image of the Eastern portion
           of Pulau Sebatik. East Pillar is located at the edge
           of the vegetation on Lat 4° 10′ signified by the white
           line on this graphic. *Note* Indonesia's jetty just south
           of and parallel to stated Latitude. *Source* Google Earth
           Map, accessed 19 May 2024 23

Fig. 1.8   Depiction of demarcated boundary on Pulau Sebatik 25

Fig. 2.1   An adaptation of a satellite-derived enhanced image
           of the Sunda Shelf. *Source* V. L. Forbes, 2014 and Malaysia
           Remote Sensing Centre, Kuala Lumpur 30

Fig. 2.2   A simplified depiction of the bathymetry of adjacent seas.
           Source Forbes and Basiron (1998, p. 31) 31

Fig. 2.3   Malaysia's adjacent seas and straits 32

Fig. 2.4   The limits of South China and Eastern Archipelagic seas 33

Fig. 2.5   Limits of Malacca Strait 34

Fig. 2.6   Limits of Singapore Strait 34

Fig. 2.7   Limits of the Natuna Sea 35

Fig. 2.8   Limits of the Gulf of Thailand 35

Fig. 2.9   Limits of the Celebes (Sulawesi) Sea   36
Fig. 2.10   Limits of the Sulu Sea   36
Fig. 2.11   Nature and quality of the adjacent seabed   37
Fig. 2.12   A monthly depiction of the predominant wind flow   38
Fig. 2.13   Prevailing wind pattern: February and August   39
Fig. 2.14   Sea-surface temperature: January   40
Fig. 2.15   Sea-surface temperature: July   41
Fig. 2.16   Generalised barometric pressure over the adjacent seas:
              January   42
Fig. 2.17   Generalised barometric pressure over the adjacent seas: July . . .   43
Fig. 2.18   Sea-surface currents over the adjacent seas: January   44
Fig. 2.19   Sea-surface currents over the adjacent seas: July   45
Fig. 2.20   Wind Roses for selected coastal stations of Peninsula
              Malaysia   46
Fig. 2.21   Wind roses for selected coastal stations: Labuan, Sabah,
              and Sarawak   47
Fig. 2.22   East Pillar demarcating a boundary, P. Sebatik   48
Fig. 2.23   Location of East Pillar (red star), P. Sebatik   49
Fig. 2.24   Generalised coastal morphology of Malaysia   50
Fig. 2.25   Tidal regime along the Malaysian littoral zone   51
Fig. 2.26   Highest and Lowest tidal levels at Standard Ports, Malaysia   52
Fig. 2.27   High-tide level during Spring Tide (bore tide event)
              at Pulau Indah. *Source* Author; image captured on 17
              September 2017, 18:04:48 h, MyST   53
Fig. 2.28   Low-tide level during Spring Tide (bore tide event)
              at Pulau Indah. *Source* Author; image captured on 18
              September 2017 at 11:50:28 h (GMT + 8)   53
Fig. 2.29   The levels of the Tides of significance to navigation
              and the nautical charts   54
Fig. 3.1   Thematic map of an economic intent. Offshore oil and gas
              wells and permits off Peninsula Malaysia. *Source* Author   68
Fig. 3.2   General-purpose sketch map. *Source* Author   69
Fig. 3.3   Statement of map projection and scale   74
Fig. 3.4   Satellite-derived image with elevation data. *Source*
              Aimmee A. Clothier and V. L. Forbes, May 2023   78
Fig. 3.5   The terrestrial political boundary between Malaysia
              and Thailand (The boundary symbology thus: ----)
              as portrayed on a nautical chart intended for use in marine
              navigation   78
Fig. 3.6   The terrestrial political boundary between Malaysia
              and Thailand as portrayed on an aeronautical map shown
              by line symbology thus: dash and two dots   79
Fig. 3.7   Delineation of the geomorphic margin. *Source* Forbes
              (1995:8)   80

Fig. 3.8   Constraints in determining the extended limits
            of the continental shelf   81
Fig. 4.1   An index map depicting arrangements of graphics
            for Sect. 4.7 of this Chapter   89
Fig. 4.2   Maritime jurisdictional zones. The inner limits of a natural
            continental shelf underlays the EEZ and may include
            the CZ. *Source* Author   90
Fig. 4.3   Coastline and river mouths. *Source* Author   93
Fig. 4.4   Bay closing lines as a Territorial Sea datum. *Source* Author   94
Fig. 4.5   Delineation of a state's straight baselines   95
Fig. 4.6   Hypothetical maritime boundaries between states
            with opposing coasts   96
Fig. 4.7   Hypothetical maritime boundary between adjacent states.
            Source Author   97
Fig. 4.8   Malaysia's actual maritime boundaries and a unilateral
            claim   98
Fig. 4.9   Partial maritime limits between Malaysia and Thailand 100
Fig. 4.10   Malaysia's maritime limits in the vicinity of Perak
            and Jarak Islands 102
Fig. 4.11   Malaysia's maritime limits in the southern sector
            of Malacca Strait 103
Fig. 4.12   Malaysia's maritime limits at the western approaches
            to Singapore Strait 105
Fig. 4.13   Partial delineation of maritime limits within Singapore
            Strait 107
Fig. 4.14   Partial maritime limits at the eastern approaches
            to Singapore Strait 108
Fig. 4.15   Middle rocks and adjacent marine features 110
Fig. 4.16   Middle rocks the site of Abu Bakar Marine Base. It
            was inaugurated on 1st August 2017. The rocks are
            about one kilometre south of Pedra Branca. *Source* Author 111
Fig. 4.17   Pedra Branca and Horsburgh Lighthouse, Natuna Sea.
            Source B. D. Proudfoot and present author 112
Fig. 4.18   Malaysia's maritime limits off the southern Malay
            Peninsula 113
Fig. 4.19   Malaysia's maritime limits off the northern Malay
            Peninsular 114
Fig. 4.20   Malaysia's maritime limits: vicinity of Vanguard Bank,
            South China Sea 115
Fig. 4.21   Malaysia's maritime limits: vicinity of the Spratly
            Archipelago 116
Fig. 4.22   Malaysia's maritime limits off Sabah and Sarawak 117
Fig. 4.23   Malaysia's maritime limits: southwestern sector of South
            China Sea 118
Fig. 4.24   Malaysia's maritime limits off Brunei Bay 119

Fig. 4.25   Malaysia's maritime limits off northern coast of Sabah   120
Fig. 4.26   Malaysia's maritime limits off the east coast of Sabah   121
Fig. 4.27   Cooperative approaches to sustainable development
            of resources   122
Fig. 4.28   The 2009 joint submission made by Malaysia and Vietnam.
            Source Executive summary of joint submission, 2008; see
            Note 12   124
Fig. 4.29   The outer limit of the continental shelf claimed
            by Malaysia. *Source* Executive summary of Malaysia's
            claim, 2017. Available UN CLCS   125
Fig. 4.30   Malaysia's claimed extended continental shelf in the South
            China Sea. *Source* Author; MYS_EN_DOC-01_281117
            available at UN website   126
Fig. 4.31   Defined EEZ limit in the Celebes Sea between Indonesia
            and the Philippines   130
Fig. 5.1   Air traffic systems over Malaysian airspace and adjacent
            seas   136
Fig. 5.2   Selected flight paths over Malaysian airspace   137
Fig. 5.3   Location of commercial aircraft at a specific date and time
            in 2021   139
Fig. 5.4   Defined corridors for Malaysian-flagged ships
            in the Natuna Sea   140
Fig. 5.5   Defined transit corridors: Sulawesi and Sulu Seas.
            Malaysia's landmass is depicted in brown, and The
            Philippines coloured as blue. *Source* Author's sketch map,
            2023   141
Fig. 5.6   Approximate routes of submarine communication cables.
            Source Sketch map by Leeanna B. Clothier and
            V. L. Forbes, 23 May 2023   142
Fig. 5.7   Submarine cables for domestic and international
            transmissions   143
Fig. 5.8   Named submarine seamounts within Malaysia's maritime
            limits   145
Fig. 5.9   Vessel traffic system within the straits of Malacca
            and Singapore   146
Fig. 5.10   A daily average flow of marine transportation
            in the Malacca Strait   147
Fig. 5.11   Location of Buana Prosperity 1, west of Port Dickson,
            Malaysia. *Source VesselFinder*, at 1421 h, MyST, 7th July
            2021 < vesselfinder.com >   151
Fig. 5.12   Density of shipping within the seas of South and Southeast
            Asia. *Source VesselFinder*, at 1337 h, MyST, 7th July 2021   152
Fig. 5.13   Shipping within the Straits of Johor and Singapore. *Source
            VesselFinder*, at 1325 h, MyST, 7th July 2021   152

Fig. 5.14   Marine traffic density in the vicinity of Peninsula Malaysia.
            Source VesselFinder, at 1330 h, MyST, 7th July 2021   153
Fig. 5.15   Vessels in the vicinity of Port Klang, Malaysia. *Source
            VesselFinder*, at 1332 h, MyST, 7th July 2021   154
Fig. 5.16   Vessels in the seas adjacent to Sabah and Sarawak,
            Malaysia *Source VesselFinder*, 1334 h, MyST, 7th July 2021 . . .   154
Fig. 5.17   Vessels in the vicinity of Brunei Bay and Labuan,
            Malaysia. *Source VesselFinder*, 1335 h, MyST, 7th July
            2021   155
Fig. 5.18   Colour code legend for the above illustrations (*On the way
            equates to 'Underway'- moving through the water*)   155
Fig. 5.19   Marine electronic highway and vessel traffic separation
            scheme. Marine casualties within the seas and straits
            of the Southeast Asian region can be found in literature
            and reports issued by international, national and regional
            agencies and authorities   157
Fig. 5.20   Maritime Districts of the Malaysian Maritime Enforcement
            Agency   160
Fig. 5.21   Submarine exercise area off Sabah and Sarawak, Malaysia   162
Fig. 5.22   Submarine exercise area off Peninsula Malaysia   162
Fig. 5.23   Image of lighthouse at Kuala Selangor, Selangor. This
            structure was built in 1907 and is 27 m tall. Elevation
            of light facility is 73 m. *Source* Author's collection   164
Fig. 5.24   Pulau Angsa Lighthouse. Built in 1887; elevation 36 m   165
Fig. 5.25   Original lantern of Angsa Light and the author. *Source
            Author's collection*   166
Fig. 5.26   Location of landfall lighthouses of Malaysia   167
Fig. 6.1   Protected zones around the Johor Group of islands   172
Fig. 6.2   Protected zones around Pulau Perhentian group   173
Fig. 6.3   Protected Zones around the Pulau Payar and Pulau Tioman
            Groups   174
Fig. 6.4   Locational map for Fig. 6.2   175
Fig. 6.5   Locational maps for islands mentioned in Fig. 6.3   176
Fig. 6.6   Protected area around Pulau Redang Group; and Labuan
            Island   177
Fig. 6.7   Marine Parks off the East coast of Sabah   178
Fig. 6.8   Image of fish at Tunku Abdul Rahman M.P.   179
Fig. 6.9   Tunku Abdul Rahman Marine Park   180
Fig. 6.10   Mangrove ecosystem in T.A.R. Marine Park. *Sources*
            Figs. 6.8, 6.9 and 6.10 courtesy of Mr. Vee Chek   180
Fig. 6.11   Marine protected areas off the west coast of Sabah   181
Fig. 6.12   Areas of concern: threat of pollution to coastal zone   183
Fig. 6.13   Tracking an oil spill in the Malacca Strait   184
Fig. 6.14   Sewage discharge and biochemical oxygen demand (1994)   186

Fig. 6.15   Vulnerable marine ecosystems: seagrass beds at nominated
            locations 188
Fig. 6.16   An area of mangroves, Malaysia. *Source* Courtesy of Ms.
            Sue Suri Hati 192
Fig. 6.17   Vulnerable marine ecosystems: mangrove forest and coral
            islands 192
Fig. 6.18   Marine ecosystem: coral and fish. *Source* Courtesy of Ms.
            Sue Suri Hati 194
Fig. 6.19   Commercial fishing ports and fishery infrastructure 195
Fig. 6.20   Known turtle nesting sites 196
Fig. 6.21   Offshore exploration tenement blocks granted in the past
            (c.2009) 197
Fig. 6.22   Offshore tenement blocks granted in Sabah and Sarawak
            in the past 198
Fig. 6.23   Tenement holders and operators 199
Fig. 6.24   The Malaysia/Thailand JDA (Joint Development Area)
            2009. *Source* Author, 2009 201
Fig. 6.25   Revised permits within the JDA 202
Fig. 6.26   Brunei's blocks 'J' and 'K' 203
Fig. 6.27   Offshore hydrocarbon fields and pipelines, Sarawak 204
Fig. 6.28   A cluster of platforms operating in Malaysian waters.
            Source Mr. Mohd Muzakir 204
Fig. 6.29   Tectonic structure of South China Sea seabed 205
Fig. 6.30   Schematic portrayals of lateral and vertical maritime
            jurisdictional zones 206
Fig. 7.1   Satellite-derived imagery of the West Johor Strait
            and of PTP (top, left of centre) and Tuas (Singapore). This
            is an area of rapid development and busy with shipping
            movements at the two ports. *Source* An extract of graphic
            courtesy of Google Earth Map 214
Fig. 7.2   Major ports of Malaysia 216
Fig. 7.3   Port Langkawi and Telok Ewa 219
Fig. 7.4   Butterworth deep water port and Penang port 220
Fig. 7.5   Satellite-derived imagery of George Town and Butterworth,
            Malaysia. Note the reclamation work off T. Tokong.
            Source An extract of Google Earth Map 221
Fig. 7.6   Limits of Port Klang, Malaysia 222
Fig. 7.7   The Three Ports of Klang 223
Fig. 7.8   South Port, Port Klang 224
Fig. 7.9   North Port and West Port, Port Klang 225
Fig. 7.10   Limit of Kuala Linggi International Port 226
Fig. 7.11   Reclamation work in progress off Malacca. *Source* Google
            Earth Map 227
Fig. 7.12   The Ports of Tanjong Bruas and Anchorage, Malacca Port 228

Fig. 7.13 Port of Tanjong Pelapas (PTP), Johor. *Note* Development
           of *Forest City*'s location (blue star) depicted in Fig. 7.22
           in an area just above the scale bar and the existing
           coastline—centroid Lat.1° 20′ N, Lon.100° 34.5′ E 230
Fig. 7.14 Satellite-derived image of reclamation works (orange
           arrows) in West Johor Strait. Three features belong
           to Malaysia and the fourth, Tuas, relates to Singapore 231
Fig. 7.15 Delineation of Johor Port limits (western section) 232
Fig. 7.16 Port of Pasir Gudang, Johor 233
Fig. 7.17 Delineation of Johor Port limits (eastern sector). Yet to be
           confirmed: 30 June 2024 234
Fig. 7.18 Ports of Kuantan and Kuching-Pending, and Biawak
           Centres 235
Fig. 7.19 Ports of Kemaman and Kota Kinabalu 236
Fig. 7.20 Bintulu Port and Bintulu Fishing Port 238
Fig. 7.21 Port of Labuan 240
Fig. 7.22 Sarawak: Limits of Ports. *Source* Present author 241
Fig. 7.23 Ports of Sabah 242
Fig. 7.24 Prescribed navigational approaches to ports of Sabah 244
Fig. 7.25 Indonesia's Archipelagic Straight Baselines and Designated
           Sea Lanes 247
Fig. 7.26 Reported Acts of Armed Robbery and Piracy, 2003–2007,
           off Peninsula Malaysia 251
Fig. 7.27 Reported acts of Armed Robbery and Piracy, 2003–2007,
           off North Borneo coast 252
Fig. 7.28 Acts of Armed Robbery, Hijacking and Hostage Taking,
           2008–2021 253
Fig. 7.29 Reported incidents of actual and attempted acts of armed
           robbery in Singapore Strait. *Source* Data from *Annual
           Report* of ICC-IMB, 2023, graphic by V. L. Forbes, 2024 254
Fig. 7.30 Incidents on ships in Singapore Strait. *Sources* ReCAAP
           and IMB 255
Fig. 8.1   Marine traffic in the vicinity of Malaysia. Image captured
           at 19:10:1 on 10 April 2024 (MyST). *Source VesselFinder* 264
Fig. 8.2   Marine traffic within Singapore Strait. Image captured
           at 19:08:20 on 10 April 2024 (MyST). *Source VesselFinder* 265
Fig. 8.3   Marine traffic within central Malacca Strait west of Port
           Klang. Image captured at 19:11: 35 on 10 April 2024 (my
           ST). *Source VesselFinder* 266

List of Tables

Table 1.1   Summary of Malaysia's Territorial sea basepoints   19
Table 4.1   Geographical coordinates of the outer limit of continental
             shelf as claimed   128
Table 4.2   Turning Points of EEZ boundary in Sulawesi Sea   129
Table 4.3   Azimuth and distances between points   129
Table 5.1   Discovery of submarine seamounts within Malaysia's
             Maritime limits   145
Table 5.2   Statistics maintained by KLANG VTS, 2010–2020   149
Table 5.3   Operational areas of MMEA   159
Table 6.1   Legislation and statistics relating to MPAs   171
Table 6.2   Islands within the Johor Group Marine Park   173
Table 6.3   Islands within the Perhentian Group Marine Park   173
Table 6.4   Pulau Seri Buat and Pulau Sembilang   174
Table 6.5   Islands of the Pulau Payar Group   175
Table 6.6   Islands of the Pulau Tioman Group   176
Table 6.7   Islands of the Pulau Redang Group   177
Table 6.8   Islands of the Labuan Group   177
Table 6.9   Threats through human activities to selected seagrass
             beds in Malaysia   189
Table 6.10   Threats to corals and coral reefs in Malaysia   193
Table 6.11   Number of fishers by state and zone, 2023   207
Table 6.12   Number of fishing vessels by state and zone, 2023   207
Table 7.1   Cargo and passenger jetties of Malaysia   217
Table 7.2   Port of Telok Ewa   218
Table 7.3   Break bulk terminal, Port Klang   225
Table 7.4   Dry bulk terminal, Port Klang   225
Table 7.5   Liquid bulk terminal, Port Klang   225
Table 7.6   Berthing facility   229
Table 7.7   Storage and warehousing   229
Table 7.8   Berthing facilities at Port of Kota Kinabalu   237
Table 7.9   Port Facilities at Labuan   239

Table 7.10   Private Jetties at Labuan   239
Table 7.11   Geographical coordinates of prescribed Sea Lanes
              to Ports of Sabah   243
Table 7.12   Sections, number of line segments and distances   245
Table 7.13   Number of base lines categorized by length of line   246
Table 7.14   Reported attempted attacks from 2003 to 2007   250

Chapter 1
Mapping to Enhance the Nation's Blue Economy

Abstract The importance of a blue economy to a coastal and island nation should not be understated nor underestimated. Accruing the maximum benefits within sustainability guidelines and practices is feasible if the data is recorded and maintained. Maps and other graphics are of fundamental value to the basis for theories, measurements, and analysis; and are useful for assisting those who engage in planning (in the marine and terrestrial contexts), and applications of marine cadastre, natural hazard assessment, resource development and environmental protection. This chapter alludes to Malaysia's long-term development plans, its regional partnership, and importantly the nation's ambitions to spur national growth by taking advantage of its geographical location at the crossroads of regional and international maritime trade.

Keywords Maritime state · Blue economy · Maritime limits · Territorial sea basepoint · Grey zone

1.1 Introduction

The study of geography is about more than just memorising places on a map. It is about understanding the complexity of our world, appreciating the diversity of cultures that exist across continents. And in the end, it is about using all that knowledge to help bridge divides and bring people together.

Barack Obama, www.politico.com 24 May 2012

Maps are of fundamental value to the basis for theories, measurements, and analysis and are useful for assisting those who engage in planning (in the marine and terrestrial contexts), and applications of marine cadastre, natural hazard assessment, resource development and environmental protection. The geological, geophysical, environmental, climatic, and hydrological components of Earth science research are fundamentally global in their physical scope, impacts, and interactions and to the implications of the economic benefits that could accrue to a nation's wealth. Malaysia is a focal point of the Indo-Pacific region. It is a member of the Association of South-East Asian Nations (ASEAN).

V. L. Forbes, *Malaysia's Maritime Jurisdictional Limits*,
https://doi.org/10.1007/978-3-031-78783-6_1

The Andaman Sea, Straits of Malacca and Singapore, Gulf of Thailand, South China Sea, Sulu Sea, and Celebes Sea (Sulawesi Sea) collectively wash the shores of Malaysia. The coasts and seas semi encompass the nation's social, economic, security, cultural and natural parameters, which are interlinked and influenced by internal as well as external economic and geopolitical factors. These factors are dynamic and continuously changing, providing goods and services and in turn being affected by their utility.

A Special Interest Group of the Academy of Sciences of Malaysia took the initiative to publish, in 2022, a *Position Paper on the Blue Economy: Unlocking the Valus of the Oceans*, in which the Academy offered three recommendations to the Government of Malaysia based on gaps identified in the Blue Economy ecosystem "that are critical for ensuring effective management of ocean and marine resources in the region, in partnership with other countries in the ASEAN region" (Page 25).

Azam A.H.M. and others (July 2023) opined that Malaysian policymakers on realising the importance of ocean-related industries incorporated the concept of a Blue Economy in the Twelfth Malaysia Plan. However, there are challenges and hence careful planning at national and local government levels must be undertaken.

The global ocean-related risks such as sea level rise due to climate change, marine plastic pollution, land-based and vessel-sourced oil-spills, and illegal, unreported, and unregulated (IUU) fishing, acts of armed robbery and piracy, and acts of actual and perceived terrorism are of persistent concerns and require the undivided attention of law-enforcement agencies and authorities.

Illegal anchoring within the nation's maritime jurisdiction limits, however defined; human trafficking; and unlawful importation or transshipment of drugs and money laundering are of major concerns of Malaysia's enforcement agencies. Indeed, law enforcement agencies were active during 2021, and later years, in apprehending drug traffickers and other incidents including acts of armed robbery in Malaysian waters (*The Star*, 2021; and other regional news media). Such illegal activities have continued in the following years and are well documented in the electronic and print media on a regular basis.

During mid-May 2023, Malaysian maritime authorities detained four ships, allegedly including two Singaporean-registered vessels, for anchoring illegally in eastern Johor waters bringing the total number to 28 since the beginning of 2023 (Zunaira Saieed, *The Star/Asia News Network* 22nd May 2023). Such an activity is an offence under Section 491B(1)(L) of the *Merchant Shipping Ordinance* 1952.

On the morning of Tuesday, 4th July 2023, armed robbers boarded a Malaysian-flagged tug and barge, in Malaysian waters, somewhere between Brunei and Malaysia, and stole some scrap iron. The attackers boarded the barge from a sampan (a flatted-bottom boat). The barge was allegedly unmanned, the nine crew on the tug were unharmed, and the incident was over within one hour (*The Star*, 5th July 2023). The consequences of the event were not available when completing this study.

A total of 100 incidents of armed robbery against ships, barges and tugs took place in Southeast Asian seas during 2023 according to the Information Sharing Centre (ISC) of the Regional Cooperating Agreement on Combatting Piracy and Armed Robbery Against Ships in Asia (ReCAAP) (*The Straits Times*, 14 January 2024;

ReCAAP *Annual Report* 2023). These incidents encapsulate the gist of this book's narrative: security at sea, safety of navigation, ensuring that maritime jurisdictional limits are enforced and enabling legislation is implemented within the zones and effectively managing the blue economy and enhancing ocean governance.

Sabah's ongoing dusk-to-dawn sea curfew which ended on Thursday 3rd August 2023 was extended for another two weeks to 18th August 2023. The curfew, which was extended for the 216th time under Section 31(4) of the *Police Act* 1967 (as amended) was first implemented on 16th July 2014, according to *The Star* (5th August 2023, p. 11). The curfew's extent covered an area of up to three nautical miles (M) off the east coastal districts of Sandakan, Beluran, Kinabatangan, Lahad Datu, Kunak, Semporna and Tawau (1 nautical mile (M) = 1.852 kms).

The extension of the curfew was justified due to the existing threats from cross-boundary criminals including kidnap for ransom (KFR) groups. The action was required to deter intrusion by criminals and terrorists from the islands of southern Philippines into Malaysian territorial waters, however defined. It was also intended to oversee the safety of international researchers and/or foreign tourists who visit the islands of Sabah. The curfew was enforced following a spate of kidnappings. KFR groups as well as the Abu Sayyaf organisation continually attempt to enter Malaysia's territorial sea and threaten chalet owners and their staff and fishers operating on islands in the Celebes and Sulu Seas adjacent to the coast of Sabah.

The issues of maritime security in the Sulu Zone are thoroughly analysed and identified in contributing articles published in Hamzah and Forbes, 2019 and Hamzah, Leong and Mokhtar, 2019, respectively.

1.2 Structure of This Study

This study comprises eight chapters. This chapter contains a narrative concerning the importance of the 'blue economy' to Malaysia, and the country's adoption of the concept in May 2023. The narrative continues with brief commentaries on regional partnership and of monitoring illegal activities within Malaysia's maritime jurisdictional limits. An analysis is presented towards the end of this chapter of the declaration of Malaysia's Territorial Sea Basepoints signed on 2nd August 2022; and a commentary relating to the signing on 8th June 2023, by the Heads of Governments of Indonesia and Malaysia, of documents on matters, in part, relating to finalisation of maritime boundary delimitation.

The geographical setting and the ocean basin environment are cartographically portrayed in 28 illustrations which appear in Chap. 2. The information has been generalised for the purpose of this study. More detailed studies and observations of climatic and weather phenomena are to be found in numerous publications originating from Malaysia and other international sources and in educational—primary, secondary, and tertiary—material.

Chapter 3 of the volume is devoted to an overview of literature of geographical and legal concepts related to charts and maps and other graphics, of which there are eight, that are employed for the purpose of delimitation of a maritime boundary. Brief commentaries are offered on aspects of various types of maps and charts and an elaboration on the natural and legal extended continental shelf.

Chapter 4 focuses on the concepts of defining, delimiting, and delineating not only international political maritime boundaries but also other jurisdictional zones. A discussion on maps and mapping terminology is undertaken herein. Other maps, diagrams, satellite imagery and photographs are included in this chapter to illustrate concepts mentioned in the discussion. Commentary for each of the 31 illustrations is presented in this section, The suite of maps in this chapter is at a scale suitable for portraying, in the present format, the international maritime boundaries negotiated between Malaysia and its neighbours and other actual and potential jurisdictional limits.

The importance of safety of navigation—aeronautical and marine—cannot be overstated. For this reason, a series of 26 illustrations in Chap. 5 depict various jurisdictional limits and zones in the air and water columns—over, on, and under—that Malaysia claims in accordance with international law, as do other coastal and island States in their respective adjacent seas if they abide by the international rule of law.

The themes chosen for the 30 illustrations in Chap. 6 of this volume are linked to the protection of the marine environment and the potential marine biotic and mineral resources that exist in the water column, on the seabed and in the substratum of the adjacent seas. Marine Protected Zones and Marine Parks around the islands are designed to provide protection for species and habitats within a marine park, whilst allowing for ecologically sustainable utilisation of the space. Multi-managed use zones provide for a full range of commercial and recreational activities provided they are sustainable and consistent with the overall objectives of the Malaysian representative system of marine protected areas.

In Chap. 7, the 30 illustrations presented therein focus on basic infrastructure of selected commercial ports of Malaysia. Four maps in this section whilst not specifically focusing on Malaysia, nevertheless, have an impact on that country's maritime jurisdiction and issues. One map depicts Indonesia's proclaimed archipelagic base points and straight baseline system and delineates the archipelagic sea lanes designated by Indonesia and approved by the International Maritime Organisation (IMO). Three maps in this section portray a spatial/temporal distribution of actual and attempted acts of armed robbery and piracy in the regional seas, of past decades, as reported to the relevant authorities and agencies.

A Conclusion is presented in Chap. 8, followed by extensive Annexures that contain a catalogue of primary documents including the provisions of the 1982 Convention pertaining to international jurisdictional maritime limits. A Select Bibliography completes the study.

The very nature and format of this study—thematic in content—would suggest that all the maps cannot be uniform although there is a semblance of uniformity in colour, design, and cartographic style as deemed appropriate for this volume. All the maps, prepared by the author, specifically for this volume, and progressively updated, are oriented with NORTH point at the top of the map.

The negotiated maritime boundaries are depicted as a straight line or a series of lines, in black, joining Turning Points whose geographical coordinates are defined in the text of the Treaties negotiated by the parties to the Agreements, which are reproduced in the Annex II. The Turning Points and Terminal points are numbered and/or lettered in accordance with their prescribed designation in the respective Agreements/Treaties.

On this note, it is important to state that Malaysia deposited at the UN, its list of geographical coordinates of Territorial Sea basepoints on 22nd August 2022. A brief discussion on this action is offered in this Chapter, below.

Where relevant, the Territorial Sea baselines, as depicted, are shown in black with open triangles whose apex point landward. The outer limit of a State's Territorial Sea is portrayed as an arc of 12-nautical mile (M) radius from the coastline or the outermost seaward point of an island, reef, or Low-Tide Elevation (LTE) or straight base line employed by the State. Some exceptions exist on the maps as explained within the relevant chapters.

Submarine cables are shown in grey wavy lines and pipelines (in black) linking the hydrocarbon fields (black squares) to the onshore processing facilities are depicted as lightweight black lines. Areas within which general navigation is restricted are marked by a pecked line (black) to show the extent of the zone.

For the generic naming of sea areas, with limited exceptions, English language has been used for the names of 'Oceans' and 'Seas' to conform with the title. For other areas, such as 'Straits,' 'Bays,' 'Channels,' 'Gulfs' and 'Troughs' English language has been used when the areas are surrounded by more than one country and the national language has been used when the area is surrounded by only one country. Exceptions to these rules may occur when common usage indicates otherwise, for example, in the use of the name Penang Island or Pulau Pinang or Pulau Penang.

1.3 Malaysia: A Maritime State

Malaysia is termed a 'coastal State' by the provisions of the 1982 United Nations Law of the Sea Convention (the 1982 Convention). Malaysia is also considered a 'zone-locked country' in the context of the law of the sea in that it cannot claim a full entitlement to an Exclusive Economic Zone (EEZ), and it is a developing nation in economic terms. Whilst Malaysia's terrestrial surface area is about 328,550 km^2 its marine area is about 614,170 km^2 that comprises an Inland Water of 97,310 km^2; a Territorial Sea of 63,670 km^2; and an EEZ of nearly 453,190 km^2. The area of its natural continental shelf is approximately 476,762 km^2; however, its extended (legal) continental shelf could be larger in the South China Sea, when submissions made in 2009 and 2018, are approved by the UN Commission on the Limits on the Continental Shelf (CLCS).

Peninsula Malaysia accounts for about 40% of country's total land area, Sabah for nearly 22.5% and Sarawak approximately 37.5%. Sabah's approximately 4400-km coastline includes about 320,000 hectares (ha) of highly productive mangroves, and ecologically diverse coral reefs located along the coast and around its many islands. The state's coastal waters are likewise home to rich seagrass and seaweed beds. Sarawak's coastline is about 750 km along portion of the north coast of Borneo Island.

In the context of economy and finance, Malaysia recorded approximately Malaysian Ringgits (MYR) 264.6 billion (USD 59.9 billion) of approved investment during the period to March 2023 (The rate of exchange on 25 May 2023 was 1US$ = MYR4.59185).[1] An updated version issued on 22 February 2024, inferred that total approved investments of RM329.5 billion during 2023—an increase of 23% from 2022. The approved investments were for the manufacturing, services, and primary resources sector. These approved investments involved 2110 projects that were expected to create about 45,000 job opportunities.[2]

Malaysia is a country deemed to have a 'developing economy' by which, in recent years, has successfully transformed itself from an exporter of raw materials into a diversified economy—in short, it is an upper-middle-income, export-oriented economy. Malaysia ranks 55th out of 157 countries (World Bank, 29 Nov. 2022). The Malaysian economy demonstrated a rapid annual economic growth in 2022 at a pace of 8.7%. Malaysia recorded a GDP of over RM1110.1 billion by mid-2022. The Gross National Income was RM1080.0 billion. Exports recorded strong progress which was strengthened by rising world commodity prices and buoyant growth in manufacturing exports.[3]

The foreign trade component of Malaysia grew by 14.8% to RM 718.7 billion with gross exports and imports increasing by 11.8% and 18.7% respectively. The growth in foreign trade performance of foreign direct investment (FDI) was RM 19.3 billion in the fourth quarter of 2022.[4] However, in line with the pessimistic global outlook of a 2.9% growth in 2023, offered by the International Monetary Fund (IMF) and the World Bank (WB), Malaysia would probably experience a similar situation. Malaysia's balance of payments for goods and services were 117.6 and negative 36.3 billion respectively by mid-2022.

Given the difficulties experienced during the first ten months of 2021, the trade surplus extended sharply to USD 202.5 billion from USD 145.3 billion in 2022. Growth was driven mainly by domestic demand. Net trade contributed negatively to the Gross Domestic Product (GDP) with exports shrinking by 3.3% and imports falling by 6.5%. On the production side, activity eased for all sectors: services 7.3%; manufacturing 3.2%; agriculture 0.9%; mining 2.4%; and construction 7.4%. The GDP was expected to be around 5.0% growth in 2023 which would be driven by domestic demand.

Whereas, [5] Foreign Direct Investment (FDI) was generally directed to the manufacturing sector, investments from Malaysian companies were directed towards services and primary divisions. According to the Malaysian Investment Development Authority (MIDA) the nation's strategic position offers the Southeast Asian region an alternate supply chain hub for manufacturing operations. These activities will position Malaysia for greater integration into the global supply chain especially if the nation takes advantage of developing its blue or ocean economy.

The fishery industry in 2015 supported about 64,000 local fishers. This number had declined due to the cost of fuel, reclamation works, the after-effects of COVID-19 pandemic and possibly the reduction in fish population in the regional seas. A significant problem is that of the sovereignty dispute in the South China Sea. The fishing industry contributed nearly 1.5% to the nation's GDP to the value of over RM 13 billion per annum. Fish as a food source provides over 60% of protein for the population of Malaysia. Largely driven by an increasing population in search of new and sustainable sources of food, economic growth and technological advances, the adjacent seas of Malaysia have come to represent a new economic frontier for operators of different interests across the public and private sectors.

1.4 Malaysia Adopts the Blue Economy Concept

A Blue Economy is aimed at the sustainable development of ocean and water resources. Realising the benefits that a Blue Economy to a nation lead to better and new job creations, achieve higher rates of economic growth, and secure biodiversity and sustainable development obligations (Colgan et al. 2021). Achieving a balance between competing uses and users will be crucial for attaining a sustainable ocean economy, consistent with Malaysia's ongoing international policy commitments and effective ocean governance management. According to the Academy of Sciences Malaysia, the Blue Economy contributes 23% to Malysia's GDP (Academy of Sciences Malaysia, 2022, page 25).

To increase the public's awareness of the value of Malaysia's Blue Economy to the nation and the importance of defined maritime jurisdiction on a suite of maps and related graphics will assist the administrative duties and implementation of enacted legislation for the sustainable development of marine biotic and mineral resources within Malaysia's Exclusive Economic Zone and on and within its legal and natural continental shelves.

The concept of a Blue Economy is of a sustainable ocean economy that emerges when the economic activity is in balance with the long-term capacity of ocean ecosystems to remain resilient and healthy. A vibrant Blue Economy ecosystem is predicted to potentially increase the contribution of marine and ocean resources from 21.3% to 31.5% of the GDP from 2020 to 2030 which is close to RM 1.4 trillion contribution to the economy of Malaysia.

Malaysia needs to develop a thorough understanding and appreciation of marine and coastal resources and create new tools and effective legislation to meet and contain the challenges and to reap the benefits of a Blue Economy. Implementation of sustainable policies to revitalise the economy and spur the recovery—an aspect of ocean governance—are the vital actions that are required. A coordinated regulatory and policy platform for each of the sectors encompassed by the Blue Economy should be followed to create a conducive, favourable, and sustainable business environment.

The maritime sector of Malaysia contributed about 40% of the nation's gross domestic product (GDP) in 2020 and an estimated employment of about four million persons. The fisheries' sector provided about ten per cent. The National Oil and Gas production company is the largest contributor at about 14.5%. The share of the offshore energy has limitations and hence sustainable developmental policies need to be implemented. Significant issues come to the fore that include the care and health of the marine ecosystems, pollution control and management of marine biotic and mineral resources.[6]

The Malaysian Government has adopted the concept of a Blue Economy and Blue Strategy. The UN infers that the Blue Economy is one that aims at the improvement of human well-bring and social equity, while significantly reducing environmental risks and ecological scarcities. Key issues, such as pollution control, marine ecosystem health, and sustainable fisheries are guiding priorities in achieving the United Nation's 17 Sustainable Development Goals (UNSDG).[7] Striking a balance between achieving economic growth and maintaining ocean health, requires proper planning, and management of maritime space. Sustainability in this context is the ability to maintain at a certain rate or level the economic growth of a natural resource to preserve an ecological balance.

A collective effort by all agencies is required. Indeed, the Malaysian National Hydrographic Centre (NHC) has collaborated in joint surveys within Malaysian and international waters where the nation's sustainable economic growth will align with maintaining the ocean's health.

1.4.1 Blue Economy Concept

The World Bank defines the 'Blue Economy' as one that endeavours to promote economic growth, social inclusion and the preservation or improvement of livelihoods—employments and income—whilst at the same time ensuring environmental sustainability of the oceans and coastal zones which may include waters, estuaries, and river systems. The Blue Economy concept has essentially evolved from the broader green movements and as a growing awareness of the threats imposed on ocean ecosystems by human activities such as overfishing, habitat destruction, pollution, and the impact of climate change.

The Blue Economy concept also focuses on development of the existing ocean sectors to further generate employment, promote entrepreneurship in new areas of economic activities, facilitate the inter-connectedness of the regional economy, and contribute to sustainable development and climate change mitigation.

The full implementation of the 1982 Convention and related UN marine-related instruments and international conventions were important measures toward ensuring the sustainable development of the world's oceans. Employment and investment opportunities have been created for industries to advance to more environmentally sustainable practices and for new and innovative developments. For example, offshore wind-generated energy has been a success story for many coastal states who have seized the opportunity to utilise this renewable energy, albeit the protests by activists (Mehhilef & Chandrasegaram, 2011).

In 2023, the Government of Malaysia, supported an objective to achieve the renewable energy target of 20% by 2025. Two distinct locations were identified that are characterised by favourable wind conditions for the harnessing of wind energy. They are off the eastern coast of Peninsular Malaysia, extending from the eastern side of Johor State to the southern region of Terengganu (in the south-western sector of the South China Sea/Natuna Sea) and off the northwest coast of Sabah in the South China Sea. Further feasibility studies are required to determine the potential of wind-generated turbines to serve as valuable references and offer recommendations for future offshore wind development initiatives in Malaysia (Li et al., 2024).

Ocean resources are diverse. There are ecologically and economically valuable fisheries as well as marine mammals and sea birds. Coastal and estuarine wetlands are vital as fish hatcheries. Marine parks and special marine protected zones are vital in ensuring the blue economy is sustainable. In 2023, offshore hydrocarbon reserves are the most valuable ocean resource as measured by dollar value of annual production.

Malaysia's maritime security is greatly challenged by non-traditional threats, such as illegal trafficking of goods and humans; illegal, unregulated, and unreported fishing (IUU fishing); piracy, terrorism; threats to the marine ecosystem, like climate; as well as shipping and land-based pollution. The Government has over the years enacted at least 15 pieces of legislation and *Orders* to manage its maritime space. Enforcement of the law is entrusted to more than ten Ministries and 31 maritime-related Agencies. Such an array of legislation and agencies may be perceived as overlapping jurisdictions, which may be professed as an inefficient use of resources. It is acknowledged that much needs to be done and many challenges to overcome and collaboration should be accepted and welcomed to seize the opportunities and progress to an international maritime nation.[8]

1.5 Regional Partnership

The Partnerships in Environmental Management for the Seas of East Asia (PEMSEA) in its 2018 Report identified the following nine key industries of the 'Blue Economy' as: Aquaculture and Fisheries—small- and large-scale operators; Ports, Shipping and Marine Transport—integrating sea and inland waterways; Tourism, including eco-tourism, Resorts and Coastal Development; Oil and Gas Industries; Coastal Manufacturing; Seabed Mining; Renewable Energy; Marine Biotechnology; and Marine Technology and Environmental Services. The Blue Economy concept is that of the sustainable utilisation of ocean resources for economic growth, improved livelihoods and employment while preserving the health of the of the ocean ecosystem.[9]

Initiatives that could be undertaken include developing a Blue Economy profile and conducting pilot studies of the ocean resources and marine cadastre to help define and/or refine Malaysia's conception of a blue economy and promoting the use of ocean economy data in marine planning at the national level to facilitate further engagement by Malaysia with other countries in the region on related areas.

Ocean-based sectors contribute significantly to the economic growth of Malaysia. The contribution of the ocean economy to the country's GDP was in the vicinity of US$359 billion by the end of 2021. According to financial models and analysts' expectations the GDP was presumed to trend around US$380 billion in 2022, and in the following year to about US$440 billion. The trend was impressive for the developing country especially after the Covid-19 Pandemic. The estimated GDP per capita is just over US$10,600. However, Malaysia's economy shrank by 4.5% year on year during the third quarter of 2021, reversing from a record high growth of 16.1% during the second quarter possibly because of the Covid-19 Pandemic and the ongoing restrictions on business activities.

A promising activity of the Blue Economy for Malaysia is that of port, shipping, and marine transport which comprises the maritime industry. Nearly 90% of Malaysia's exports and imports are by maritime transport. Since 2010, ports of Malaysia have recorded an average growth rate of three per cent in compound cargo output. The ports of Klang Complex and Tanjung Pelepas (PTP) have ambitious plans to increase their cargo-handling capacities during the period to 2040 and beyond.[10]

The maritime industry appreciates the fact that it is continuously under pressure to meet the demands of the commercial marketplace and carbon emissions regulations established by the International Maritime Organisation (IMO). The recorded value of total trade by mid-2022 was RM 2374.4 billion; the incomes derived from exports and imports were nearly equal amounting to RM 1290.1 and 1084.1 billion respectively.[11]

Various studies focusing on the valuation of ecosystem services have been conducted to justify the interest in biodiversity protection on economic grounds. For instance, total economic valuation studies have been carried out in the major national marine protected areas, with the values recorded to be RM 8.7 billion. The total value of ecosystem services for the country is projected to be about US$17.7 billion.[12]

Major priority areas identified to boost the national economy and spur growth include fisheries and aquaculture, ocean energy, ports and shipping, ship building, oil and minerals exploitation, renewable energy research and development and sustainable tourism activities through various platforms at the local, national and regional levels.[13]

1.5.1   The 2018 Industry 4.0 Policy

Humankind is currently amidst a technological transformation that is fundamentally altering the way humans live and work. The First Industrial Revolution (1IR) started with the advent of steam and waterpower, enabling mechanisation of production processes. The Second Industrial Revolution (2IR) was driven by electric power and mass manufacturing techniques. Information technology and automation ushered in the Third Industrial Revolution (3IR). The Fourth Industrial Revolution (4IR), presently, is accompanied by the range of technologies, for example, Artificial Intelligence (AI), that are fogging the distinction among physical, digital, and biological spaces.

In response to 4IR, the Government of Malaysia launched its Industry 4.0 Policy on 31st October 2018 to highlight the prominent role the country could play in the global economic and manufacturing environments. The Policy has made the country a strategic partner for global manufacturers, and a primary hub for the high technology sectors. A key beneficiary of the nation's Industry 4.0 Policy is the machinery and equipment (M&E) sector which is progressing rapidly. Malaysian manufacturers are gaining from cutting-edge technologies such as the industrial Internet of Things (IoT), robotics process automation, Artificial Intelligence (AI), Cloud and edge computing, big data analytics, and many more.[14]

Some Industry 4.0 technologies have been in use for several years; however, new developments have given the M&E sectors renewed impetus. The proliferation of 5G telecommunications in coming years will be significant, enabling equipment and machines to make more timely, critical decisions, and allow businesses to operate plants and factories remotely with limited human activity. However, due diligence is required to block potential cyber-attacks to equipment and business activities including disruptions to logistics and supply chains. The 5G technology is expected to be rolled out in Malaysia by 2023.[15] The 5G wireless technology is designed to deliver higher multi-Gbps (Gigabytes per second) peak data speed, ultra-low latency with more reliability and supported by a massive network capacity and increased availability.

Multinational corporations in this field are widening their global reach to capitalise on growing demand for manufactured goods, while local businesses offer niche services to support overseas partners. However, it is vital that the nation's manufacturing and equipment sector receives support from the Government of Malaysia and the various States' administrators.

1.6 Monitoring Activities Within Malaysia's Maritime Limits

The employment of the term 'limit' in this study infers a boundary or bound, as of a country, area, or district. A coastal state would legislate jurisdictional limits for varied purposes. For example, there are harbour and port limits that are defined in official documents and delineated on maps and charts; safety limits around offshore oil rigs and work platforms and marine parks. Jurisdictional limits for Internal Waters, Territorial Sea, Contiguous Zone, Exclusive Economic Zone, Natural and Extended (Outer) Continental Shelf are defined in international law. Limits for commercial aircraft for reporting flight paths and other flight information are defined in international conventions. In the maritime context, limits are set for operations by government agencies, fisheries, and marine conservation and pollution control.

Malaysia has several naval craft of varying size, and naval personnel in place to police its coastlines. However, senior officials have stated that they were examining ways to deploy more personnel and assets at strategic locations to prevent Islamic State (IS) members from entering Malaysian territory. Malaysia's national police force had strengthened its efforts to monitor the coastal and cross-border movements of Malaysian supporters of IS (Islamic State—a political/terrorist group), who had travelled to Syria or Iraq to join the extremist group and who may be trying to return home and plot terrorist attacks (Gunaratna, 2018).

The Malaysian Maritime Enforcement Agency (MMEA) has bolstered the presence of ships and officers working along Malaysia's different coasts. The Agency sought the government's permission to enhance coastal security in other ways. It constructed two radar systems at Pulau Perak at the northern approaches to the Malacca Strait and Pulau Tengah (Middle Rocks) at the eastern approaches to the Straits of Malacca and Singapore. MMEA performs a myriad function using cutters (patrol boats) and small boats on the water and fixed and rotary wing (helicopters) and UAVs (unmanned air vehicles) to monitor activities offshore. The service has approximately 5000 officers and personnel who are considered civil servants despite having military ranks and insignia.[16]

The plans for increased security along the coasts and borders have taken on greater urgency considering arrests in Indonesia, on 21st July 2015, of six suspected IS-linked militants who, according to police in that country, were planning to launch a cross-border rocket attack on Singapore from the nearby Indonesian island of Bantam (This author was delayed, with other tourists, for several hours in Singapore's Immigration and Quarantine Complex at the Second Crossing whilst awaiting border clearance, during late afternoon, on this day).

Examples of the surveillance of maritime space and monitoring of illegal activities include the apprehension of a boat with about 40 polystyrene boxes that contained crabs that were estimated to be worth about US$3000; detaining nine vessels for illegally transferring catches from one vessel to another; since January 2021, the MMEA detained 101 ships for various offences; and on 23rd September 2021, three fishing boats were detained in the Langkawi region for trespassing within the country's waters and were subsequently disposed by being sunk as artificial reefs around Pulau Susu Dara. The Agency is also involved in actions of detaining ships that illegally anchor in Malaysia's maritime jurisdiction, for example, the bulk carrier *Pacific 07*, that was allegedly anchored about 21 nautical miles southwest of Sekinchan, Selangor at 10 p.m. on 4th May 2023.[17] The problem became acute during the COVID-19 Pandemic when some coastal states in the region prohibited ships from entering certain port limits thereby creating grey maritime zones.

1.6.1 'Grey Zone'

A 'grey zone' in the geopolitical context may be considered an area, in this instance, maritime space, where maritime jurisdiction has not been delineated on a chart or map, nor clearly defined by narrative and/or an agreement wherein illegal activities—actual or potential—are contemplated. Operations within such zones are of maritime security concern. This is particularly the case for Malaysia in at least four marine areas. Two areas are within the Straits of Singapore—at the eastern and western approaches of the Strait; the others are in the southern sector of the South China Sea; and in the north-west sector of the Celebes Sea (Sulawesi Sea), which is briefly discussed at Sect. 1.10, below.

Battling or tackling any illegal activity calls for a cautious approach by administrators and those personnel tasked with implementing international law and national rules and regulations. Diplomacy and delicate negotiations are probably the first and wisest approach. However, having clearly defined maritime boundaries and no territorial dispute does help the situation.

A marine area claimed by Malaysia through its Exclusive Economic Zone and natural and extended continental shelf rights, in accordance with the provisions of the 1982 *UN Convention on the Law of the Sea* (the 1982 Convention) Parts V and VI within the southern sector of the South China Sea, is subject to a territorial dispute mainly with China (PRC).

Between 13 and 18th May 2023, the MMEA reported that it had detained four ships, including two Singapore-registered vessels, for anchoring illegally in eastern Johor waters. The vessel detained on 13th May was a Singapore-registered ship, a liquified petroleum gas (LPG) vessel seized about 17.7 nautical miles east of Tanjung Sedili Besar in Kota Tinggi district. The second vessel was detained about 13.2 M east of Tanjung Siang. The other two ships were Indonesian-registered tankers. These offences fall under Section 491B(1)(L) of the *Merchant Shipping Ordinance* 1952. Indeed, since the beginning of 2023, 28 vessels, including these four ships, had been seized for illegally anchoring in Malaysian waters.

On 30th May 2023, Malaysian authorities detained a China-registered bulk carrier for anchoring illegally in Malaysia's waters—presumably within the nation's territorial sea/contiguous zone limits without seeking permission to do so.[18]

The MMEA in recent years, noted that it had detained a Chinese-registered ship that was suspected of looting from other sunken vessels and the two British World War II shipwrecks, namely the HMS *Prince of Wales* and the HMS *Repulse* that the Japanese Air Force sunk on 10th December 1941 in the south-west sector of the South China Sea off the coast of Malay Peninsula. The barge on which the 'loot' discovery was made was allegedly registered in Fuzhou, China and was anchored without a Malaysian Marine Department permit, in waters off southern Johor state.[19]

1.7 Caveat

Conveying as accurate as possible image of the landscape in a static series of maps, like this volume, is only feasible for a single point in time of the day that the manuscript was ready for delivery to the publisher, which in this instance was 31st August 2023 (with subsequent updates on certain facts up to 23rd July 2024). The reader/user of this publication will appreciate that in a world of constant change—natural, geopolitical, and socio-economic—there will be a difference between the information in this volume and that in reality. Considering the potential and major implications, reclamation projects call for a comprehensive assessment of environmental impacts ranging from the direct to the residual, and from the quantifiable to the intangible. Often authorities, land developers and funding agencies are reluctant to divulge the true extent of the proposed reclamation project(s) for varied reasons which are not elaborated in this study.

Satellite-derived imagery of spatial and temporal scales offer ample evidence of the changing nature of coastal geomorphology in many parts of the world and in Southeast and East Asia. Reclaiming portion of the sea that is adjacent to the coast, or re-shaping geography, is a human activity that can transform an entire coastal zone and marine environment in a relatively short time. Whole new islands can be created in this way, and, in the process, landscapes, eco-systems and livelihoods of the coastal communities could be altered, and maritime jurisdiction extended. In Asia, land reclamation has become a contentious issue, with Cambodia and Malaysia banning sand exports, while Jakarta has suspended its reclamation project, and a plan to build an artificial island in Hong Kong drew fierce criticism.

Destroying coral reefs and marine habitat to transform a reef system into artificial islands to the extent of justifying extended territorial jurisdiction by several littoral states has been witnessed in the South China Sea basin between 2010 and 2023.[20]

The Malaysian Government is firmly committed to protecting its maritime areas in the South China Sea which includes its sovereignty, sovereign rights, and interests in accordance with international law and its stated territorial claim.

1.8 Reclamation Projects off the Coast of Malaysia

Reclamation works have been carried out along Malaysia's coast since the 1950s, however, it was only in the 1990s that such drastic changes captured much public attention, and mostly, due to the conceptual proposals of several large-scale reclamation plans along the coast of Peninsula Malaysia facing the Straits of Malacca and Singapore, in locations stretching from Pulau Penang to Johor Straits—East and West branches. Figure 1.1 offers an example of a project being undertaken off the coast of Tanjung Piai, West Johor Strait.

Large-scale reclamation allows more flexibility in city planning, but also lets local governments engage more ambitiously and aggressively with the business of land-transactions. These island-making projects are designed to boost state coffers but also the money lenders. They represent a colossal misappropriation of resources at a time of intensifying housing unaffordability and social injustice. The expanse of these projects is clearly visible in satellite imagery in spatial and temporal scales although their existence is often downplayed by local authorities and agencies when challenged for verification, based on personal experience and knowledge.

Fig. 1.1   Reclamation work in 2023, off Tanjong Piai, West Johor Strait, Malaysia. *Source* Google Earth Map. Scale bar depicted at lower right-hand corner of graphic

Another example is off the south coast of Penang Island, where construction of three artificial islands, referred to as Penang South Island, was proposed in 2023. The collective area of the islands are estimated to be 1182 ha. These islands were allegedly needed to provide housing and economic opportunities for an expanding population.

Fishers and environmentalists have challenged the government and businesses for proposing the Penang South Reclamation (PSR) project. The arguments range from ruination of lives of residents to damage the coastal zone. The area was rich in prawns and fish resources that sustained the employment of 6000 fishers. Permanent environmental degradation will occur when the islands are completed and populated. Many fishers have rejected any compensation offered, as well as the Environmental Impact Assessment (EIA) report, which conservationists say does not reflect the potential damage or propose adequate mitigation measures.[21]

A negative aspect of this and other projects around the island is the scale of the dredging and reclamation work over more than a decade deemed to cause massive and long-term environmental destruction. Land reclamation is a long offensive to any marine ecology and the fishery industry that depends on it. The reclaimed islands will bury existing fishing areas while deteriorating the surrounding marine water quality. Coastal communities who rely on the marine and coastal area for their livelihood will experience an irreversible negative impact. The fishers of Tanjung Tokong were banned from entering what was once an extremely productive fishing ground but there are now fewer fish in the waters near their village. The mud from the project suffocates the crabs and they are dead long before they get tangled in the fisher's net.

Malaysia has two other major reclamation projects underway: *Melaka Gateway*, a deep-seaport and cruise terminal that is part of China's massive Belt and Road

Initiative infrastructure plan, and *Forest City* in Johor near Singapore (discussed in Chap. 7), aimed at foreign investors with proper design and construction methods applied to dredging and reclamation, and pollution prevention and mitigation measures to minimise environmental impact. Reclamation has hugely benefited Penang, with parts of the Bayan Lepas industrial zone, as well as heritage clan jetties built on reclaimed land.[22]

However, there are concerns that such megaprojects were getting out of control and doing irreparable harm to the environment and put an end to land reclamation projects that were not spearheaded by central government.[23]

Sabah's coastal shallows—coastal zone—have traditionally been home for seafaring communities whose free movement across porous international maritime boundaries had long been tolerated. Large communities—sea villages—have been established along the shores at Kota Kinabalu (see Fig. 1.2), Kudat, Sandakan and Tawau and other towns and offshore islands, for example, Pulau Gaya, of Sabah. However, following a few violent clashes, kidnapping of locals and tourists, the Government of Sabah has plans to resettle these water villagers. Such villages are regarded as havens for the illegal migrants that crisscross the *ill-defined or porous* maritime boundaries. Figure 1.3 illustrates the simplicity of the lifestyle of the sea-gypsies, or *Baju Laut*, where public infrastructure needs more coordination, especially clean, fresh water, and basic education for children and comfortable lifestyle.

With the aid of satellite-derived imagery enhanced for the purpose of geospatial analysis, charts, maps, and other graphics combined with digital technology that planning and implementing enabling legislation, rules and regulations are possible. Assurance of territorial integrity and security in the maritime domain is possible when

Fig. 1.2 A sea village in Sabah, Malaysia. Image courtesy of Mr. Vee Chek, June 2023

Fig. 1.3 Lifestyle at a sea village sea in Sabah. Image courtesy of Mr. Vee Chek, June 2023

the territorial sea baselines have been declared and published, and where feasible, delineated on large-scale charts and other graphics. Importantly, the State's territorial sea datum must be defined and available in the public domain.

1.9 Malaysia's Territorial Sea Basepoints: An Analysis

The legal basis for Malaysia's maritime claim and its jurisdictional limits is linked to defined Territorial Sea basepoints which must be given due publicity and portrayed, where feasible, on nautical charts or maps at an appropriate scale. On 2nd August 2022, pursuant to Article 16:2 of the 1982 Convention, Malaysia deposited a list (M.Z.N. 159.2022) of geographical co-ordinates of the Basepoints (Refer: Annex XII). Straight baselines connect these basepoints.[24] An analysis of the Malaysia's Basepoint declaration is offered in this section of the chapter.

The alignment of declared baselines appear to mirror the 'Inferred Baselines' delineated on the 1979 Map—*Peta Baru*—portraying Malaysia's Continental Shelf boundary—a unilateral claim that created adverse protests from Malaysia's near neighbours. However, the then Government of Malaysia opined that its claim was legitimate and proceeded to negotiate its maritime boundaries where it was relevant. These boundaries are discussed in Chap. 4.

In mid-June 2023, whilst compiling this manuscript there were at least three areas that required urgent attention to finalize the maritime boundary dispute. The present author had been assured, on several occasions since 1998, that negotiations are ongoing. Perhaps some of the technical hiccups may be resolved with the publications of the 2022 Baseline coordinates. The 2022 baselines mark the outer limit of what is considered as *Internal Waters* deemed as the offshore waters landward of the Territorial Sea straight baselines.

Within the northern sector of the Malacca Strait this outer limit extends to Pulau (P. Pu.) (Island) Jarak and Pulau Perak. The islands, Perak and Jarak, are located 100 and 35 nautical miles (M), respectively, westward off the west coast of Peninsula Malaysia (Fig. 1.4). One nautical mile (M) is equivalent to 1.852 kms (km). The defined straight baselines in this northern sector of the Malacca Strait creates a vast expanse of 'Internal Water', in legal parlance, the concept of which is described in Chap. 4, below.

Likewise, off the coast of Sarawak and Sabah's northwest coast the straight baseline system creates vast areas of 'Internal Waters'. The straight baseline segments off the coast of Sarawak are of excessive lengths and distant, for example, 30 M from the adjacent coast. The straight baselines off the east coast of Sabah are in accord with the provisions of the 1982 Convention.

Basepoint SM1 is located on Batu Puteh off the west coast of Peninsula Malaysia and Basepoint 95 is SM80 located on Pengkalan Kubor in Lat 6° 14′ N, Lon 102° 05′E. Within the Straits of Johor, the defined Malaysian basepoints are Numbered from SM22/1, located at Pulau Merambong, to SM59, which is at Tanjung Pengelhi. The straight baselines connecting the successive points are of short distances.

Basepoint 96 is labelled SWK1 which is located at Tg. Datu and Basepoint 111, numbered as SWK 15, is pinned to Beting Sunda. Basepoint 112 is SBH01 located

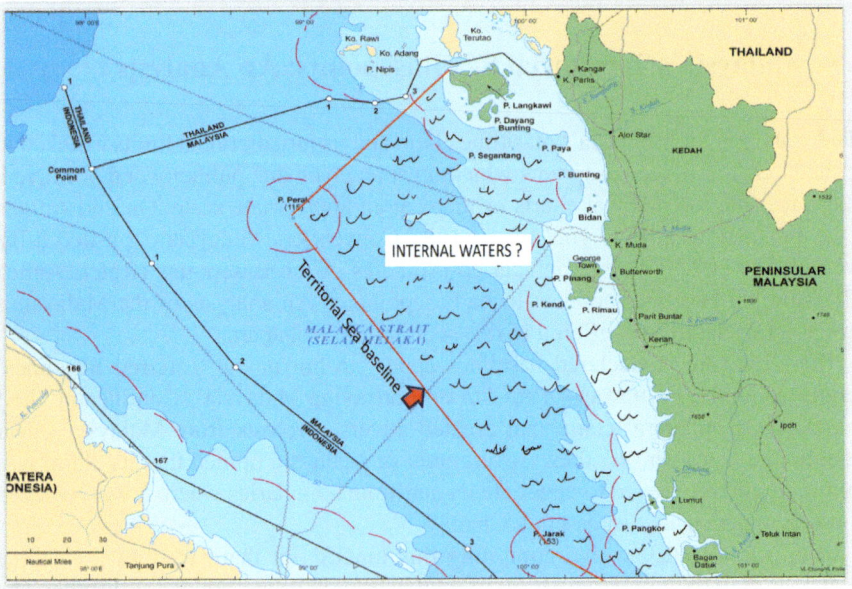

Fig. 1.4 Territorial sea straight baselines connecting P. Langkawi, P. Perak, and P. Jarak. *Source* Author; North is at the top of the map as indeed with all the maps in this volume

Table 1.1 Summary of Malaysia's Territorial sea basepoints

A tabulated summary of the basepoints and the relevant straight baselines follows:

• Total number of defined basepoints	158
• Total number of Straight baselines	155
• Straight baselines encompassing Pen. Malaysia	94
• Straight baselines enclosing Sarawak coastline	15
• Straight baselines encompassing Sabah's coast	46
• Longest length—points 11 to 12-P Perak to P Jarak 124 M (nautical miles (M)) **(1 km = 0.539957 M) or (1 M = 1.852 km)**	
• 24 Straight baseline segments are each more than 24 M	
Category of straight baseline segments (M)	
• 0–0.99	118
• 10–10.99	2
• 11–19.99	6
• 20–20.99	5
• 30–39.9	6
• 40–49.9	1
• 50–59.9	4
• 60–69.9	4
• 70–79.9	3
• 80–89.9	1
• 90–99.9	2
• 100–109.9	2
• 110–119.9	0
• Over 120	1

at Kuala Sg Padas Damil and Basepoint 158 is SBH43 pegged to Pu. Sebatik border. This boundary post is discussed in Chap. 4 of the present volume.

The term often used for an exaggeration for exceptionally long TS baselines is 'rogue baselines' employed in the literature on the determination of maritime boundary delimitation (Table 1.1).

1.10   Joint Statement of 8th June 2023 on Boundary Delimitation

On 8th June 2023, the Honourable Prime Minister of Malaysia and His Excellency, President of the Republic of Indonesia issued a joint statement relating to their bilateral discussions on a range of issues of mutual interests. Both leaders stressed their commitment to further strengthen relations between the two countries to promote comprehensive peace and security and stability in the region for the prosperity of their citizens.[25]

The leaders witnessed the signing of the following instruments:

- *Agreement between the Government of Malaysia and Government of the Republic of Indonesia on Border Crossing (BCA).*
- *Agreement between the Government of Malaysia and the Government of the Republic of Indonesia on Border Trade (BTA).*
- *Memorandum of Understanding (MoU) between the Government of Malaysia and the Government of the Republic of Indonesia on Cooperation in Investment Promotion.*
- *Memorandum of Cooperation (MoC) between the Government of Malaysia and the Government of the Republic of Indonesia on Mutual Recognition of Halal Certification for Domestic Products.*
- *Treaty between Malaysia and the Republic of Indonesia Relating to the Delimitation of the Territorial Seas of the Two Countries in the Southernmost Part of the Straits of Melaka (SOM Treaty); and*
- *Treaty between Malaysia and the Republic of Indonesia Relating to the Delimitation of the Territorial Seas of the Two Countries in the Sulawesi Sea (Sulawesi Sea Treaty).*

The signing of the last two named items, namely the Strait of Melaka (Malacca) Treaty (SOM Treaty) and the Sulawesi Sea Treaty was a significant milestone after twenty years of negotiations. The two leaders expressed their commitments to expedite their respective internal processes towards realizing the simultaneous entry into force (ratification) of the Treaties.

On the agenda will be the resolution of the land boundary issues on Pulau Sebatik, Outer Boundary Post (OBP) on Pulau Sebatik, OBP Sungai Sinapad-Sesai and new West Pillar on Pulau Sebatik to AA Pillar, as well as issues relating to the intertidal area in the east of Pulau Sebatik and the gap from the low-water line to Point M by June 2024.

Signing the Agreements on 8th June 2023 symbolized the harmonious relationship and contemporary friendship between the two countries that provides a robust foundation for future maritime boundary negotiations. The maritime boundary negotiation teams from Indonesia and Malaysia have been tasked to commence, as soon as possible, negotiations on all the remaining and outstanding maritime boundaries between both countries simultaneously.

The Leaders of the respective countries emphasized their dedicated efforts towards effective implementation of a MoU in respect of the Common Guidelines concerning the treatment of fishers by Malaysia's MMEA and the corresponding agency from Indonesia. Both parties urged for closer cooperation and continuous open channels of communication by the authorities on both sides to safeguard the safety and to protect the livelihood of the fishers from both countries.

It is doubtful that the details of maritime boundary alignments referred to above will be in the public domain before this volume is published. However, the present author will be tracking developments on these issues and of the sovereignty claims in the South China Sea and Sulawesi Sea as they impact on Malaysia maritime space.

In the instance of the Sulawesi Sea, Fig. 1.5 is a 'working document' of the author's interpretation of a potential alignment of a Territorial Sea, Contiguous Zone and partial EEZ boundary between Indonesia and Malaysia in the vicinity of the politically shared island of Sebatik, north-west Sulawesi Sea (Celebes Sea). The illustration, it must be stated, has no authority, however, it is compiled on facts, namely, the defined basepoints of Indonesia's Archipelagic straight baseline system (mentioned in a later chapter of this study) and Malaysia's Territorial Sea Baseline declaration of 2022.

The Territorial Sea basepoints of Malaysia, relevant to this narrative are Basepoints SBH 39, 40, 41, 42 and 43. Basepoints 39 and 40 are located at Pulau Ligitan; SBH 41 is at Pulau Sipidan; SBH 42 is at Batuan Tangan; and SBH 43 is at the land/sea interface on the east coast of Pulau Sebatik.

The basepoints for Indonesia, in the context of this study are TD 36 and TD 37. The former is in the vicinity of SBH 43 but not coincident—there is a difference of seconds of arc value in Latitude and Longitude (which are listed below). The latter is located at Batuan Unarang, a feature whose sovereignty has been disputed. TD 36A and TD 36B are not depicted in this illustration. Both marine features are allegedly on Low-Tide Elevations. This author cannot verify this fact for lack of access to a large-scale chart—authorized or otherwise and confirmation from the respective authorities in Indonesia and Malaysia.

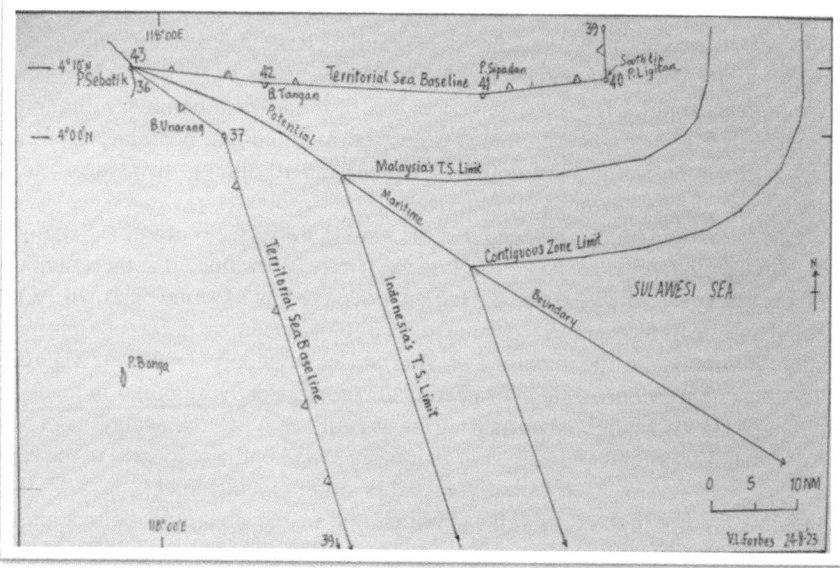

Fig. 1.5 A working document depicting potential alignment of a potential maritime boundary. *Source* Author

The Territorial Sea boundary projected from a 'Common Point' on the east coast of Pulau Sebatik (in the vicinity of Lat. 4° 10′ N Lon 117° 54′ E) is determined on the equidistance principle to the 12-M limits of each country and continued as a Contiguous Zone boundary and extended further south-eastward as a partial EEZ boundary. A Common Point on the East Coast of Pulau Sebatik is yet to be finalized.

Within the context of Fig. 1.5, the relevant Territorial Sea basepoints for Malaysia are:

39	P. Ligitan (i)	04° 08′ 43.5″N 118° 53′ 42.9″E
39A	P. Ligitan (ii)	04° 08′ 13.1″N 118° 53′ 33.6″E
40	P. Ligitan (iii)	04° 08′ 03.2″N 118° 53′ 16.0″E
41	P. Sipadan	04° 06′ 13.2″N 118° 38′ 00.9″E
42	BatuanTangan	04° 08′ 24.4″N 118° 10′ 40.8″E
43	P. Sebatik	04° 10′ 00.5″N 117° 54′ 31.8″E

The relevant Territorial Sea basepoints for Indonesia in Fig. 1.7 are:

36	P. Sebatik	04° 10′ 10.0″N 117° 54′ 29″E
37	B. Unarang	04° 00′ 38.0″ N 118° 04′ 58″E

Indonesia's Basepoints 36A and 36B are not depicted on the map due to scale limitations; Point 38 is officially not used; and Point 39 is outside the southern limit of Fig. 1.5.

An interesting point that illustrates the potential conflict that arises when neighbouring—adjacent coastal states—have differing views as to the location of their land/sea interface point for the commencement of a maritime boundary alignment is depicted in Fig. 1.6. The negotiated terrestrial international boundary on Pulau Sebatik is along the parallel of Latitude 4 degrees and 10 min (4° 10′) North (Refer Annex III, Sabah Document of 20 June 1891).

A boundary stone (pillar) is located in the sand (hard surface) of Pulau Sebatik is the eastern terminal point of the land boundary. It may be assumed that this boundary would project seaward across the drying area depicted as *M* (mud) and where the projected line meets the sea surface would offer a commencement point for a maritime boundary alignment or specifically a Territorial Sea boundary. However, Malaysia and Indonesia have defined their respective Territorial Sea basepoints. Indonesia's basepoint (+TD36) is slightly further northwards of Lat 4° 10′ N, Malaysia's basepoint (+SBH43) is slightly southeast of Indonesia's, and both are located at the edge of the drying area. A proposed compromised position is indicated by the '?+'. The three points are highlighted in yellow. Which one of the three will be recognised as the Common Point for the commencement of the alignment of the international maritime boundary? Interestingly, Indonesia's defined basepoint is north of the recognised international boundary and the low-water mark as depicted on an authorised nautical chart.

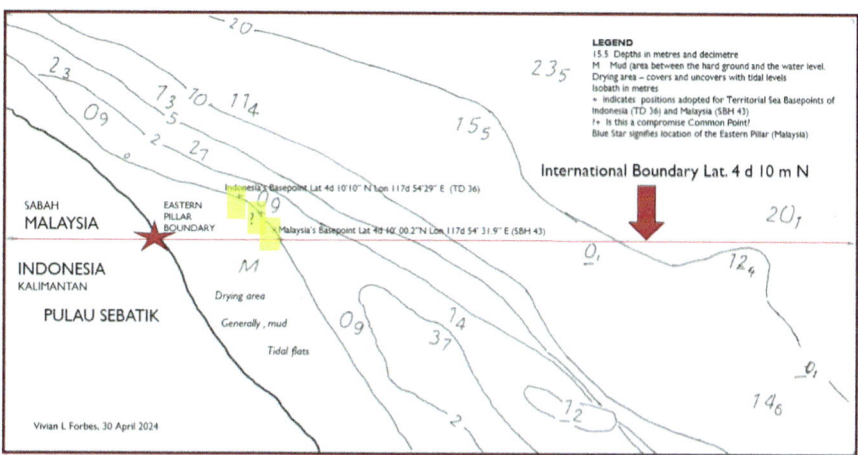

Fig. 1.6 Proposed territorial sea basepoints on Pulau Sebatik

Fig. 1.7 An extract of Google Earth image of the Eastern portion of Pulau Sebatik. East Pillar is located at the edge of the vegetation on Lat 4° 10′ signified by the white line on this graphic. *Note* Indonesia's jetty just south of and parallel to stated Latitude. *Source* Google Earth Map, accessed 19 May 2024

The isobath marked '0' (metres) represents the Low Water Mark (LWM) or Lowest Astronomical Tide (LAT) level as recognised and adopted by the National Authority.

The geographical coordinates of the following features in Fig. 1.7 are NW corner 04° 10′ 02″ N. 117° 54′18″ E; NE corner 04° 10′ 02″ N. 117° 55′ 01″ E; SE corner 04 09 45 N 117° 55′ 01″ E; SW corner 04° 09′ 45″ N 117° 53′ 35″ E. The agreed Malaysia/Indonesia terrestrial boundary on Pulau Sebatik is along Lat 04° 10′ 00″ N. Eastern limit of vegetation on boundary is at Lon 117° 54′ 03″ E. The Eastern end of white 'boundary' line is 117° 54′ 22″ E.

The difference in Longitude between 'A', Malaysia's basepoint (117° 53′ 51″ E) and 'I' of Indonesia (117° 54′ 10″ E) in this graphic is 14 s of arc value. The 1st bend of northern jetty 04° 09′ 57″ N 117° 54′ 04″ E. The 2nd bend of the northern jetty is at 117° 54′ 13″ E. The Eastern edge of the southern jetty 04° 09′ 48″ N 117° 54′ 40″ E.

As stated above, it is highly unlikely, given the prolonged period of boundary negotiations between Malaysia and Indonesia, that the details of maritime boundary alignments referred to above will be in the public domain before this volume is published.

However, between June and November 2019, the Joint Committee that was formed to demarcate the boundary line between the two nations that claim portions of Pulau Sebatik established 148 new boundary pillars consisting of three 'A' type Pillars; 10 of 'B' type Pillars; and 135 of 'C' type Pillars that stretched across the island, about 24 kms in width, along parallel of Latitude 4° 10′ N. This was the designated Parallel of Latitude that was stipulated in the 1891 Treaty between the United Kingdom and The Netherlands.

On the agenda of the 8 June 2023, mention was made of a potential resolution of the land boundary issues on Pulau Sebatik, Outer Boundary Post (OBP) on Pulau Sebatik, OBP Sungai Sinapad-Sesai and new West Pillar on Pulau Sebatik to AA Pillar, as well as issues relating to the intertidal area in the east of Pulau Sebatik and the gap from the Low-Water Line to Point M (depicted in Figs. 1.6 and 1.8) by June 2024. Confirmation of the boundary demarcation and survey and finalization of the work was mentioned by the Ministry of Natural Resources and Environmental Sustainability at Labuan on 4 March 2024. The survey marks included East Pillar, PB002 near Kampung Sungai Haji Kuning, on the east coast of the island; New West Pilar, on the west coast of the island; and boundary mark AA2 in Sungai Sikapal, Tawau.[26] The approximate position of West Pillar is Lat. 4° 10′00″ N, Lon. 117° 40′37″ E. The azimuth of the boundary line depicted in Fig. 1.8 is approximately 269° from East Pillar to West Pillar—a generally East–West international terrestrial boundary alignment.

A maritime boundary alignment between Indonesia and Malaysia, within the Strait of Sebatik, to the west of the island awaits delimitation. The potential boundary will have a westward projection from West Pillar.

The present author and other scholars will be tracking developments on these issues and of the sovereignty claims in the Singapore Strait, the South China Sea, and the Sulawesi Sea as they impact on Malaysia's maritime space and jurisdictional limits.

1.11   Summary

The narrative in this Chapter sets the tone for discussion on the remaining contents in this volume, namely Malaysia's maritime jurisdictional limits and its reliance on a blue economy and a blue strategy. As a maritime nation, Malaysia needs to take

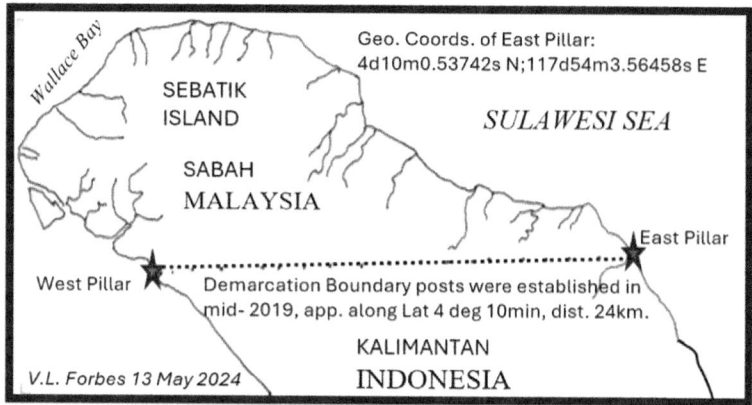

Fig. 1.8 Depiction of demarcated boundary on Pulau Sebatik

full advantage to benefit from the marine biotic and mineral resources within its jurisdictional limits in accordance with international law and in keeping with its obligations and rights as a member of the international community.

An initial self-imposed deadline of 31st August 2023 was set for completion of this manuscript. It was an auspice occasion as it marked Malaysia's National Day (66 years of *Merdeka*). Good fortune permitted the inclusion of an announcement of a Joint Statement relating to six Agreements between the Governments of Indonesia and Malaysia that focuses on border crossing, border trade, cooperation in investment and delimitation of the Territorial Sea in the southern sector of the Malacca Strait and in the Sulawesi Sea.

Malaysia's priority is to reclaim its status as a consequential actor in global efforts to promote sustainability, while bolstering domestic cooperation to ensure its constructive efforts at home are reflected in its international image. The 17 global objectives of the UN's Sustainable Development Goals are a crucial part of Malaysia's long-term development plans and is embedded in its five-year Twelfth Malaysia Plan (2021–2025).

A narrative with maps alludes to the geographical setting and ocean basin environment in Chap. 2. Chapter 3 offers an overview of geographical and legal themes and a brief discussion on charts and maps. Chapter 4 presents an appraisal of Malaysia's maritime jurisdictional limits in its adjacent seas. The following chapter focuses on marine traffic and maritime safety and includes issues relating to the airspace above Malaysia's authority. In Chap. 6 the highlight is on Malaysia's marine parks, protected zones and harvesting of marine biotic and exploitation marine mineral resources. The penultimate chapter contains a narrative and series of maps relating to the commercial ports of Malaysia. The Conclusion is contained in Chap. 8.

Notes

1. Currency Converter https://www.xe/com, accessed 25 May 2023.
2. Malaysian Investment Development Authority (MIDA) (2023) 'Malaysia Attracted RM264.6 billion in Investment for 2022…' mida.gov.my/media-release… accessed 23/5/2023).
3. R. Biswas 'Malaysia Records buoyant GDP Growth in 2022' S&P Global web page, 2023.
4. Malaysia's Ministry of Finance Press statement of 11 February 2023 online version.
5. Malaysia's GDP Annual Growth Rate. www.tradingeconomics.com/Malaysia/gdp-growth.
6. Hanizah Idris (2022) *Realising Blue Economy Potential in Malaysia, Opportunities and Challenges*, Taylor, and Francis eBooks.
7. 'Malaysian Government to adopt blue economy', *The Star*, 24 May 2023, online version. See also Paul Hallwood's *Economics of the Oceans* (2014), London: Routledge.
8. Mani Juneja and others, 'Contextualising Blue Economy in Asia–Pacific Region', The Energy and Resource Institute, March 2021, Konrad Adenauer Stiftung E.V. and RECAP, HK. Blue Economy and Blue Finance Towards Sustainable Development and Ocean Governance, Edited by Peter J. Morgan and others, Asian Development Bank Institute, Japan, 2022.
9. PEMSEA Report 2018.
10. Future plans for Port Klang and Port of Tanjung Pelepas announced by the Government of Malaysia. *The Star*, 27 March 2024, online version.
11. IMO's carbon emissions—IMO website; The Malaysian Economy in Figures, 2022 produced by the Malaysian Ministry of Economy.
12. Total Economic Value of Marine Diversity, Malaysia Marine Parks. wdpa.s3.amazonaws. Fact sheet. See also Academy of Sciences Malaysia, 'Position Paper on Blue Economy…', 2022, page 30.
13. For trends in Malaysia's GDP see World Bank. data.worldbank.org/… and Department of Statistics, Malaysia https://open.dosm.gov.my, accessed 30 June 2023.
14. National Policy on Industry 4.0. Malaysia.gov.my/portal/content/30917.
15. Government Policy on Industry 4WRD, 12 July 2019. pmo.gov.my/ms/2019/07…
16. Malaysia's Maritime Enforcement Agency (MMEA). mmea.gov.my/…
17. 'Malaysia: MMEA detains bulk carrier 'Pacific 07' over illegal anchoring', Manifold Times. manifoldtimes.com/new/Malaysia-mmea-detains-bulk-carrier… 9 May 2023.
18. Z. Saieed, 'Grey Zones', *New Straits Times*, 22 May 2023.
19. 'Malaysia finds 100 old artillery shells on Chinese barge…' *Associated Press*, 31 May 2023, apnews.com/article/Malaysia-British-shipwrecks- china-….
20. Reef Check, Malaysia the Value of Our Marine Ecosystems *Fact Sheet*. reefcheck.org.my.
21. Rina Chandran "Boon or bane? Malaysian island reclamation plan divides residents, 23 February 2021, Reuters. reuters.com/articles/us-malaysia-environment-islands…; see also, Forbes, V.L. 'Artificial Islands in the South China Sea…', 2015: 30–53; and James Clark 'Penang South Islands—Three Islands land reclamation project', 2023, available online https://futuresouthesatasia.com/penang-south-islands/.
22. Sahabat Alam Malaysia 'Impacts of Coastal Reclamation in Malaysia' (no date). foe-malaysia.org/wp-content… (accessed 30 June 2023); see also https://www.reuters.com/art icles/us-malaysia-environment-islands-trfn-idUSKBN2AN031/.
23. Yvonne Tan 'Land Reclamation in Malaysia', 17 January 2022. climatetracker.asia….
24. Malaysia's Territorial Sea Baseline declaration. M.Z.N. 159.2022, UN LOS.
25. Malaysia Ministry of Foreign Affairs Press Statement of 8 June 2023.
26. The International Court of Justice's *Case Concerning Sovereignty over Pulau Ligitan and Pulau Sipadan* (Indonesia/Malaysia), 17 December 2002, General List 102, available online. Sections 36 to 58 offers arguments by the Parties to the Dispute and their interpretations of the provision of Article IV of the Treaty, which appears in Annex III of the Annexures, below. The reader may also be interested in reading 'The Unsettled Agreement of the 2019 Re-Demarcation of the Indo-Malaysia Boundary Line's Impact' by Amin Nurdin, A. Sudjito and I.M.A. Arsana, in *Indonesian Journal of Multidisciplinary Science* 2(9): 3058–3069 and

'Examining the Negotiation Model of the Disputed Boundary between Indonesia and Malaysia on Sebatik Island' by Amin Nurdin, Atmar Sudjito and I,M,A. Arsana, *Indonesian Journal of International Law*, Vol. 21, No.2, Article 7, pp. 399–434, January 2024. Malaysia's Ministry of Natural Resources and Environmental Sustainability at Labuan confirmed, on 5 March 2024, that the boundary demarcation on Pulau Sebatik was progressing and should be completed by June 2024 (*Bernama*).

Chapter 2
Geographical Setting and Ocean Basin Environment

No water, no life. No blue, no green.
Dr Sylvia Earle. Founder of Mission Blue.
The World is Blue: How Our Fate and the Oceans are One.

Abstract The landmass that comprises Malaysia sits atop the Sunda Shelf and is almost surrounded by sea and straits of Southeast Asia. The narrative of this Chapter focuses on the geographical setting and the ocean basin environment. A suite of maps portrays the limits of the gulf, seas, and straits; the nature and quality of the seabed; the climate and marine meteorology over the regional seas; and the coastal morphology and tidal regime of Malaysia. Whilst the limits of the individual bodies of water (seas and straits) have no political significance the information is included in this study solely for the convenience of the user to gain an understanding of the extent of the regional seas and the locality when referring to maritime issues whether they are of a geopolitical intent or socio-economic context.

Keywords Sunda shelf · Limits of the sea · Climate · Marine meteorology · Marine environment

2.1   Introduction

In this chapter, Figs. 2.1, 2.2, 2.3, 2.4, 2.5, 2.6, 2.7, 2.8, 2.9, 2.10, 2.11, 2.12, 2.13, 2.14, 2.15, 2.16, 2.17, 2.18, 2.19, 2.20, 2.21, 2.22, 2.23, 2.24, 2.25, 2.26, 2.27, 2.28, and 2.29 inclusive, depicting the geographical setting and ocean basin environment, offer the reader a foundation for a greater understanding of the generalised climatic conditions and marine environment that prevail over the adjacent, semi-enclosed seas that abut the Malaysian coastline.

A few vital points sourced from the *Fact Sheets* that accompanied the Sixth IPCC (Intergovernmental Panel on Climate Change) Report, 2021, will assist in setting the discussion of Chap. 2. For the section focusing on South-East Asia and its adjacent seas, in the context of climate change, there is a *high confidence* that future regional warming will be slightly less than the global average. However, observed mean

© The Author(s), under exclusive license to Springer Nature Switzerland AG 2025
V. L. Forbes, *Malaysia's Maritime Jurisdictional Limits*,
https://doi.org/10.1007/978-3-031-78783-6_2

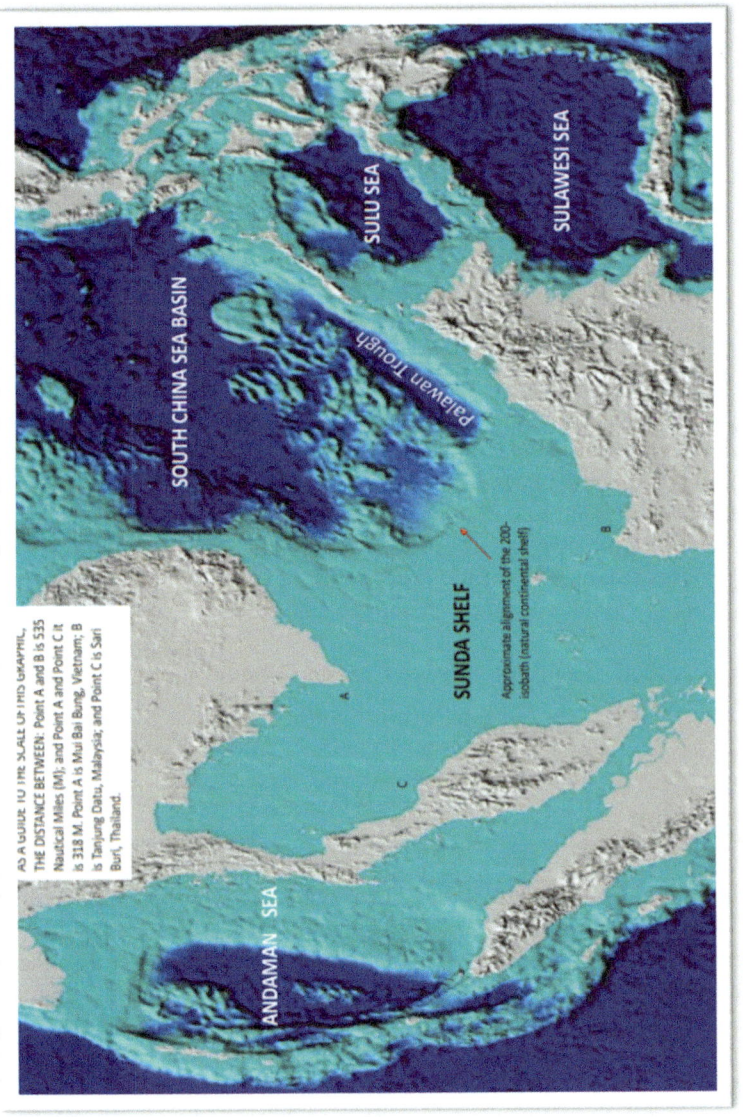

Fig. 2.1 An adaptation of a satellite-derived enhanced image of the Sunda Shelf. *Source* V. L. Forbes, 2014 and Malaysia Remote Sensing Centre, Kuala Lumpur

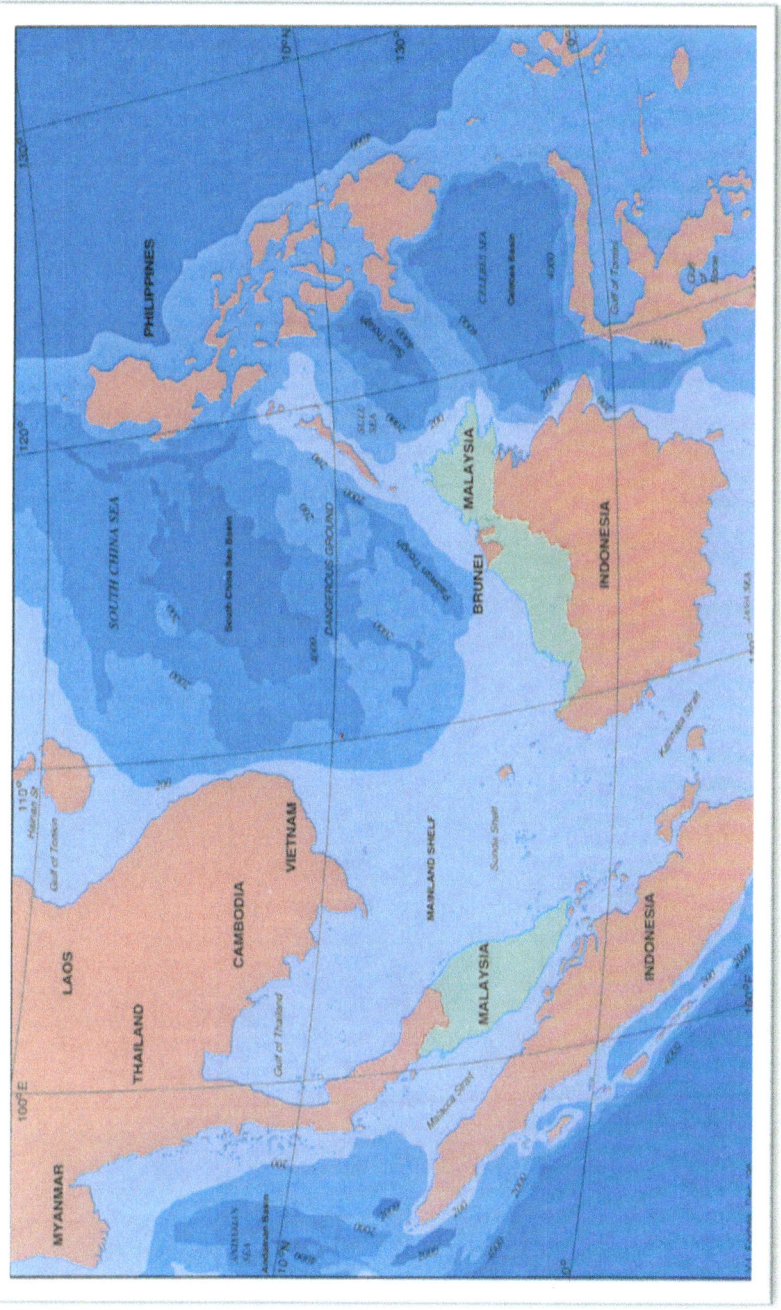

Fig. 2.2 A simplified depiction of the bathymetry of adjacent seas. *Source* Forbes and Basiron (1998, p. 31)[2]

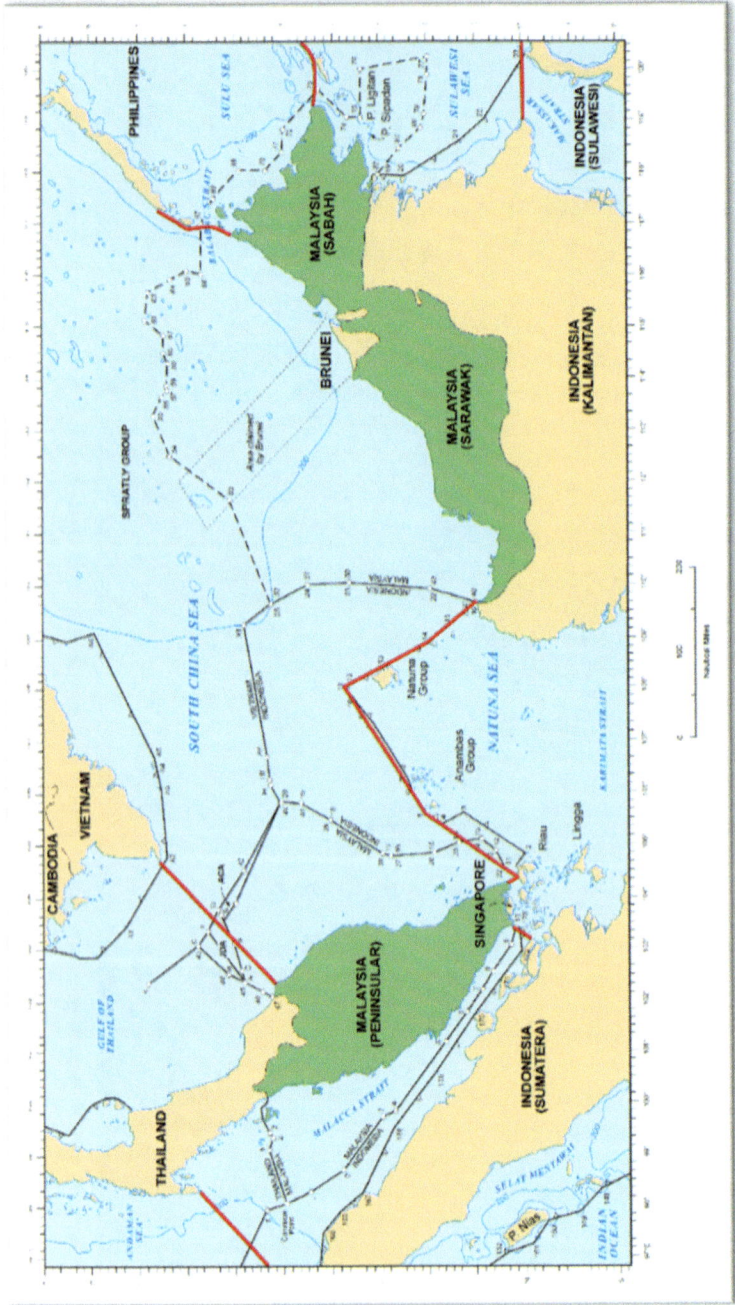

Fig. 2.3 Malaysia's adjacent seas and straits

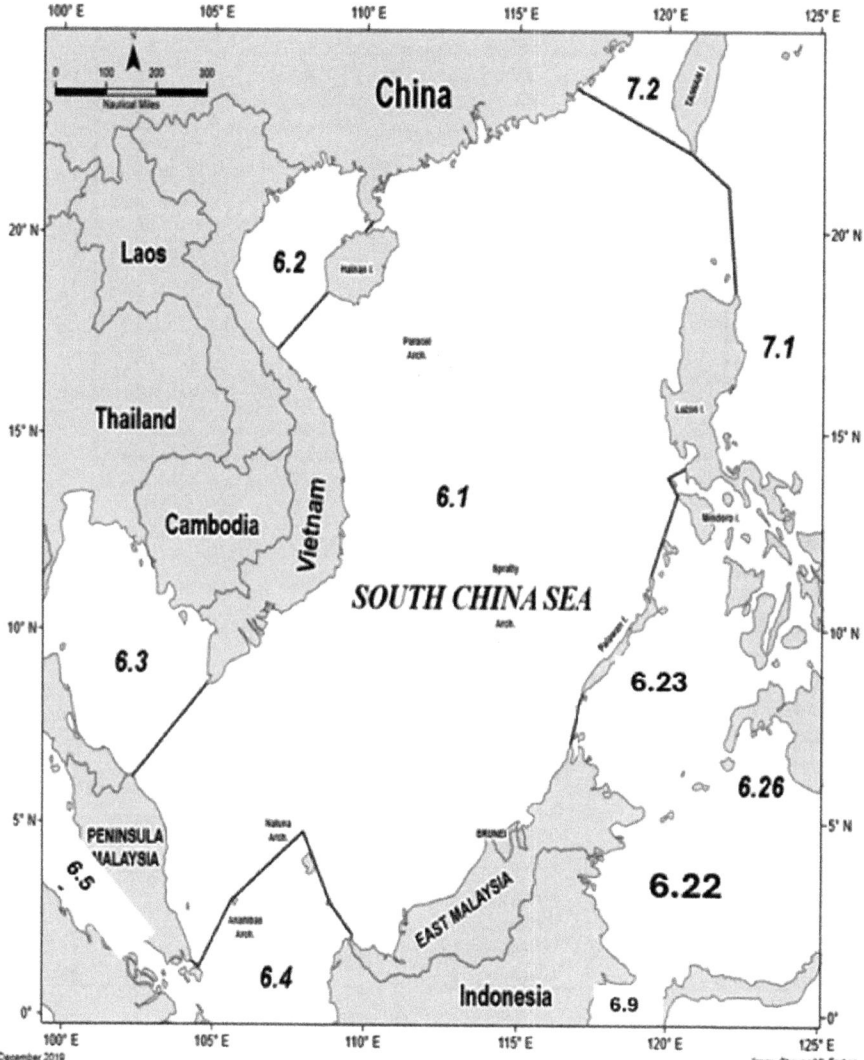

Fig. 2.4 The limits of South China and Eastern Archipelagic seas

rainfall trends are not spatially coherent or consistent across datasets and seasons. The prediction is that rainfall will increase in northern parts of the region and decrease in the Maritime Continent, stated with *medium confidence*. Sea level rise rates vary locally and around Malaysia it may equate to about 3 and 5 mm per year currently, depending on the location along the vast coastline.[1]

Compound impacts of climate change, sea level rise, land subsidence and local human activities will lead to higher flood levels and prolonged inundation in the Mekong Delta is *confidentially forecasted*. Although there has been no significant

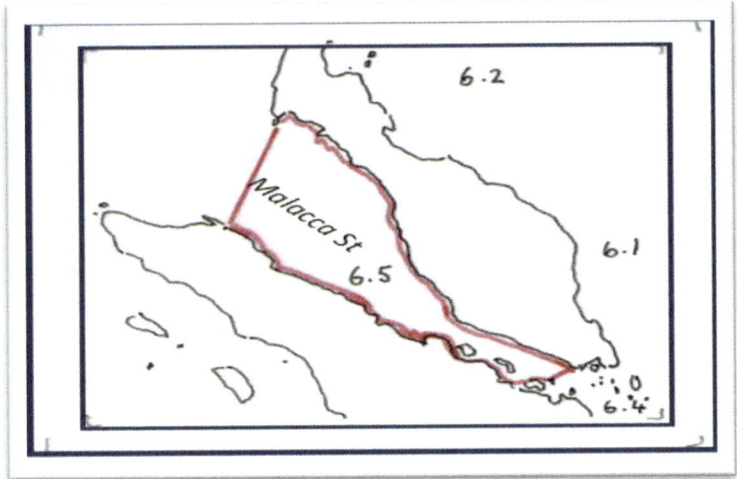

Fig. 2.5 Limits of Malacca Strait

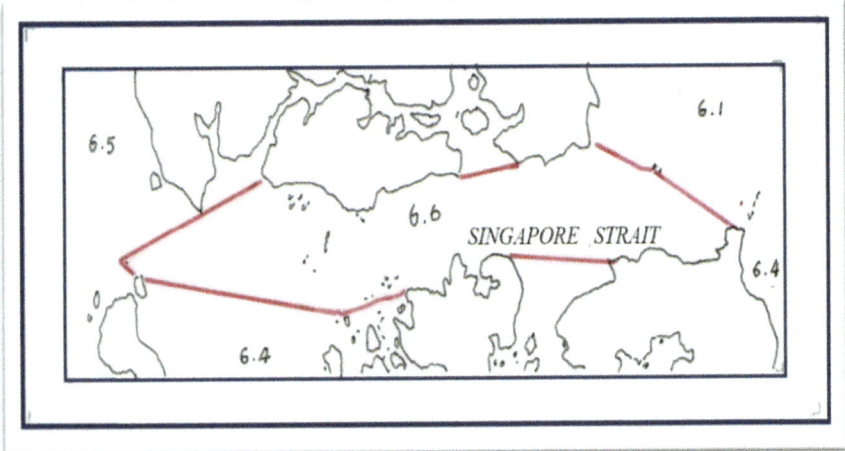

Fig. 2.6 Limits of Singapore Strait

long-term trend in the overall number of tropical cyclones, fewer but more extreme tropical cyclones have affected the region. Malaysia is not within the sphere of influence of cyclones, although heavy rainfall may be expected along the east and west coasts of Peninsula Malaysia.

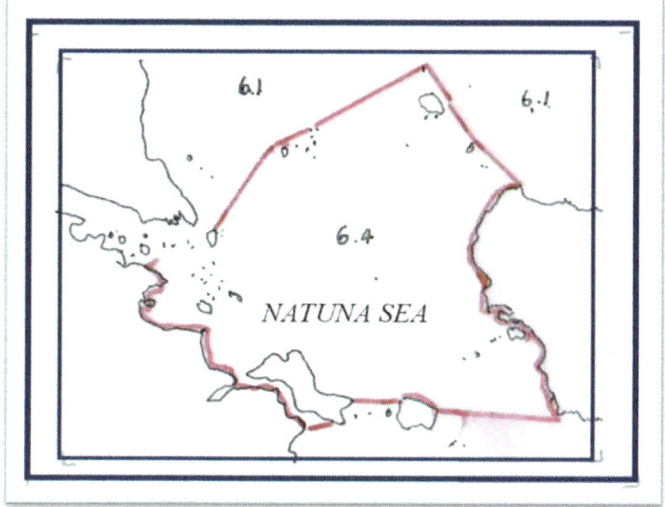

Fig. 2.7 Limits of the Natuna Sea

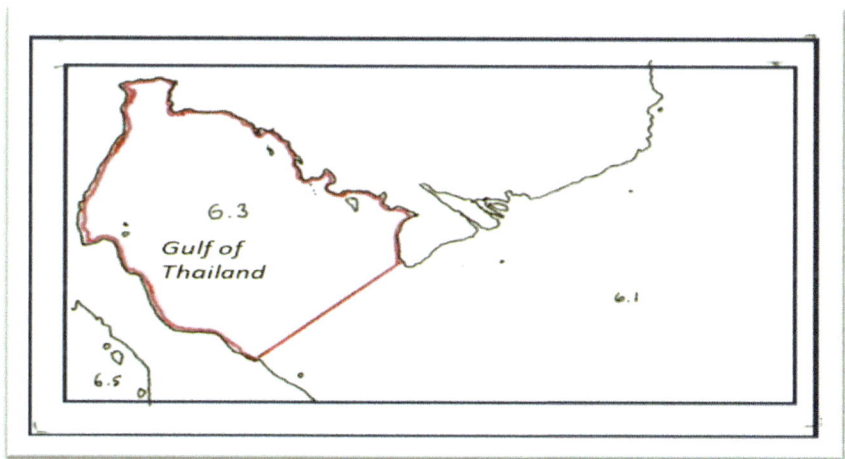

Fig. 2.8 Limits of the Gulf of Thailand

2.2 Sunda Shelf and Limits of Adjacent Seas

Figure 2.1 is an enhanced satellite-derived image of part of maritime Southeast Asia. The landmass of the mainland and some of the islands of the Indonesian and the Philippines archipelagos, depicted in the grey scale, conveys an impression of the topography of the land, namely the mountain ridges, relatively higher ground, and

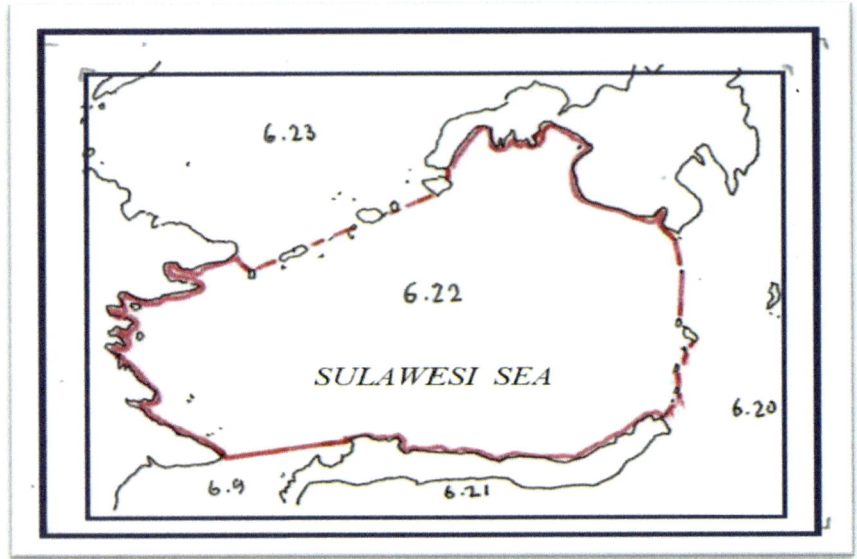

Fig. 2.9 Limits of the Celebes (Sulawesi) Sea

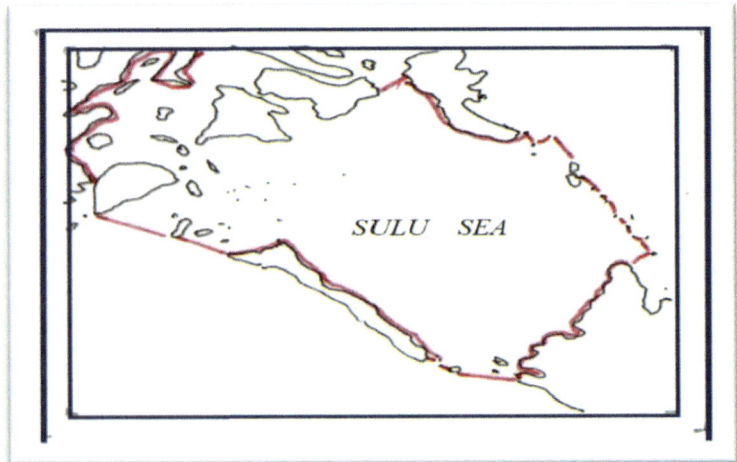

Fig. 2.10 Limits of the Sulu Sea

the coastal plains. The natural continental shelf and continental margin are shown in a light blue tinge and deeper waters in dark blue. A distinctive edge indicative of the limiting 200-m isobath in the South China Sea basin is apparent in this graphic. The Palawan Trench is almost parallel to the north-west coast of Borneo Island. Palawan

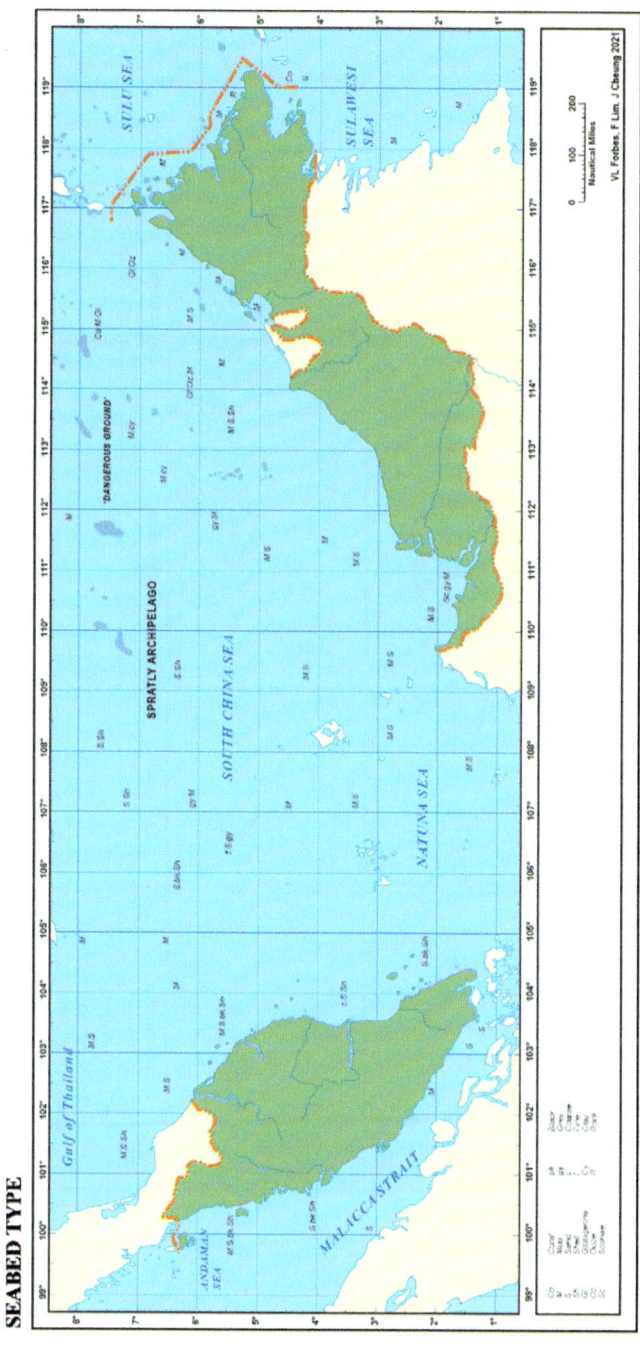

Fig. 2.11 Nature and quality of the adjacent seabed

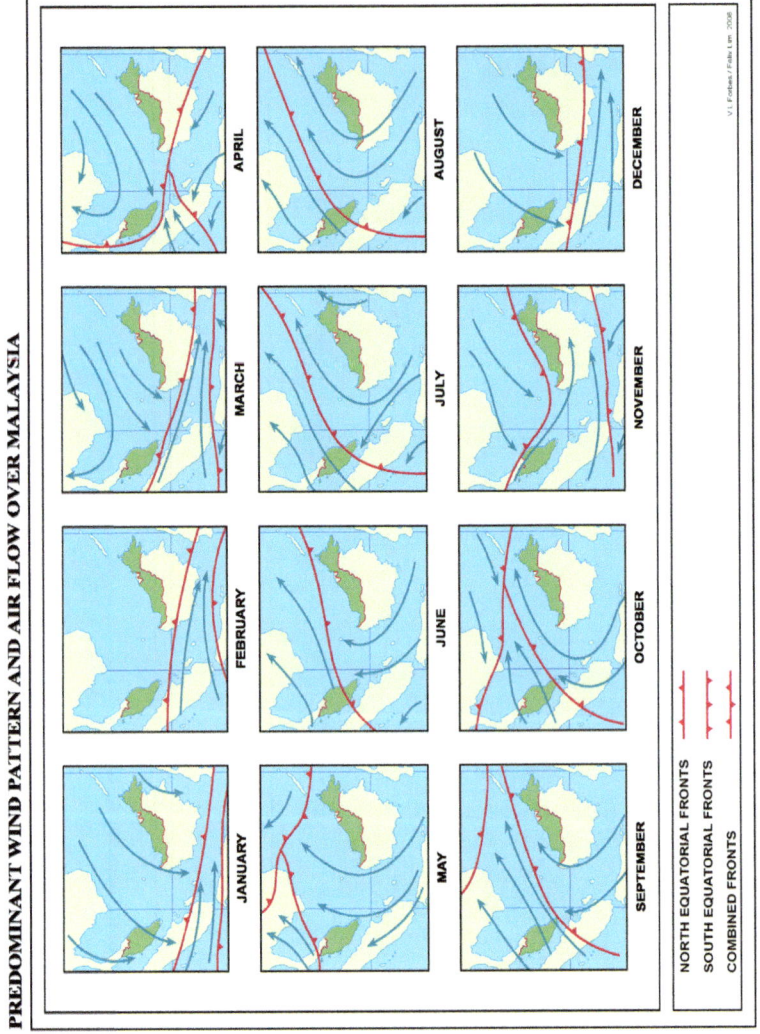

Fig. 2.12 A monthly depiction of the predominant wind flow

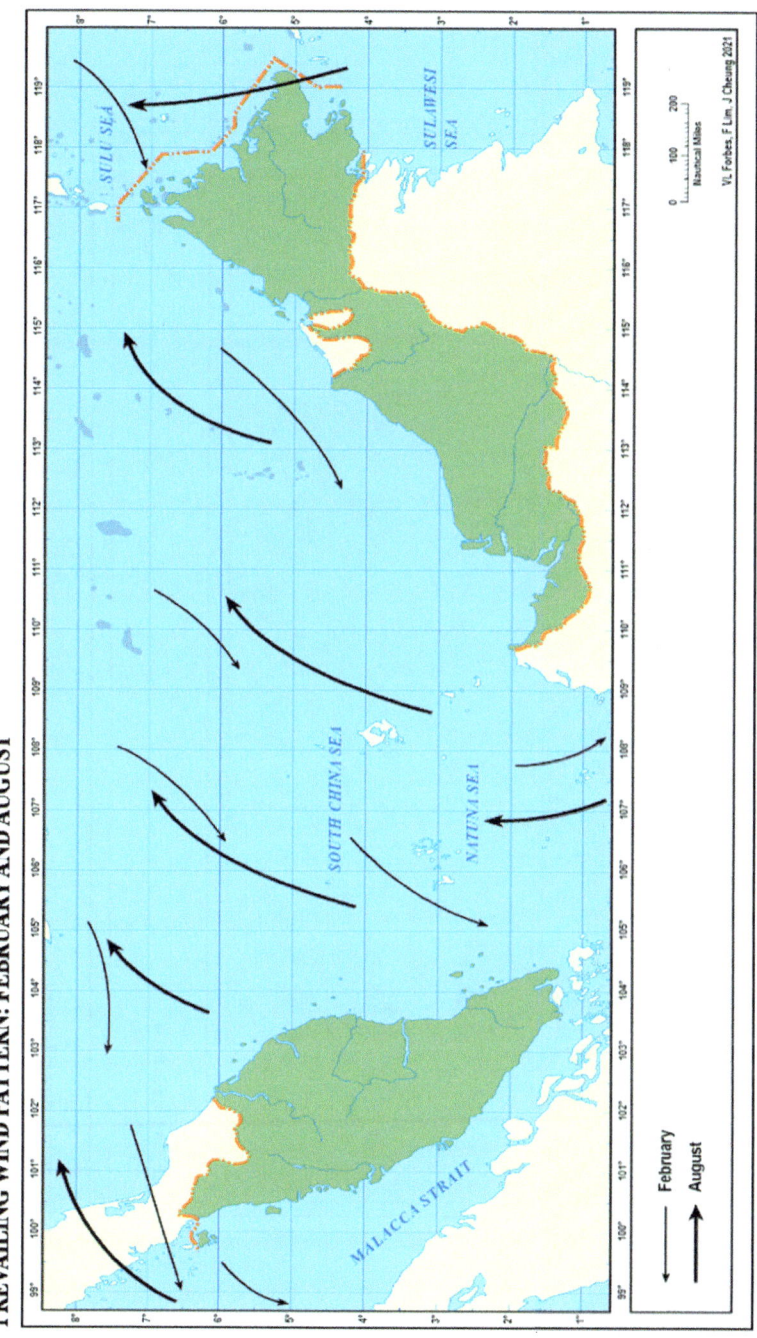

Fig. 2.13 Prevailing wind pattern: February and August

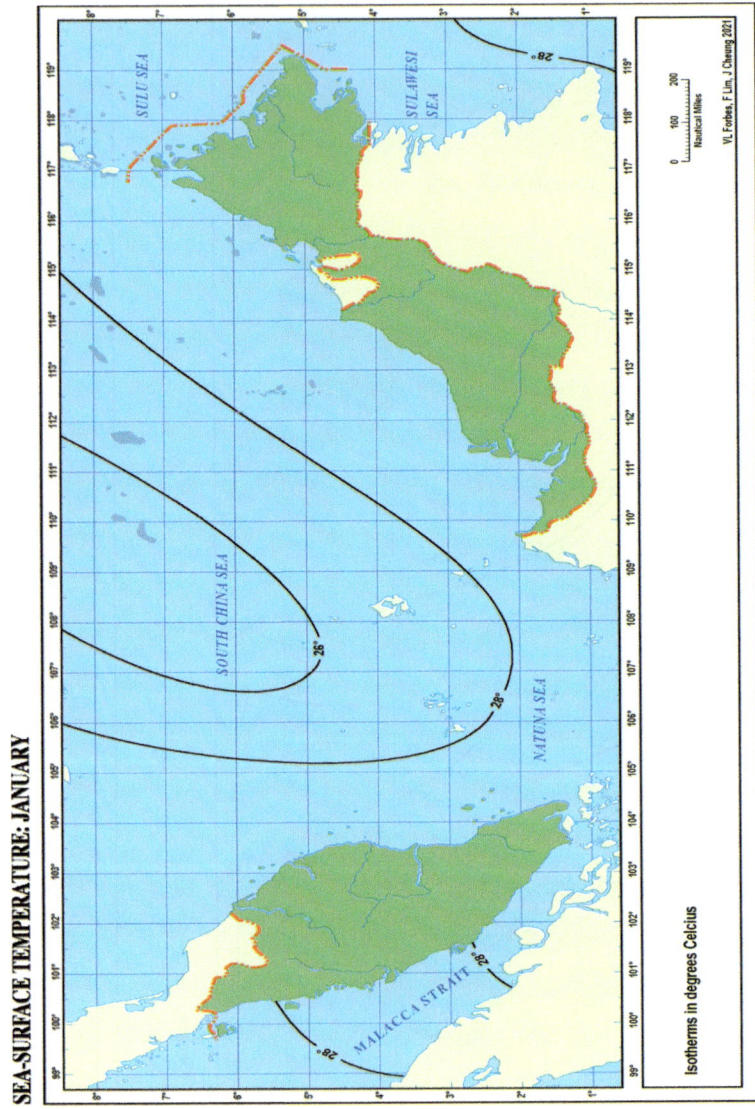

Fig. 2.14 Sea-surface temperature: January

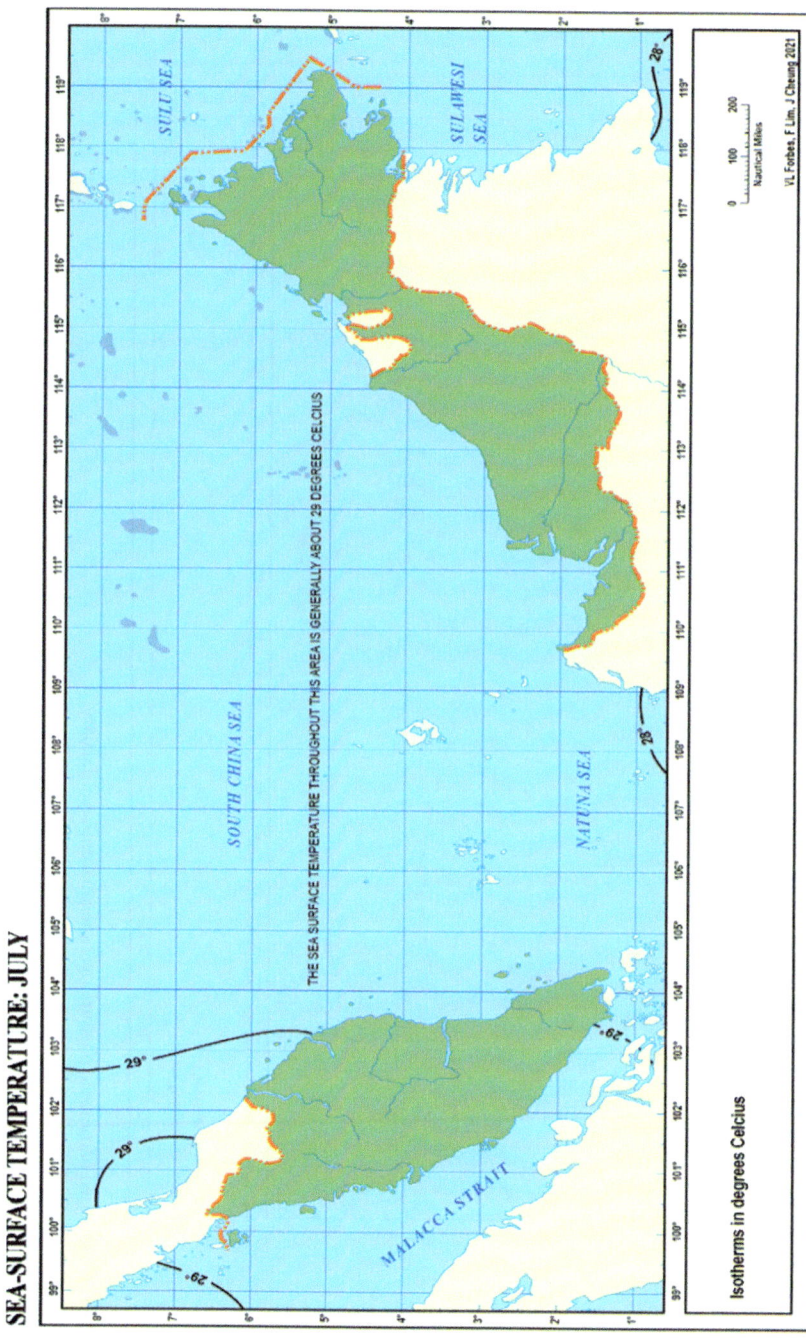

Fig. 2.15 Sea-surface temperature: July

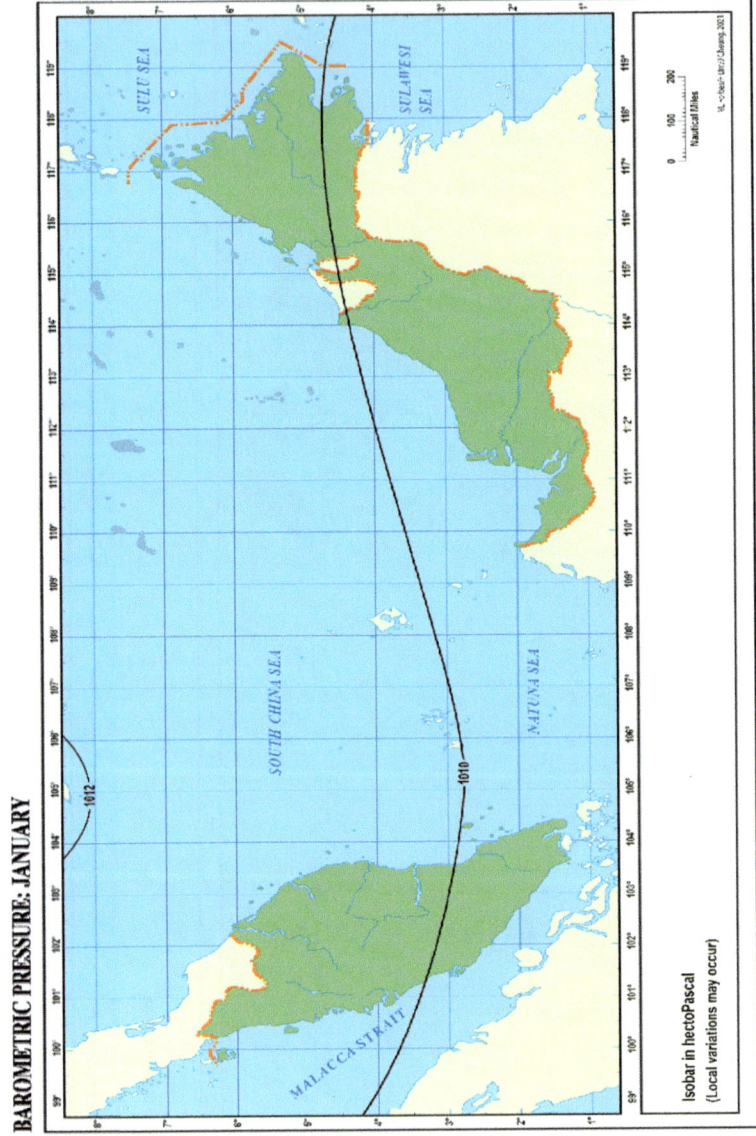

Fig. 2.16 Generalised barometric pressure over the adjacent seas: January

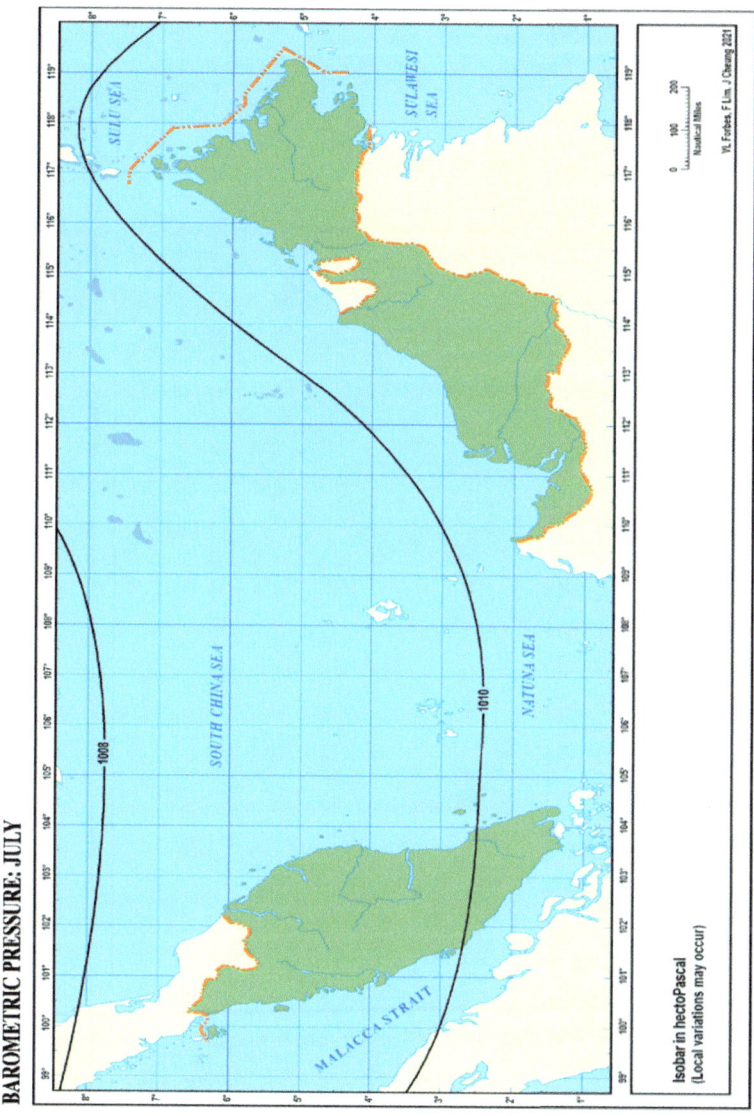

Fig. 2.17 Generalised barometric pressure over the adjacent seas: July

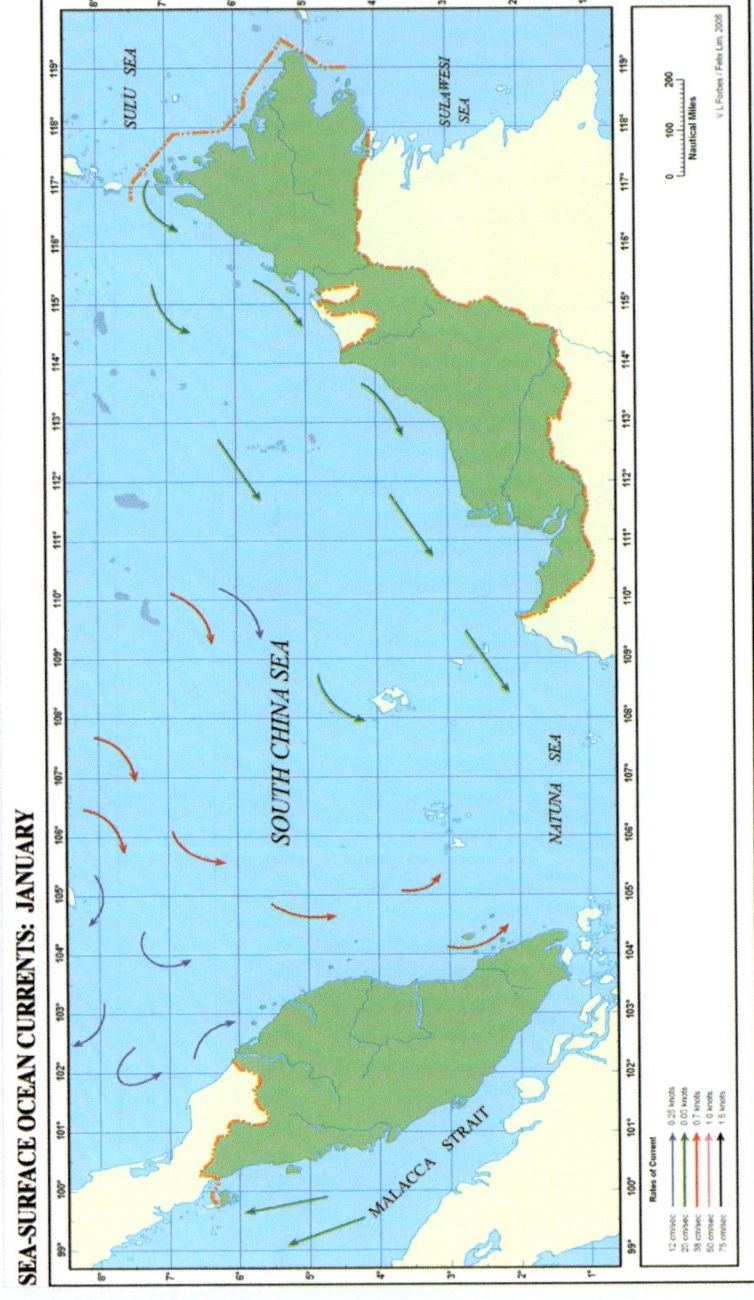

Fig. 2.18   Sea-surface currents over the adjacent seas: January

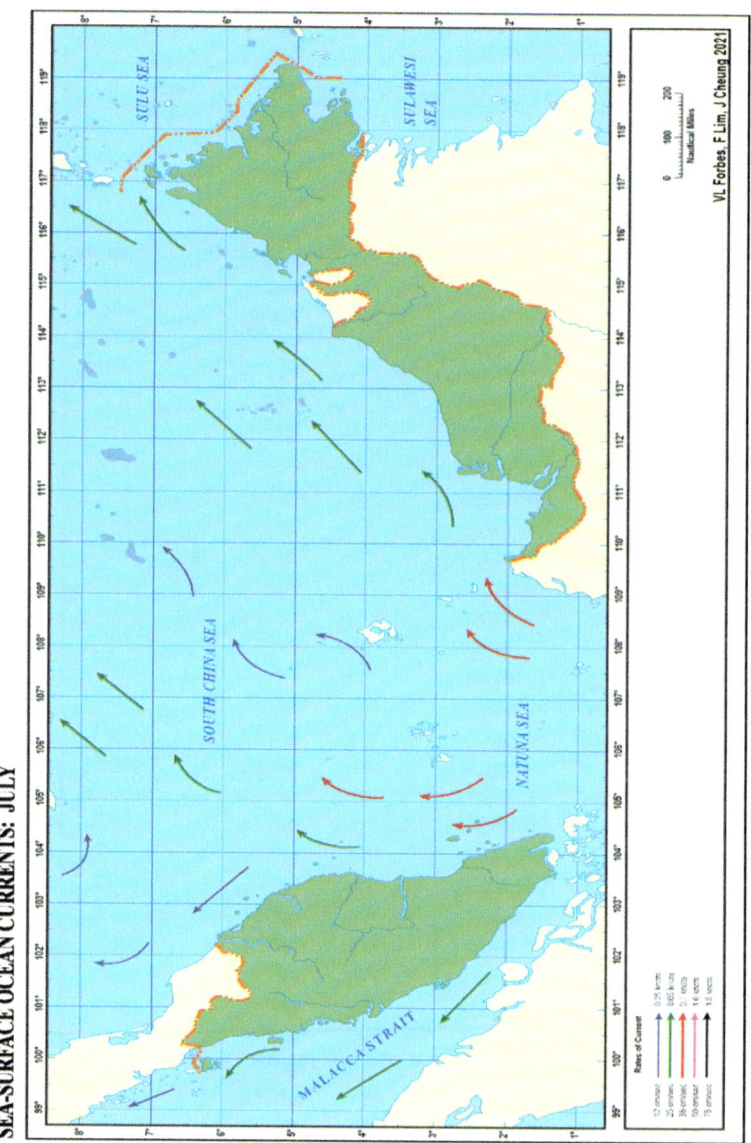

Fig. 2.19 Sea-surface currents over the adjacent seas: July

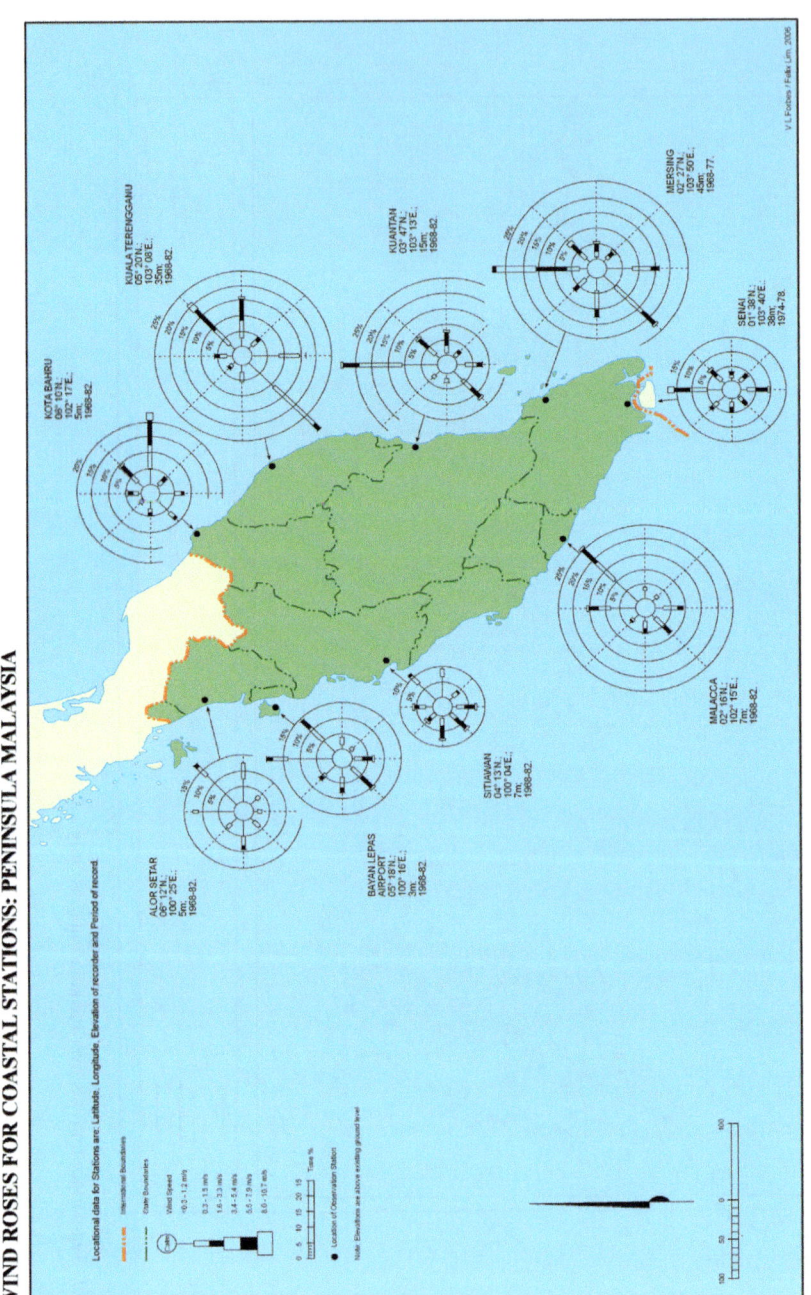

Fig. 2.20 Wind Roses for selected coastal stations of Peninsula Malaysia

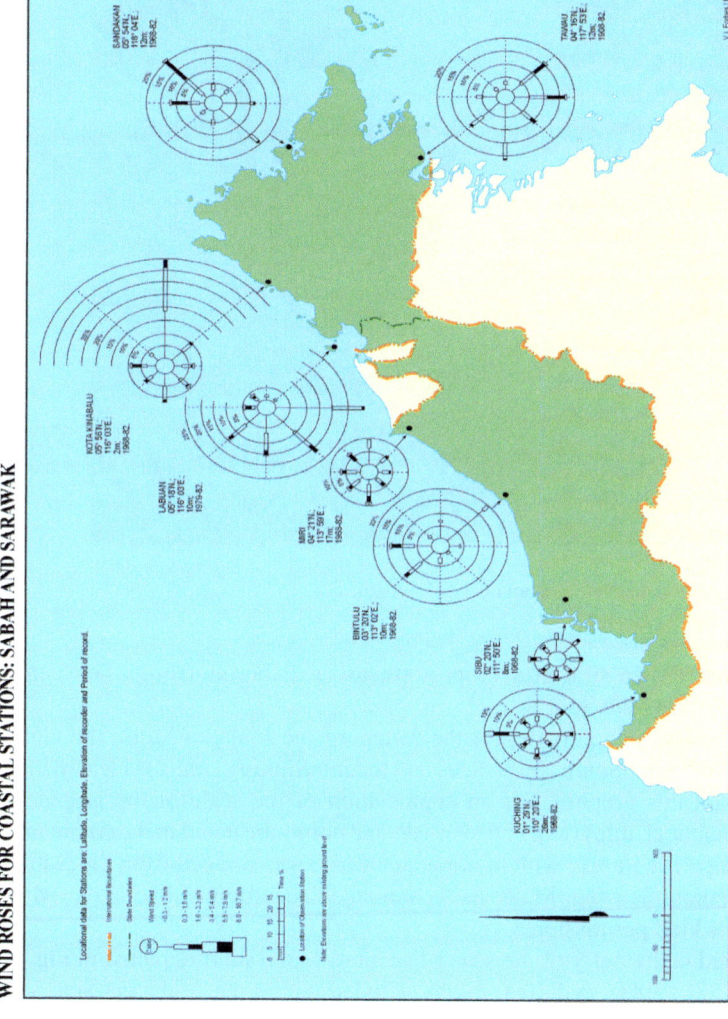

Fig. 2.21   Wind roses for selected coastal stations: Labuan, Sabah, and Sarawak

Fig. 2.22   East Pillar demarcating a boundary, P. Sebatik

Island is depicted as a slim grey strip, northeast of Palawan Trough between the South China Sea and the Sulu Sea.

The Sunda Shelf which underlies the waters of the Malacca Strait, the Gulf of Thailand, the Java and South China Seas and the interlinking straits is clearly shown and labelled on this graphic. For an appreciation of the scale of the graphic the distances, in nautical miles (M), between selected points are noted in the commentary box on the map. The image, with minor adaptations, was sourced from the National Centre for Remote Sensing, Malaysia, is annotated accordingly, and is reproduced here with their kind permission.[2]

A simplified bathymetry of the seas adjacent to Malaysia is depicted in Fig. 2.2. Isobaths for 200-, 2000- and 4000-m are included within the Andaman Sea, South China Sea, Celebes (Sulawesi) Sea and Sulu Sea basins are portrayed in Fig. 2.2. A greater portion of the Java Sea, Malacca Strait, the Gulf of Tonkin, Gulf of Thailand are underlaid by the Sunda Shelf.

The Mainland and Sunda Shelves, which lie between Vietnam and Borneo, slopes gently towards its edge. The Saigon, Conson, Mekong, Sarawak and Yingge Basins

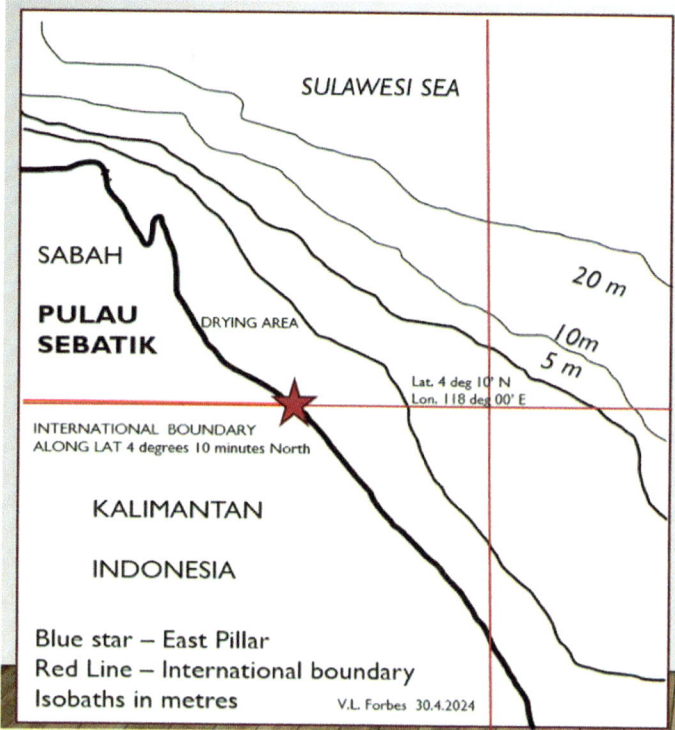

Fig. 2.23 Location of East Pillar (red star), P. Sebatik

are on these shelves. The width of this shelf varies: for example, north of the Natuna Archipelago and north of Point Sirik, Sarawak, it is about 150 M; off Bintulu, Sarawak it narrows to about 60 M; in the vicinity of Lat. 13° N off the east coast of Vietnam it is a mere 40 M in width; whilst southeast of Yulin, southern coast of Hainan Island it is about 45 M.

The limits of the seas and straits, shown as bold red lines on the following illustrations (Figs. 2.3, 2.4, 2.5, 2.6, 2.7, 2.8, 2.9, and 2.10, inclusive) in this chapter, have no political significance whatsoever. The illustrations are included in this study solely for the convenience of the user to gain an understanding of the extent of the regional seas and the locality when referring to maritime issues whether they are of a geopolitical intent or socio-economic nature. In other words, issues and implications that relate to human, political and physical geography, science, resource potential, economics, and international law.

The geographical limits of the South China Sea basin were defined by the International Hydrographic Bureau (now Organisation) in their *Special Publication* 23 of 1953. However, in a revised draft of *Limits of Oceans and Seas*, 4th Edition, of June 2002, the IHO officially recognised the Natuna Sea by approving the limits as

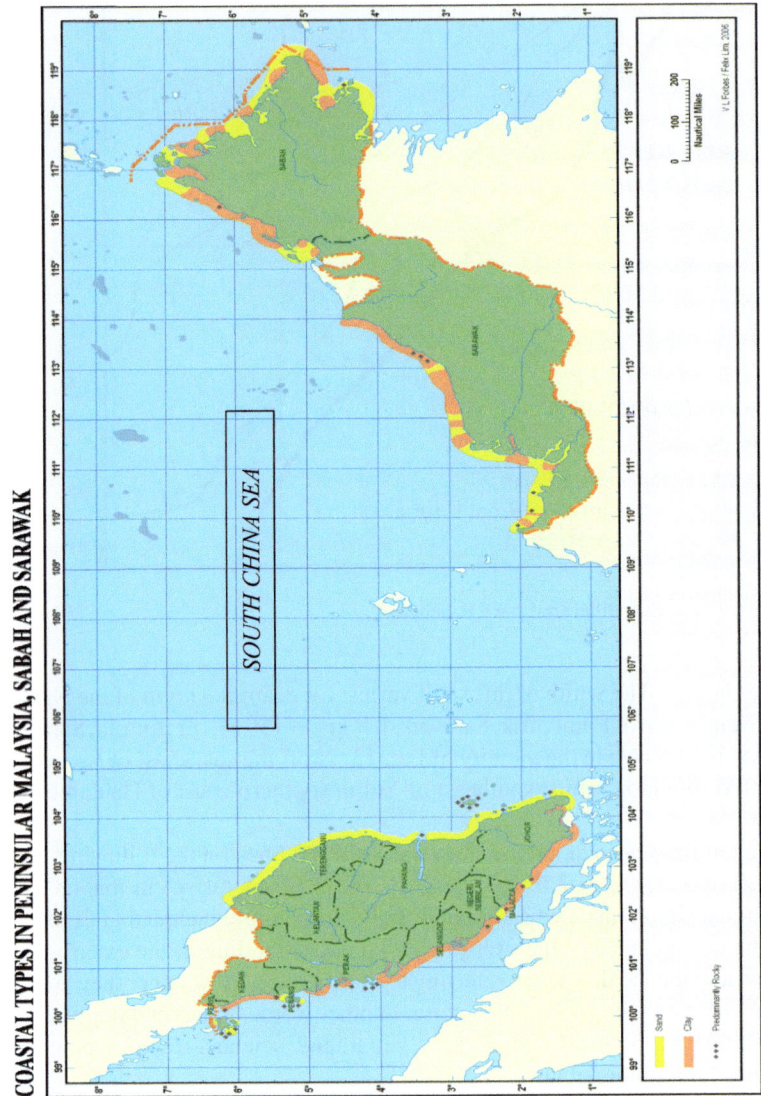

Fig. 2.24 Generalised coastal morphology of Malaysia

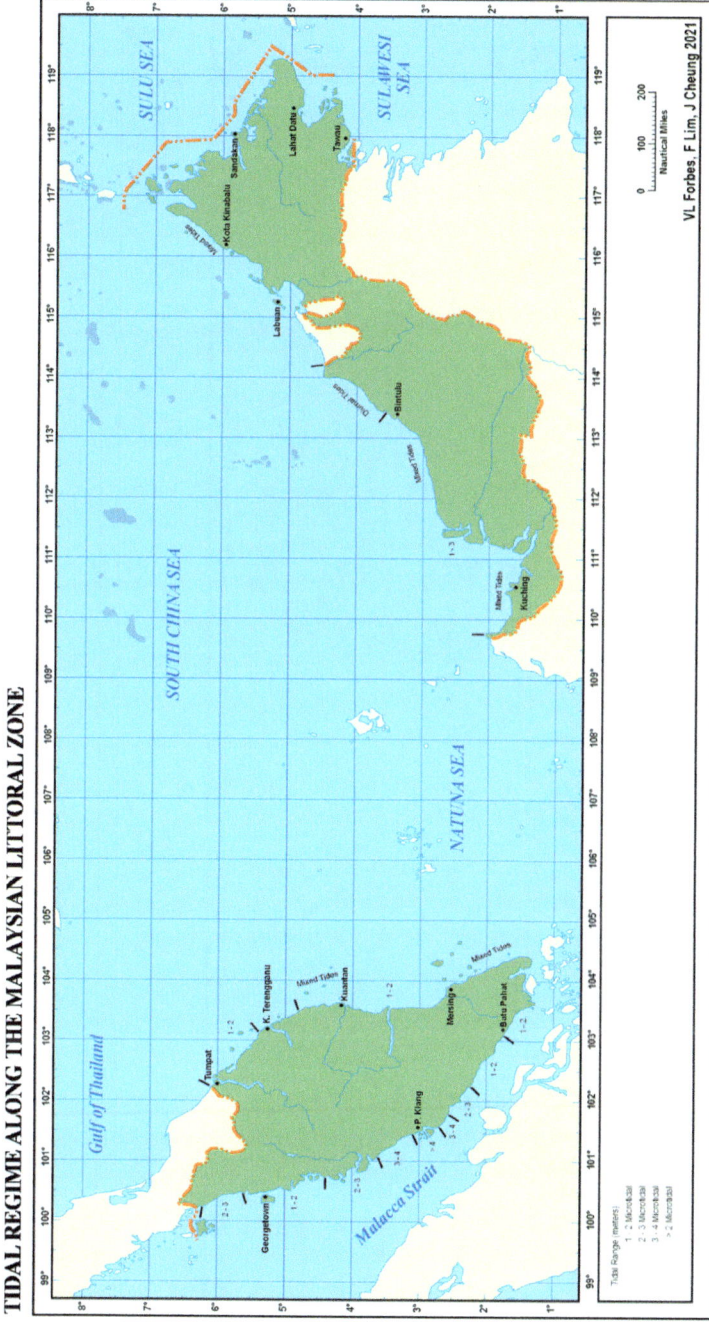

Fig. 2.25   Tidal regime along the Malaysian littoral zone

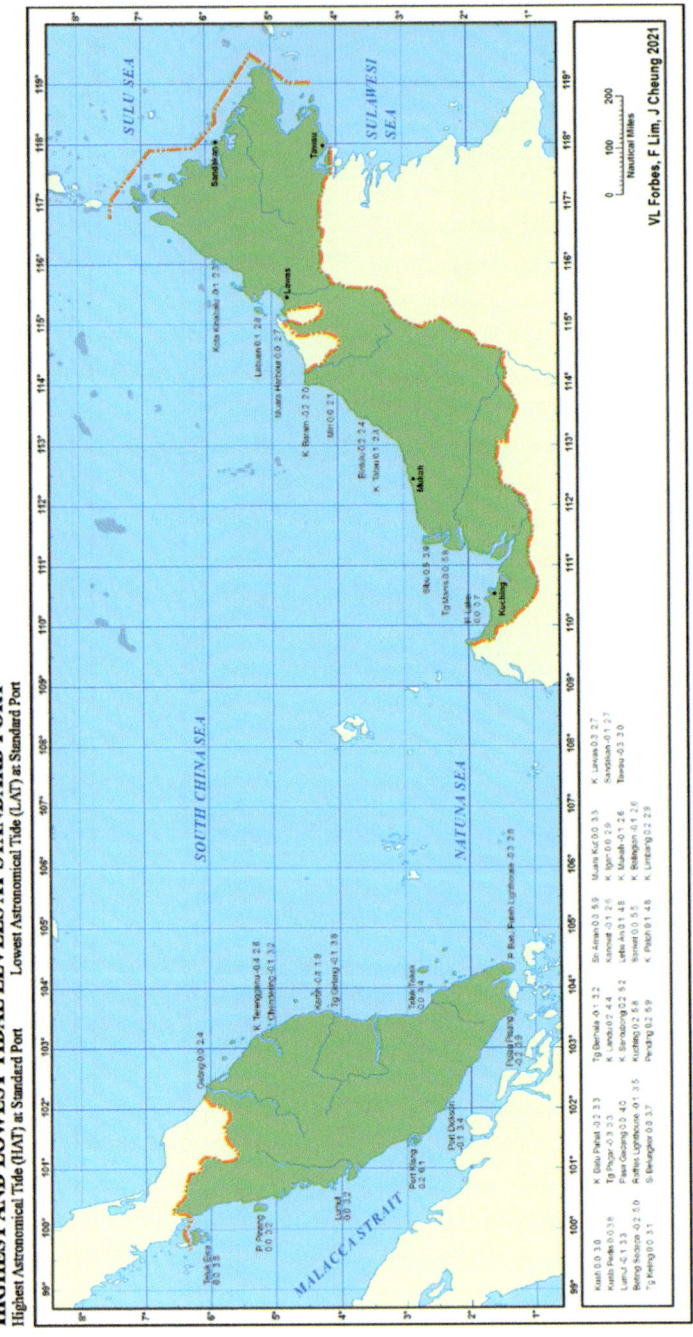

Fig. 2.26 Highest and Lowest tidal levels at Standard Ports, Malaysia

Fig. 2.27 High-tide level during Spring Tide (bore tide event) at Pulau Indah. *Source* Author; image captured on 17 September 2017, 18:04:48 h, MyST

Fig. 2.28 Low-tide level during Spring Tide (bore tide event) at Pulau Indah. *Source* Author; image captured on 18 September 2017 at 11:50:28 h (GMT + 8)

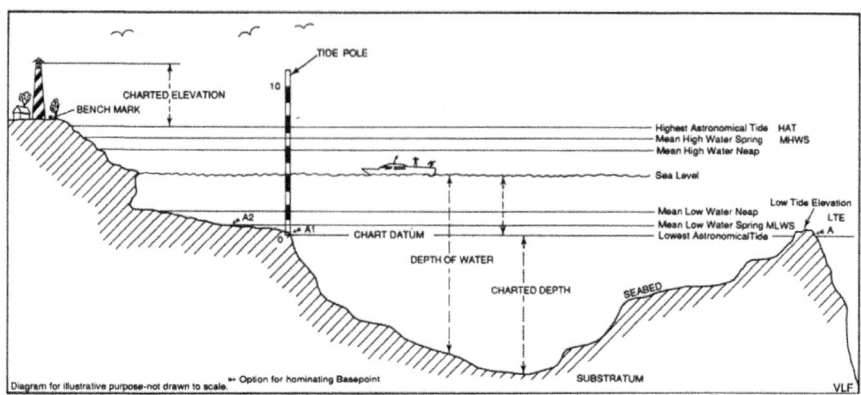

Fig. 2.29 The levels of the Tides of significance to navigation and the nautical charts

proposed (Fig. 2.4). The illustrations (*the sketches*) offered below within this narrative are reproduced from 'Names and Limits of Oceans and Seas' the Final Draft of the International Hydrographic Organisation's (IHO) *Special Publication No. 23*, 4th Edition, of June 2002 (as of 12th July 2021). No mention is made in this publication of the names 'East Sea' (Vietnam) and 'West Sea' (the Philippines) when these nations refer to their respective portion of the South China Sea.

To be consistent with IHO standards, for example, this study employs the name Malacca Strait and not the plural Straits of Malacca and Singapore Strait, except in the instance when referring to the collective Straits of Malacca and Singapore. Straits joining two seas have been allotted to one of the seas in accordance with the resolution of an International Hydrographic Conference held in London in 1919. The southern limit of South China Sea, IHO code number 6.1, is revised, northward, from the Bangka Belitung Islands to the Natuna Islands. A description of the geography of the littoral and of the limits of the South China Sea and adjacent Gulf, Seas and Straits are offered by Forbes (2017; and 2021, pages 1–26) and others.[3]

2.2.1 South China Sea (Code 6.1)

In summary, the common limit of the South China Sea in the north is with the southern boundary of the T'ai-wan (Taiwan) Strait (see Code 7.2). The eastern limit is coincident with the Philippine Sea limit (see Code 7.1); the southern limit of the sea is coextensive with the Sulu Sea (See Code 6.23).

The southwestern limit of the South China Sea shares a common limit with the Natuna Sea (see Code 6.4) and also corresponds with the limit of Singapore Strait (see Code 6.5). The northwest common limit is with the common limit with the Gulf of Thailand (see Code 6.3), and Gulf of Tonkin (see Code 6.2).

2.2.2   Malacca Strait (Code 6.5)

The Malacca Strait (Fig. 2.5) is situated between the southern coast of Thailand, the west coast of the Malay Peninsula and the east coast of Sumatera (Sumatra Island). The eastern limit of the Malacca Strait is along the western coast of the Malay Peninsula.

  The southern limit of the Strait is an imaginary line joining Tanjong Piai south-westward to Pulau Iyu Kecil in Indonesia. The western limit of the Strait aligns with the east coast of Sumatera Island and the northwestern limit is coincident with the southern limit of the Andaman Sea, see 5.13 and 5.14).

2.2.3   Singapore Strait (Code 6.6)

The northern limit of the Singapore Strait is bounded by the southern coasts of the Malay Peninsular and Singapore Island, the northern coasts of the islands of Karimun Kecil, Pemping Besar, Batam and Bintan (Fig. 2.6) The Johore Straits are not included in the definition used by the IHO. The eastern limit of Singapore Strait extends to Pedra Branca. The western limit of the Strait is coincident with the limit of the Malacca Strait, see Code 6.5).

2.2.4   Natuna Sea (Code 6.4)

The northern limits of the Natuna Sea are coincident with the south-west limit of the South China Sea (6.1) and encompasses the Anambas and Natuna Archipelagoes. Its eastern limit follows the western coast of Kalimantan. The limits of this sea, as defined by the IHO, do not touch the coast of Malaysia as illustrated in Fig. 2.7.

2.2.5   Gulf of Thailand (Code 6.3)

The Gulf of Thailand (Fig. 2.8), is bounded by the coasts of Peninsula Malaysia, Thailand, Cambodia, and Viet Nam and its southeastern limit is coincident with the South China Sea (see 6.1).

2.2.6   Sulawesi Sea (Celebes Sea) (Code 6.22)

The limits of the Sulawesi Sea are delineated in red as depicted in Fig. 2.9. The sea is bounded on the west by the eastern coast of Sabah (East Malaysia), to the north and east by the islands of the Philippines and to the south by the north coast of Sulawesi Island.

2.2.7   Sulu Sea (6.23)

The limits of the sea for the southeast, east, and northwest are depicted as the red lines that are aligned with the land mass, islands and reefs of the Philippine Archipelago. The western limit of the sea follows the eastern coast of Borneo Island as portrayed in Fig. 2.10.

2.2.8   Nature and Quality of the Seabed

The quality and nature of the seabed is depicted cartographically in Fig. 2.11. The oceanographic information for randomly selected spots provides a general overview of what mariners and scientists could expect when operating in the marine environment and geographical setting of the semi-enclosed seas and on the Sunda Shelf as encompassed on the map. The structure of the seabed and that of the littoral makes the region a complex mix of terrestrial and marine features.

The quality and texture of the material is self-explanatory; the nature of the seabed varies from tiny pieces of broken coral, mud, sand and shell to globigerina, ooze, and scoriae. *Globigerina ooze* is the dead shells of most foraminifera—fixed or free bottom-living animals. These planktonic foraminifera contain symbiotic chrysophycean algae, as do other planktonic rhizoflagellates. Foraminifera are valuable fossils, for the group have been so abundant and varied since early Cambrian, more than 550 million years ago, that they are important not only in paleo-ecological studies but also in the exploration for new petroleum reserves.

The composition and nature of the seabed off the east coast of Peninsula Malaysia is described in nautical guides and other oceanographic literature of the regional seas[4] (Forbes 2017, 2021).

In the revision of the 1979 *China Sea Pilot* (Hydrographer, 1982, 4th Edition, p. 19) six identified and named seamounts were listed for the South China Sea basin. By early-2024, as portrayed on Map C (above), there are many more as a result of the activities of international and regional hydrographical associations and agencies who have realised the importance of surveying and mapping the seabed and updating the bathymetric and marine information on electronic and paper-based charts and other graphics.

For example, due to the water exchange between the South China Sea and the Indian Ocean basin, strong tidal currents occur along the seabed, causing large uniform sand ripples to form in the Straits of Malacca and Singapore have been identified. The height of these ripples, or sand waves, which form at right angles to the current, are between 4 and 7 m and their wave lengths vary between 250 and 450 m. In addition, there are other large, long ridges running parallel with the direction of the tidal currents. Satellite-derived SPOT imagery and other remotely sensed imagery can deduce and delineate the shifting sandbanks in the Malacca Strait.

2.3 Climate and Marine Meteorology Over the Regional Seas

The climate and marine meteorology experienced over the seas adjacent to Malaysia are adequately described in various meteorological and nautical publications. A generalised predominant wind pattern and airflow over Malaysia for each month of a year is portrayed in Fig. 2.12. For each month there is also a depiction of the approximate alignment of the weather fronts in this regional focus. The fronts are termed North Equatorial, South Equatorial or Combined Fronts. Note the geographical movement of the fronts in relation to the land mass.

During late September and October annually, the monsoon reversal is experienced in Southeast Asian Seas: however, it is not until December that the Northeast Monsoon is firmly established across this region.[5]

An understanding of the mechanisms by which the atmosphere and ocean exchange heat, moisture and momentum is fundamental to human's understanding of the overall air-ocean interaction and to the further development of medium- and long-term weather forecasting techniques.

Generalised prevailing wind patterns for the months of February and August for this geographical region are symbolised by arrows to show the direction of the wind (Fig. 2.13). The static nature of the atmospheric conditions experienced over the semi-enclosed seas offer slight variation in this regional context.

Along the southeast coast of Malaysia, in the Malacca Strait and off the northwest coast of Borneo Island winds are from the northwest through northeast for about 65–90% of the time from December to March.

The distribution of surface water properties in the sub-tropic sub-area is influenced by an excess of evaporation over precipitation, cycles of heating and cooling, convective mixing, and currents. The tropic sub-area is characterised by higher water temperatures and is influenced by excess precipitation, daily cycles of heating and cooling, wind mixing, and currents.

Temperature is almost uniform throughout the year. Seldom does the difference between the highest and lowest mean monthly temperature exceed 2 °C. At coastal observation stations the temperature rises to 31 °C during the day and falls to about to about 23 °C in the early morning. The daytime heat is oppressive due to high humidity.

Given the equatorial setting of the semi-enclosed seas that abut the landmass of Malaysia, there is only a slight variation in the sea surface temperature. Figure 2.14 portrays the approximate alignment of the isotherms (lines joining points of equal value, in this case, for temperature of the sea surface) for 26 and 28 °C during the month of January.

Along the west coast of Malaysia Peninsula, the climate is hot and humid throughout the year. During the Northeast Monsoon (November to March) showers and thunderstorms develop in the afternoon and tend to drift towards the coast in the evening. From May to September (the period of the Southwest Monsoon) showers and frequent thunderstorms occur during the daytime.

The mean sea surface temperature ranges between 29 °C in May to a minimum of 27 °C in February. In general, the sea is slightly warmer than the air above it.

Figure 2.15 shows the isotherms for the values of 28 and 29 °C of the water-surface for the semi-enclosed seas in the context of this atlas for the month of July. In this month there is little variation over the whole area apart from a slight variation between the coastal waters and the open sea.

The temperature of the air near the sea surface seldom varies more than one degree Celsius above or below the sea surface temperature. Sudden temporary changes are encountered during violent rain squalls and at the passage of frontal troughs generally further north of the physical limits of the map.

There is a relatively regular seasonal change in the average monthly pressure over most of East Asian seas. The rise in pressure in the winter months of the Northern Hemisphere is due to the anticyclone which forms over southern Siberia. These two seasonal pressure systems are large enough to be the dominating influence on the coasts. There is a corresponding decrease in pressure in Spring (in the Northern Hemisphere) to give a low-pressure value of about 996 hectoPascal (hPa) over Northern India in July. The seasonal variation of pressure varies according to Latitude, being very pronounced in the North, and almost marginal towards the Equator.

Average barometric pressure for the areas over the seas shown in Figs. 2.16 and 2.17 is between 1008 and 1010 hPa as delineated by the isobars in all months with a regular diurnal range of about 3 hPa, but further north there is a seasonal variation from 1004 hPa during the summer to 1020 hPa in winter. This latitudinal reversal of the pressure gradient is responsible for the Summer and Winter Monsoonal patterns.

Average barometric pressure for the areas over the seas shown on these two maps is between 1008 and 1010 hPa as delineated by the isobars in all months with a regular diurnal range of about 3 hPa, but further north there is a seasonal variation from 1004 hPa during the summer to 1020 hPa in winter. This latitudinal reversal of the pressure gradient is responsible for the Summer and Winter Monsoonal patterns.

Apart from the irregular variations of pressure, there is a regular diurnal variation with a maxima at 1000- and 2200-h local time and a minima at about 0400- and 1600-h local time. This underlying fluctuation has an amplitude of about 3 hPa from maximum to minimum.

The movement of the surface water over the South China Sea is related, in general, to the monsoons, although the relationship is complex and not direct (Fig. 2.18). The main southwest setting current during the Northeast Monsoon (November to March).

The direction of the water movement is controlled largely by the locations of eddies which occur in the South China Sea in most months and by the flow of water to and from the Sulu Sea and through the Karimata Strait and Sunda Strait since these geographical marine features are controlled by varying oceanographic and meteorological conditions, both within and outside the area under discussion.

During the month of January, a considerable volume of water enters the South China Sea from the Sulu Sea. During February and March, with the NE Monsoon generally weakening, the main Southwest current slowly declines and by mid-March the mean rate is reduced to less than 0.5 knot. A month of transition between monsoons, the predominant flow starts to recurve during April off the coast of Peninsular Malaysia. The SW Monsoon over the southern half of the South China Sea begins in May. The flow of water is now from the Java Sea and as the month progresses a NE-going current becomes increasingly evident on the west side of the South China Sea and there is a slight increase in rate of the current.

Over the greater part of the area the currents are weak; the mean rates over most regions in most months are less than 0.5 knot. During the months when the Monsoons are fully developed, January and July, the mean rates increase to one knot or slightly more; however, no currents have been reported of more than two knots.

In June, the SW Monsoon extends over the whole South China Sea and from this month to the end of September the current sets between North and East as illustrated in Fig. 2.19. Within the Malacca Strait the predominate direction trends North-westward. The Malacca Strait is relatively shallow, the greater part of the area having depths of less than 80 m and therefore the main movement of water is from tidal influences. Though the predominant direction in the Malacca Strait is north-westerly, currents from all direction have been observed and the percentage frequency of the predominant flow is never high.

The Northeast setting current in the Southwest Monsoon (May to September) generally flows along the west side of the South China Sea. In July, during the height of the Southwest Monsoon, there is an absence of eddies in the South China Sea. The flow between the South China Sea and Sulu Sea becomes more complex in August, when a clockwise pattern of currents is observed to prevail around Mindoro.

Conditions are usually favourable for the development of sea breezes during most of the year. The sea breeze begins around mid-morning and reaches a maximum in the afternoon, fading away at sunset. The resultant speed in any locality depends on a good deal on the Monsoon direction so that a combined effect can produce a fresh breeze of about 20 knots or just a light and variable wind when the components are opposed. The decisive factors are the morphology of the coast and the nature of the adjacent countryside and the amount of solar radiation reaching the ground. The sea breeze extends some 10 and 20 M both inland and out to sea and, in favourable circumstances, may extend even farther.

Wind Rose patterns for nine selected coastal stations located on Peninsula Malaysia are depicted in Fig. 2.20. The name, geographical coordinate values of

Latitude and Longitude, elevation in metres of the recorder and the time frame of recording the observations are given for each station.

Each Wind Rose portrays the direction, speed, and the percentage in time for each cardinal and inter-cardinal point of the compass. The map also delineates the approximate alignment of the boundaries for each of the States in the Federation.

Peninsula Malaysia is just a portion of Continental Southeast Asia and as such experiences maritime conditions. Within the peninsular there are extensive lowlands and high mountain ranges. The climates of Continental Southeast Asia are generally controlled to a significant extent by the system of the Asian monsoons. Near the Equator where conditions vary but little from the mean, average values are representative of the actual conditions. Local conditions may also vary from one coastal town to another.

Each Wind Rose portrays the direction, speed, and the percentage in time for each cardinal and inter-cardinal point of the compass. The map also delineates the approximate alignment of the boundaries for each of the States in the Federation.

The land breeze forms on clear quiet nights but normally the speeds are less than the daytime sea breeze. Usually, the speed is in the range of 4 to 8 knots, but where the land slopes steeply to the coast the land breeze gets an additional energy.

Gales are infrequent. Over the greater part of the area depicted on the two maps only about one per cent of ship observations record winds of 30 knots or over. The frequency increases to five per cent in the northern approaches to Malacca Strait from June to September.

Tropical Revolving Storms very rarely reach the area covered on these maps. Only a couple of TRS have been experienced in the extreme north of these maps in the vicinity of the Malaysia/Thailand eastern terrestrial boundary.

Figure 2.21 has Wind Roses for eight stations. Four are located along the coast of Sarawak, one on Labuan and the remaining three along the coast of Sabah. The name, geographical coordinate values of Latitude and Longitude, elevation in metres of the recorder and the period of recording the observations are given for each station.

2.4   Coastal Morphology and Tidal Regime of Malaysia

2.4.1   The West Coast of Peninsula Malaysia

The coast between Ko Phuket, Thailand and Pulau Pinang, Malaysia, about 180 M south-southeast, is generally low with some hills near the mountainous island of Langkawi. Islands and islets project widely from the coast, fringed by a coastal bank. The Butang Group is wooded and uninhabited, except for a small fishing village. Pulau Langkawi is about 15 M wide, mountainous, and densely wooded.

To the southwest of Pulau Langkawi lies Pulau Perak, a barren white rock about 115 m elevation. The rock lies nearly midway in the northern sector of the Malacca Strait. The island of Pulau Pinang is separated from the mainland by a strait that varies in width from 1.5 to 7 M wide which provides a sheltered anchorage. The northern part of Pulau Pinang is mountainous.

Further south, there is a practically continuous strip of mangrove forest which varies in width from 0.5 to 8 M between Tanjung Piandang and Tanjiung Batu. The port of Lumut is approached through River Passage and the Sungai Dinding. There is a commercial port and a naval base at Lumut.

2.4.2 The East Coast of Peninsula Malaysia

The East coast of Peninsula Malaysia between Sungai Kelantan and Tanjung Penawar, 310 M SSE, is characterised by low swampy areas with numerous rivers discharging into the sea. Coastal ridges and hills extend to the coast at isolated points. In general, the off-lying islands are quite high, wooded, and good landmarks for coastal navigation.

Sungai Kelantan has two main entrances. The Sungai Kelantan Delta is low and featureless and is backed by numerous lagoons and waterways. The port area of Tumpat is sheltered by a curved sandpit. A shoal exists about 2.5 M NNW of the port.

The coast between Sabak and Sungai Besut, 25 miles SE, is low, sandy, and bordered by coconut trees. Depths of less than 11 m extend up to 4.5 miles offshore, shoaling gradually shoreward. The coast between Sungai Besut and Tanjung Merang, 30 miles SE, is low and bordered by sandy beach. Kuala Setiu Bharu, an inlet along the coast, lies 17 miles SE of Sungai Besut. Hinterland along this coast, a range of mountains, with many peaks, attaining its greatest elevation at Gunong Lawit, 1519 m high, about 17 miles SW of Kuala Setiu Bharu.

The coast between Tanjung Merang and Kuala Terengganu, 16.5 miles SE, is low, but at Batu Rakit the elevated land approaches the coast. The coast from abreast Pulau Kapas to Tanjung Dungun, 28 miles SSE, is flat, covered with jungle and scrub, and backed by many low hills. A range of wooded hills is located 3.5 miles inland, in the S part of this coast. A rocky headland, with an elevation of 203 m is in the vicinity of Tanjung Chenering. Parts of the coast is low and featureless.

2.4.3 Off-Lying Islands and Dangers

Between Latitudes 2° and 6° N a series of distant offshore islands, islets, rocks, and reefs exist, non-continuous, that in parts, semi-mask the East coast of the Peninsular. Many of the islands have an elevation more than 70 m, for example Pulau Kapas is densely wooded and has an elevation of 124 m.

2.4.4   The Coast of Sarawak

The northwest coast of Borneo, from Tanjong Datu to Tanjong Sempang Mangayau, which lies about 415 M northeast of Tanjong Datu, comprises Sarawak, Brunei, and a portion of Sabah. The coast is relatively regular in configuration; however, it is indented by two large bays. The larger of these bays is located between Tanjong Datu and Tanjong Sirik, a headland about 110 M to the east-northeast. The other is Brunei Bay, located about 145 M southwest of Tanjong Sampangmangio.

Several rivers discharge into the sea along this section of coast, with many of them being navigable for some distance inland by small craft. Much of the coast is fronted by tidal mud or sand flats, however, it a relatively free of fringing reef. The coastal land is relatively low in elevation, however, about 25 M inland a mountain range extends in a northeast direction and terminates at Mount Kinabalu, which has an elevation of 4100 m.

Labuan Island stands at the entrance of Brunei Bay. The port and town of Victoria is located on Labuan Island. Kuching, Bintulu, and Miri are cities/towns of Sarawak, and Kota Kinabalu (formerly, Jesselton) and Kudat are the principal towns on or near the coast of Sabah.

Mobile oil drilling rigs may be encountered off the northwest coast of Sabah and Sarawak, between the 200 m isobath and the coast.

2.4.5   The Coast of Sabah

Along the southeast coast of Sabah, the land bordering the area from Dent Haven to Sandakan Harbour, about 75 M, is relatively high, rugged, and fronted by numerous islands, islets, rocks, reefs, and other dangers. Near Dent Haven, the land on the northeast side of Darvel Bay is low and densely vegetated. Some hills of moderate height rise to the west of Dent Haven. A few scattered hills of moderate elevation appear further inland.

The coast between Sandakan Harbour and Tanjong Naruntung, about 94 M north-west, is indented by several large bays and fronted by numerous above and below-water dangers that extend up to 20 M and more offshore in places. Some high peaks are conspicuous and serve as useful navigational landmarks in the hinterland.

The coast between Pulau Ahus and Batu Tinagat, about 27 M in a north-northeast direction, is indented by several rivers. Several large islands lie in the approaches to these rivers and Cowie Harbour. The port and town of Tawau stands on the north side of the entrance of Cowie Harbour. Pulau Sebatik, a large island about 20 M in length and a maximum width of eight miles, lies on the south side of the channel. The northern portion of the island falls under the sovereignty of Malaysia, the remainder of the island belongs to Indonesia. The dividing line is the parallel of Latitude 4° 10′ N which is about 24 km in length (Refer to Figs. 1.6, 1.7, and 1.8 relating to the demarcation of the boundary and location of East Pillar in Fig. 2.22 and 2.23).

Further eastward of Pulau Sebatik and north of Lat. 4° 10′ N are the islands of Ligitan and Sipidan. The former has an elevation of about 9 m and partially vegetated and located on the south part of the main reef. The channels within the reef complex have not been extensively examined and hence caution is offered to mariners attempting to navigate within the vicinity of the reef complex. Pulau Sipidan, an oceanic island, lies to the southwest of P. Ligitan. P. Sipidan has an elevation of about 50 m. The island stands on the northwest side of a steep-to reef that lies about 7 M south of Pulau Mabul.

The predicted tide levels experienced along the coast of Pulau Sebatik during the Spring Tides could range from a low of 0.3 m to high of 3.1 m, whilst at Neap Tides it could range from 1.0 to 1.7 m.

A generalised classification of the coastal morphology of Malaysia, as the map scale permits, is depicted in Fig. 2.24. Two types are offered on the map, clay, and sand.

Figure 2.25 purports to show the tidal regime along the East and West coasts of Peninsula Malaysia and along the coasts of Sabah and Sarawak. Tides are mainly diurnal, sometimes semi-diurnal throughout the year but in places mixed regimes exist.

The overall tidal set (tidal flow) in the Malacca Strait is to the Northwest, however, from May to September there is a tendency for Southeast sets to prevail in some North and central parts, but the predominance is very slight. On the average, between 50 and 60% of all current observations in the Malacca Strait are 0.5 knot or less. A small portion of these observations exceed two knots. In the Northern part of the Malacca Strait, the general directions of the tidal currents are Southeast and Northwest. The South-easterly stream reaches maximum rate about one hour prior to High Water and the Northwest current reaches maximum rate about one hour before Low Water.

2.4.6 Highest and Lowest Tidal Levels at Standard Ports

The levels of the Highest Astronomical Tide (HAT) and Lowest Astronomical Tide (LAT) for selected ports of Malaysia are depicted in Fig. 2.26. The concept of the LAT is of relevance in the context of the 1982 Law of the Sea Convention. The water level along the coast at LAT is an accepted datum from which the seaward limit of the suite of maritime zones that a State may claim may be measured from. This datum is not to be confused with Chart Datum, a concept employed for stating the depths that are recorded on nautical charts.

Tidal bores may occur in certain rivers in Sarawak.

Figures 2.27 and 2.28 are images captured by the author in September 2017 during the Spring Tides at Pulau Indah. The author spent a week in residence at the facilities of the National Hydrographic Centre whilst contributing to a book project relating to the history and role of the organisation in safeguarding the seas adjacent to Malaysia.

Figure 2.29 illustrates the significance of the levels of tides as it relates to the nautical charts and marine navigation which is discussed in Chap. 3 in the section in the context of marine or nautical charts and topographical maps.

2.5   Summary

The narrative in this chapter focused on the geographical setting and ocean basin environment, especially within the marine jurisdiction of Malaysia. It was amply illustrated with maps and other graphics.

The following chapter presents an overview the geographic and legal themes employed in the definition, determination, delimitation of maritime jurisdictional and political boundaries and demarcation of terrestrial boundaries. An extensive discussion is offered on nautical charts, maps and satellite-derived images that are employed in delineating the maritime limits of a state's jurisdiction.

Notes

1. Intergovernmental Panel on Climate Change, *IPCC Assessment Report*, 2021. The Report may be accessed from ipcc.ch/report…; World Bank Group and Asian Development Bank Climate Risk Country Profile: MALAYSIA, 2021 available on World Bank website; Ben Tan, 'Report: Parts of Malaysia at risk of disappearing…', *Malay Mail Online,* 18 May 2023.
2. Forbes and Basiron, 1998: p. 31.
3. Forbes, Vivian Louis 'The South China Sea: Geographical Overview', Chapter 1 (pp. 1–26) *Routledge Handbook of the South China Sea,* Edited by Keyuan Zou. 2021. Taylor and Francis, UK. 524 pages; see also Forbes, V.L. 'Re-framing the South China Sea; Geographical Reality and Historical Fact and Fiction', Universiti Brunei Darussalam Working Paper No. 33, UBD Gadong, 2017; International Hydrographic Organisation's (IHO) *Special Publication No. 23,* 4th Edition, 2002.
4. Hydrographer of the Navy, UK (1961) *Eastern Archipelago Pilot*, Vol. II, London: HMSO, 821 pp; (1971) *Malacca Strait and West Coast of Sumatra Pilot*, Taunton: HMSO, 477 pp; (1978) *China Sea Pilot*, Volume 1, Taunton: HMSO. 304 pp; (1982) *China Sea Pilot*, Vol. 11, Taunton: HMSO. 218 pp; Hydrographic Office, US Navy Dep. (1932) *Sailing Directions for Malacca Strait and Sumatra*, Washington: US Govt. Printing Office, 612 pp.
5. https://www.scribd.com/document/203628845/Pub-120-Pacific-Ocean-and-Southeast-Asia-Planning-Guide-10th-Ed-2013; (PDF) SURFACE SALINITIES IN THE STRAIT OF MALACCA (researchgate.net).

Chapter 3
Overview of Literature: Geographical and Legal Concepts

> "A geographical map insensibly charms the mind with the great and
> pleasing variety of objects that it offers and incites to further study."
> Richard Burton in Anatomy of Melancholy, 1621

Abstract In discussing issues relating to jurisdictional limits—whether on land, at sea, and above and below the sea surface—a sound understanding of the bathymetric and geophysical aspects of the geographical setting and the legal concepts of implementing policies and regulations should be supported by utilising aeronautical and nautical charts and maps with secondary statistical information and state practice. The narrative that follows focuses on the processes of defining, delineating, demarcating, and delimiting international political maritime and terrestrial boundaries.

Keywords Charts and maps · Cartography and GIS · Continental shelf · Delimitation · Maritime boundary

3.1 Introduction

This chapter offers an overview of literature of the geographical and legal concepts employed in defining, delimiting, and delineating international political maritime boundaries and in the demarcation of terrestrial border/boundary markers and frontier limits. During the negotiation processes the parties to an agreement will most likely consult charts, maps and other graphics of their respective countries to establish an initial alignment of the boundary each perceives is their sovereign territory. Delegates of each party will then debate the negative and positive points of the stance they wish to adopt. They will then meet with the other party to negotiate a settlement in what they deem will be favourable terms in the dispute resolution. This process may take several months or indeed, many years for ratification of the Agreement and entry into force as a bilateral or multilateral Treaty. There are many examples of maritime boundary agreements in this regional context. Reliance on charts, maps and satellite-derived

© The Author(s), under exclusive license to Springer Nature Switzerland AG 2025
V. L. Forbes, *Malaysia's Maritime Jurisdictional Limits*,
https://doi.org/10.1007/978-3-031-78783-6_3

and other remotely sensed data play a vital role in the understanding of maritime and terrestrial boundary alignments in the legal, political, and social contexts.

3.2 Charts and Maps

The preparation of accurate charts and maps showing coastline configurations and the bathymetry of adjacent seas has been an objective of maritime nations. Accurate charts and maps are considered essential for land and marine cadastre for effective and proper planning; in addition, to their marine and terrestrial navigational value (Forbes, 2008d: 8–13).

The introduction of an Exclusive Economic Zone (EEZ) through the provisions of the 1982 Convention, Part V, Articles 55–75, and the potential for coastal and island states to claim an extended juridical continental shelf (Part VI, Articles 76 to 85, inclusive) has increased the access to marine resource potential available to coastal and island states.[1] This in turn, increases the need for accurate geographically geo-referenced information and data which are required for the exploration, exploitation, harvesting, and management of the marine biotic and mineral resources. National resource development has been dependent on cartographic and spatial information services to assist in the process and also to promote efficiency and economy in its management.

A chart and map is intended to provide information relating to geographical and geophysical features of the Earth and the planets. The graphic—a cartographic narrative—is a scaled reduction of an area that is observed and studied. The cartographer (map maker/spatial scientist terms are inter-changeable) collates the data to be mapped and makes a judgement of what are important pieces of information that should appear on the final product—a process termed map generalisation—so as to avoid clutter of map icons. The information derived from these graphics include, but are not limited to, the elevation of the terrain, the depth of the ocean, the configuration and alignment of the coastline, and importantly, the distance between locations and points on the graphic can be deduced.

Spatial science—a conventional term—encompasses the traditional disciplines of cartography, hydrography, terrestrial surveying, photogrammetry—and encompasses remote sensing and Geographic Information System (GIS) (Forbes, 2008d: 8–13).

3.2.1 Cartography and GIS (Geographic Information SYST)

Cartography and GIS are aspects of applied graphics and are aids to analysing complex spatial relationships. It is essential that the contents of the graphic be assembled in a logical and obvious manner so that the user can easily understand the information being shown. Relevant statistics or data alone will not necessarily supply the information required.

In many resource development and spatial planning decision-making processes the incredibly useful information is often that which is obtained from studying the overall relationship of all the data. Graphics and graphical techniques can present these relationships in a way even casual observers can readily appreciate the implications (Forbes, 2004: 1–11).

Cartography, cartograms (also termed value-area map) and quantitative graphics can play key roles in socio-economic development, for example, in the exploration and exploitation of hydrocarbon reserves as portrayed in Fig. 3.1 and volume of traffic movement. However, identifying, compiling, and analysing the most up-to-date and accurate spatial information available is the first step. The next step is the selection of the appropriate cartographic techniques to display this information. These may include both traditional presentations and computer-generated compilation utilising software packages.[2] The final product may appear as an expensive multi-coloured map or a simple pen and ink drawing (Fig. 3.2).

The paper-based version of a map is generally considered 'static', whereas a map appearing on a computer monitor employing a software that can manipulate the scale change and perhaps analyse the mapped information is deemed as a 'dynamic' model. Broadly speaking, maps can be divided into two categories, the first being general purpose or reference maps, the second is special purpose or thematic maps which this volume portrays with good intentions.

Using Geographical Information System (GIS) presents considerable problems in the assembly of a suitable digital model from which solutions can be calculated and the 'dynamic' map can be displayed on the monitor or printed copy for utilising as a reference map (Forbes, 2004: 6).

3.3 General Purpose or Reference Map

A general purpose or reference map portrays the relationships of a selection of different geographical, cultural and socio-economic features in an ordered and representative context. Such graphics are carefully constructed by photogrammetric methods and satellite-derived imagery and usually produced in a series of individual sheets to specific standards, generally for ease of interpretation by the international community. Attention is paid to positional accuracy of features so that in many instances these graphics have the validity of legal documents. General purpose maps are fundamental for organising and planning national and regional development. They are considered a basic national resource and are a foundation for further development (Forbes, 2008d: 4–13).

Topographical maps are multifunctional; they are fundamental to economic and resource development of a nation or to a region. Bathymetric maps and models (two- or three-dimensional) are graphics that portray water depths and underwater topography. Uniform depths at specific intervals are usually connected by solid lines called bottom contour lines or more correctly, isobaths.

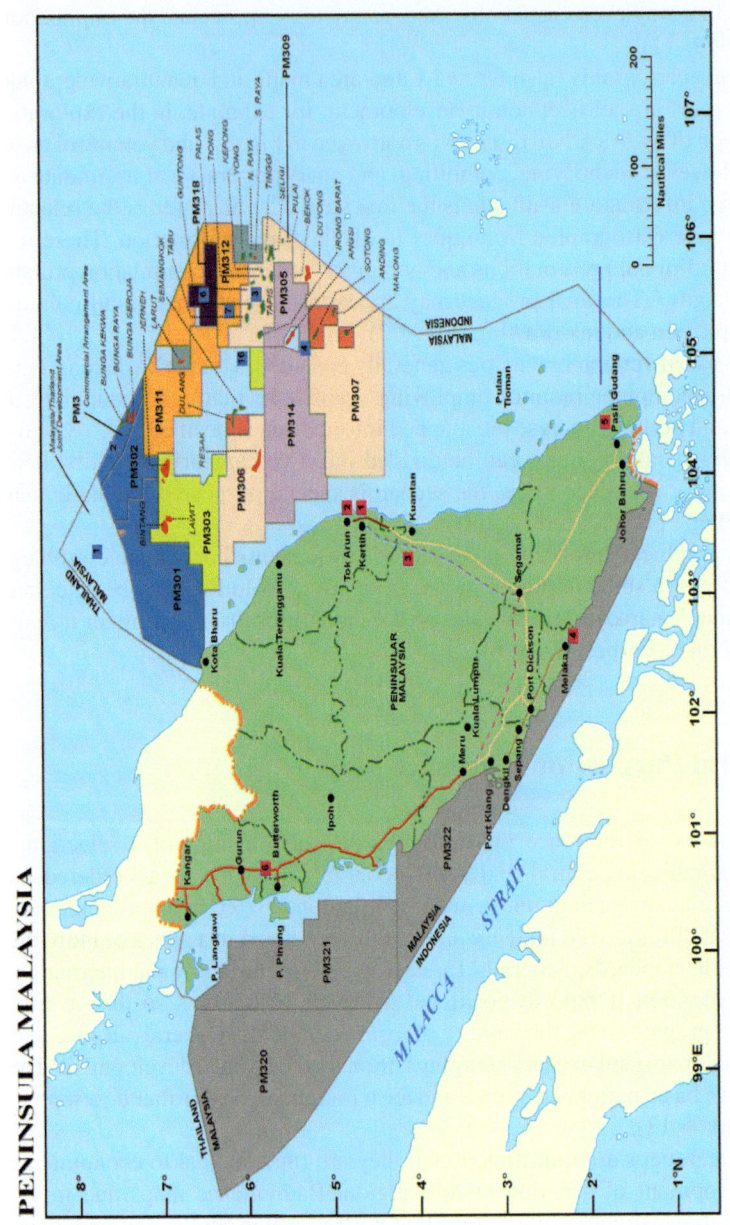

Fig. 3.1 Thematic map of an economic intent. Offshore oil and gas wells and permits off Peninsula Malaysia. *Source* Author

Fig. 3.2   General-purpose sketch map. *Source* Author

3.3.1   Special Purpose or Thematic Maps

This second broad category includes all the map variants designed to fulfil a specific purpose which can be clearly identified in advance. Special purpose mapping, for example, cartograms and quantitative information graphics is the fastest growing segment of the cartographic field because of its relevance to developmental activities. It is the discipline where a broad knowledge of design, technology and the knowledge of intended map user and ability to utilise the data are required (Tufte, 1983).

Thematic maps are composed of two major elements: the background or base map and the specific information being presented as portrayed in Fig. 3.1 which is a thematic map. Thematic maps are also called distribution maps which may emphasise single or multiple themes. The present volume comprises a suite of thematic maps that focus on maritime jurisdiction and marine biotic and mineral resources.

Nautical charts are published primarily for mariners although they serve a wider public in many related ways, for example, in planning and development of port infrastructure, laying of submarine pipelines and sub-surface communication cables. The National Hydrographic Centre of Malaysia is the designated authority for the publication of nautical charts for waters under Malaysia's jurisdiction (Forbes & Zulkifil, 2011: 4–11).

Flight information is generally displayed in bold font and size for visual impact on aeronautical charts. Map symbology may appear in magenta colour, which is ideal for use under low-light intensity in the cockpit. In the context of safety, aeronautical and marine charts require frequent revisions. The compilation and subsequent revision(s) date should be carefully noted and ideally, should appear on the map as a matter of temporal record. Such informational changes are usually published in Annual and Weekly *Notices to Airmen* and *Notices to Mariners*.

3.4 Mapping as an Aid to National Development

In most developing and developed countries topographic mapping has preceded all other types of recorded information and has formed a foundation for subsequent mapping programmes. Usually, geological surveys are compiled onto these topographic bases and become the planning tool for mineral resource and infrastructure development. This is followed by such important studies as road and railway development, improvement in regional agriculture, development of water supplies and hydro-electric power from dam construction, large-scale cultivation of new crops and other aspects of national planning required prior to undertaking major capital projects.

Other major mapping projects built on a topographical map base include tourism planning and development; census studies; forestry management; industrial plant location; land ownership (land tenure); land usage; environmental hazard understanding; and ecological studies. Topographical base-maps are also employed to depict transportation routes; archaeological and anthropological sites; investigation, control, and use of water resources; cadastral surveys; urban studies; sea defences; soil surveys; economic assessments; health investigations; irrigation systems; land reclamation; mosquito control in marshes; airport and seaport sittings; housing developments; vegetation classification; and many other themes. There have also been many schemes implemented because of military and internal security considerations and significant boundary disputes.[3] Special planning maps may be produced by a local government authority or real estate organisation or by the media to highlight a project. To the obvious economic and security aspects listed above must be added several benefits which are difficult to quantify. These would include the significance of mapping in education and a variety of recreational activities. A reading of the above list will show that many of the subject areas have a direct impact on the riverine and marine environment.

Indeed, the division of the natural world into terrestrial and marine aspects is an artificial one as each affect and impacts the other.

The economic welfare of a nation or region is directly dependent on the resources available to it and on the ability of the people to use these resources to their benefit. Inadequate knowledge often results in resources being over-exploited or even destroyed before they are deeply appreciated. These resources include all the sustainable development of minerals, soils, vegetation, wildlife, and water.

To derive benefits from a resource, it must first be identified and then managed. The map is the most efficient method of displaying the necessary resource information. Not only are charts and maps vital for recording and planning of any kind, but they are also an economic investment that produces a major, though highly diversified return. They create economic benefits by preventing a great deal of unnecessary measurement, by eliminating an enormous amount of expensive field work and by shortening the time required to assess a developmental proposal.

Once the basic frame of the survey control network for topographical or planimetric maps is in place, it can become the base on which special studies both on land and in the marine environment can be built. This is not to infer that mapping is impossible without a survey control network. Once the overview or reconnaissance phase is over, however, accurate measurements tied to survey control networks are usually necessary if major development is to take place.

Despite the benefits of national standard mapping, the current coverage of world topographic mapping is far from complete. Only about 40% of the countries of the world are mapped at the scale of 1:50,000, considered to be the most useful scale available. A study by the International Hydrographic Office in 2010 found that only 40% of the marine area covered by the Exclusive Economic Zone (EEZ) of the 39 countries it surveyed was adequately charted.[4] These values, given as examples, may have altered during the past decade.

Standardised mapping tends to have a long life. Topographic maps, for example, are typically in circulation for 15 to 20 years before they are updated. However, this statement may not apply to some national mapping agencies that have the facilities and resources to produce computer-assisted cartographic products. Many potential users and uses for such a product cannot be foreseen. Some hydrographical charts have been in continuous usage, through regular updates, for 70–100 years and their economic, social and convenience benefits can hardly be overstated. Such graphics will easily return their cost many times over during their useful life and some will make possible a return on investment and human benefit unmatched by virtually any other developmental expenditure.[5]

Maps have a catalytic effect on development and prevent the main source of haphazard planning which is the lack of easily understood information. In relation to the development and exploitation of marine resources, a mapping program can address critical information needs concerning, for instance, fisheries, oceanography, geology, navigation, coastal development, and local political realities. In the event of extreme weather conditions, charts and maps are useful tools for authorities. In addition, such important topics as education and training can be addressed.

3.4.1 Scale of the Map

All maps, aerial photographs and satellite-derived imagery are a minute representation of a portion of the earth's surface. It is their smaller than life size which is responsible for their convenience as a method for portraying the world. If these products are to be at all useful, the relationship between the size of the graphic and the actual size of the same region of the earth must be known. This fundamental concept, known as the scale of a map, is one of the most important design considerations in the field of cartography.[6]

Establishing a scale for a map is an important design decision. Among other things, the following are controlled by scale: (a) the amount of data or detail which can be displayed; (b) the size of the graphic; (c) the cost of reprographics; and d) the legibility.

Additionally, the following factors must be considered: (1) the regional extent of the information to be displayed; (2) the degree and nature of the generalization carried out; (3) the suitability of an available base map for a specific purpose; (4) the facilitation of the user; and (5) the amount of time the spatial scientist must spend on a project.

Specifically, the scale of the map is the ratio of the graphic to ground distance and its selection depends primarily on map purpose. The spatial scientist must also consider convenience and economy, striking an equilibrium between the area covered, map size and the amount of detail required. Map scales are often a compromise.

The use of the relative terms large-scale and small-scale can cause considerable confusion and must be thoroughly addressed. To understand the concept clearly, compare two maps of the same area but of significantly different scales. Select a common feature such as an airfield, a bay, or an island. The map which shows the feature drawn relatively large is the large-scale map. By contrast, the map which shows the same feature as being distinctly small is a small-scale map. As a rule: small-scale maps cover large geographical areas with modest detail while large-scale maps show greater detail and only cover a smaller geographical area.

Most maps will be a compromise between the required detail and the area of coverage. Sometimes the needs are incompatible, as when a large area must be covered but with some parts requiring detail. This may be solved by producing more than one map or by using portions of the map as insets at larger scales. The latter solution allows variation in small scales and greater detail in critical areas.

Generally, once it is calculated, the scale of a map may be shown in three different standard forms. These are representative fraction (as a ratio), scale statement (for example, one centimetre on the map is comparable to one kilometre on the ground) and graphic or bar scale. Other scale variations are sometimes used in addition to the standard forms.

Representative fractions (R.F.), also known as scale ratios, relate to the size of the map, or portion thereof, to its actual size on the ground. Thus, an R.F. of 1:10,000 means that one unit on the map is equivalent to 10,000 units on the ground. A major advantage of this system is that it is not tied to a specific measurement system; the ratio works as well in metric as in imperial or any other convenient unit of measurement.

Comparatively small numbers after the colon are associated with large-scale maps while comparatively high numbers after the colon are associated with small-scale maps. to standardise the terminology, the International Cartographic Association, suggested the following should apply:

- R.F. 1:1 to 1:25,000, are considered large-scale maps,
- R.F. 1:50,000 to 1:100,000 may be termed medium-scale maps; and,
- R.F. 1:200,000 and values greater than 200,000 are small-scale maps.

Scale statement is a notation of map distance in relation to earth distance, for example, 'one inch on the map equates to one statute mile on the ground' or 'one centimetre on the graphic equates to one kilometre' on the earth's surface. An R.F. could also be considered a scale statement since, for example, 1:1,000,000 could also be written as one centimetre on the map is to 10 kms on the ground or one millimetre on the graphic is to one kilometre on the surface of the earth. If this version of a scale is chosen, do avoid confusion by not mixing metric and imperial units in one statement.

Graphic or bar scale is the most common and the most useful method of depicting scale on a map or chart. It consists of one or more straight lines which are subdivided into units of ground distance or whatever the scale is designed to show as portrayed in Fig. 3.1. It has the considerable advantage of remaining correct even if the map is photographically or by computer-processed enlarged or reduced, which is not true for the other scale depiction variants, the R.F. and scale statement. This is especially true in the dynamic mapping of digital cartographic versions, as illustrated in all the above examples.

The cartographer must consider that the scale is designed for the user and not for the convenience of the cartographer. The subdivision units must be selected to be as even and useful as possible, whatever the R.F. For example, the common older map scale of one inch on the map represents one statute mile (imperial measurement) or R.F. of 1:63,360 should be converted by the cartographer if a metric scale is desired. By taking the original scale units of one inch and plotting them according to their metric scale equivalent, each subdivision unit of one mile would represent an inconvenient 1609.35 m. In this case a basic subdivision unit of 1000 m or one kilometre might be deemed appropriate. A calculation reveals that each unit representing 1000 m will be 1.578 cm long. This is difficult to plot but it is the effort the cartographer must make for turning out a useful and professional product.[7]

3.4.2 Map Projections

The cartographer makes use of map projections to present the three-dimensional nature of the earth's surface in the two dimensions available on a map or chart stating the map projection employed (refer to Fig. 3.3) As discussed earlier, for the purposes of medium- and small-scale graphics the basic shape of the earth can be assumed to be spherical. A small area of a large-scale map or chart can be drawn without appreciable error but for those products showing large areas, and particularly for series mapping, a projection system is vital.

Projections can be created purely graphically by projecting the earth's curved surface onto flat surfaces or developable surfaces such as cones or cylinders which can be flattened. They can also be created mathematically or by a combination of the two methods. The ideal projection would provide correct shapes, precise areas, accurate scale, exact bearings, a good overall "fit" and ease of construction. Obtaining all or even most of these properties is impossible so the cartographer must select whichever feature is the most important for a particular map, or choose a compromise projection, often one of the so called 'minimum error' types.[8]

Correct shape is a characteristic of conformal (orthomorphic) projections. It should be noted that it is only possible to keep shapes correct over small areas.

LAUTAN PASIFIK　PACIFIC OCEAN

LAUT CHINA SELANATAN
SOUTH CHINA SEA

Unjuran/Projection: Mercator
Skala/Scale 1:3 500 000 (Lat. 22° 30'N)

KEDALAMAN DENGAN METRE
DEPTHS IN METRE

KETINGGIAN DEGAN METRE
HEIGHTS IN METRE

V.L. Forbes, June 2024

Fig. 3.3 Statement of map projection and scale

Conformal projections preserve true angles and a constant scale in all directions about a given point because the parallels and meridians cross each other at right angles. This is an essential characteristic for navigational charts. Both the Mercator and the Lambert Conformal Conic are conformal projections and are widely used both for sea and for air navigational charting. As these projections preserve angles locally, they may also be used for graphics showing data based on angular measurement. These might include tidal streams, lines of gravity and magnetic variation and deflection, direction of surface-water movements, migrations, and bathymetry. Navigational charting using conformal projections has been undertaken for centuries, providing a ready source of data for use as base map information. This simplifies the cartographer's task. The maps in this volume are generally depicted on the Mercator Map Projection. This is the most convenient projection for cartographically depicting this Equatorial region.

Equal area is also known as equivalence. This property can be preserved on a map constructed from a projection such as Bonne's, but only at the expense of distorted shapes. This projection can be of excellent value for displaying spatial relationships and distributions. When the cartographic symbolisation requires an area or quantitative symbol, such as water volume movement, an equal area projection is needed although it is not applicable in this present volume.

The attainment of full equidistance, that is, the preservation of scale at all points on a projection, is impossible. On any projection the actual scale is continuously variable; it can vary from point to point and may also vary in different directions. It is possible, however, to maintain correct scale where a projection surface meets the sphere from which it is derived. Selection of those points in a careful manner can reduce scale errors to a minimum. Equal distance can be preserved on Zenithal projections. Equidistant projections are a useful compromise between conformal and equal area projections, and they are often used for general reference graphics.

The area scale changes on equidistant projections are less dramatic than those on a conformal projection and the angular errors are less than those of an equal area projection.

3.4.3   Marine or Nautical Chart and Topographical Map

A marine chart is essential for safe navigation and the practice of surveying and charting water for the purpose of navigation is known as hydrography (Forbes, 2008d: 4–13) Marine charts are important to several economic sectors including:

- commercial and military shipping,
- fishing fleets,
- aquaculture operations,
- offshore and coastal oil and gas industry and wind powered turbines,
- coastal mines and industrial plants, and,
- recreational sailing.

The primary purpose of a topographical map is to portray the topography of the land (hills, mountains, rivers and streams) within the mapped area. In addition, these maps may depict the transport network, infrastructure, land-use and built environment. The information displayed will be limited to the scale of the mapped area and the generalisation concept employed. These graphics differ in several important respects, some of which have relevance to thematic base map preparation. Nautical charts are intended primarily for marine navigation.

Marine charts, for example, are generally compiled using the Mercator and Gnomonic Projections. Topographic maps use Transverse Mercator or other map projections. The graphics generally depict a geographical coordinate system. With some conventional projections the scale varies considerably over the graphic, particularly in the higher latitudes, for example, marine charts and small-scale topographic maps giving certain areas of the coast undue emphasis.

The depiction of the coastline in the compilation for small-scale topographical maps is relatively simple because they usually require so much simplification that detail is of little consequence. For example, the coastline of the landmass, as delineated on the map, will represent the Mean Sea Level as a datum. When compiling medium- and large-scale nautical charts, however, the major difficulties facing the spatial scientists and hydrographers is to factor in the Chart Datum. Marine charts employ Low-Water Mark (LWM) or Lowest Astronomical Tide (LAT) reference level (ref: Fig. 2.28) whereas topographic maps use Mean Sea Level as a result, the shape of the coast on the graphic will differ, particularly in areas of high tidal amplitude.

In some areas of the world, the shape of the coast changes rapidly due to erosion or deposition and indeed, due to human induced alterations such as reclamation works and construction of artificial islands. These changes may be monitored by the comparison of historical and current aerial photography and satellite imagery.

The unit of measurement employed on a nautical chart is nautical miles whereas topographical and general-purpose maps use measurements in imperial and/or metric units. The marine chart may portray one or two Compass Roses, whereas the topographical map may depict the three North Points (True, Magnetic and Grid).[9]

Point, line and area symbology, generally of international standard, is commonly employed on these graphics. In the instance of choice of colours used on graphics there are a number of inconsistencies when utilising both charts and maps, for example, marshland, definitely not navigable, is likely to be coloured as land on a chart, whereas a low-lying swamp on a topographic map is likely to be coloured blue as water. The concepts of colour balance and colour contrast are essential elements to be considered in producing a graphic.

3.4.4 Geographical Information System

The form and methods of data collection are of immense importance to a marine resource mapping programme. The collection phase may include a variety of techniques from relatively simple and inexpensive data gathering exercises, such as the deployment of field personnel equipped with felt markers and base maps, to the complex operations associated with data collection from remote sensing platforms (satellites, fixed wing aircraft and helicopters). The information collected can be stored in the conventional manner on hard-copy maps, or it can be entered into the computer of a geographic information system (GIS) and stored in a digital format.[10]

Because of their complexity, earth resource analysis (for example, for fisheries) and land use planning require the investigation of the relationship of large volumes of detailed information, hence the demand for Geographical Information Systems (GIS). The essential elements of an automated GIS are Methods, People, Data, Software and Hardware.

Utilisation of these more sophisticated techniques is certainly not essential for the development of a marine resource management plan. They should be considered, however, since the more elaborate data collection, storage, analysis, and presentation procedures are now becoming readily available.

The automated production of topographical maps and hydrographical charts has changed not only the way they are made and how they are utilised. Computer-assisted map production is making it easier to produce new paper maps and to revise existing ones. National charting and mapping agencies have responded with innovative ways of compiling map data and using them for map production. Improved efficiencies in most facets of production have reduced the four to five years it used to take to produce a map by traditional methods.

The rapidity of research and development in this field can be appreciated by reviewing the proceedings of recent automated cartography symposia. Limited investment mapping systems, which consider the automated production of thematic maps, are currently available with modestly priced hardware (computer, terminals, plotters, etc.) and software (computer programmes). Widespread acceptance of computers and related technologies, including satellite-derived enhanced images has accelerated the demand for mapping information in computer-compatible form. Government agencies and private businesses now require digital mapping information for their computer-based systems.

A GIS permits automated inventory and management of geographic (spatially related) data such as that contained on maps, charts, and other graphical representations. A GIS uses a combination of hardware and software to input, store, analyse, manipulate, and output graphic and text data in a variety of formats. There are four main activities involved in this process:

- Input: User documents such as maps, charts, aerial photographs, tabular or text data are entered into the system via a digitizing tablet, through keyboard terminals or from existing digital media using several specially developed interfaces. Alternatively, a map may be scanned, rather than manually digitised, with a laser or optical scanner.
- Processing: Once data is entered into the system it can be processed, stored, analysed, synthesised, or queried to create an unlimited range of products and to answer questions regarding the data.
- Output: Production can be generated from the system for temporary viewing on a terminal (colour or black and white), or as hard copy for distribution or publication using a printer, plotter, or other hard copy devices.
- Communication: With the aid of recently developed protocols, both alpha-numeric and graphic data can now be transmitted over dedicated telephone lines with relative speed, accuracy, and scale of economy.

Traditional methods of map production are part scientific documents and part works of art. Generally, computers cannot yet create the graceful appearance of a traditional hand-drawn map from normal GIS data. Challenges include text placement, integration of different feature classes, and positioning crowded features (clustering) while maintaining clarity. Other issues are colour balance and colour contrast on graphics.

Figure 3.4 portrays an example of a graphic from satellite-derived imagery and computer-enhanced 3D model using software tools to produce spatial information useful in planning and policy making. The image is a representation of a seabed feature—the red depicting relatively higher ground than the surrounding seabed.

Figures 3.5 and 3.6 illustrate the extent of the terrestrial international political boundary between Malaysia and Thailand on two media: on a hydrographic (nautical) chart and an aeronautical map, respectively. In the former line symbology comprising a series of 'plus signs' is utilised whereas in the latter, the boundary is a continuous coloured line. Figure 3.5 emphasises the marine information whereas Fig. 3.6 focuses on the terrestrial features relevant to aviation. Note the prominence of the land boundary between the two countries as depicted on the topographic map in comparison with the nautical chart. A larger portion of the boundary is termed a land boundary and is demarcated accordingly by boundary pillars and posts and a riverine boundary that is aligned on the thalweg principle. The boundary, as agreed in a 1909 Treaty stated that the terrestrial border:

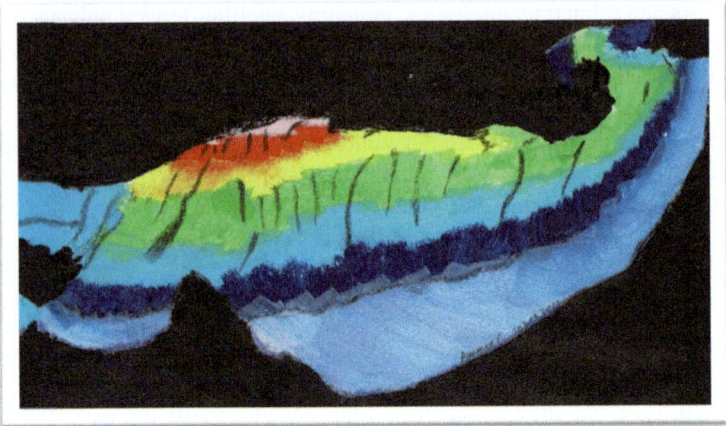

Fig. 3.4 Satellite-derived image with elevation data. *Source* Aimmee A. Clothier and V. L. Forbes, May 2023

Commencing from the most seaward point of the northern bank of the estuary of the Perlis River and thence north to the range of hills which is the watershed between the Perlis River on the one side and the Pujoh River on the other side; then following the watershed formed by the said range of hills until it reaches the main watershed of dividing line between those rivers which flow into the Gulf of Siam on the one side and into the Indian Ocean on the other; following this main watershed so as to pass the sources of the Sungei Patani, Sungei Telubin, and Sungei Perak to a point which is the source of Sungei Pergau; then leaving the main watershed and going along the watershed separating the waters of the Sungei Pergau from the Sungei Telubin to the hill called Bukit Jeli or the source of the main stream of the Sungei Golok....thence the frontier follows the thalweg of the mainstream of the Sungei Golok to the sea at a place called Kuala Tabar.

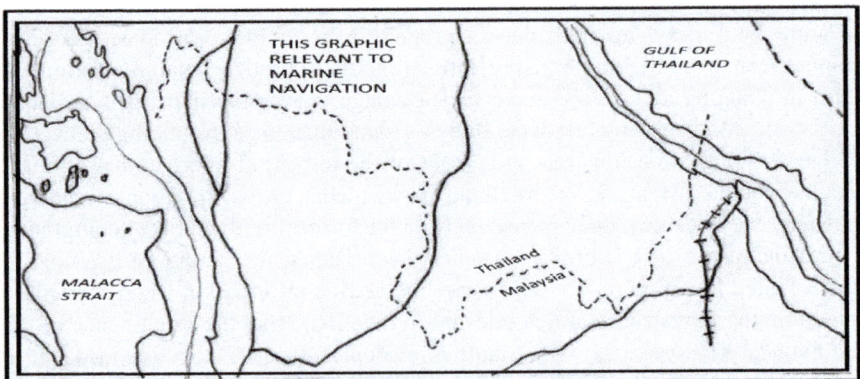

Fig. 3.5 The terrestrial political boundary between Malaysia and Thailand (The boundary symbology thus: ----) as portrayed on a nautical chart intended for use in marine navigation

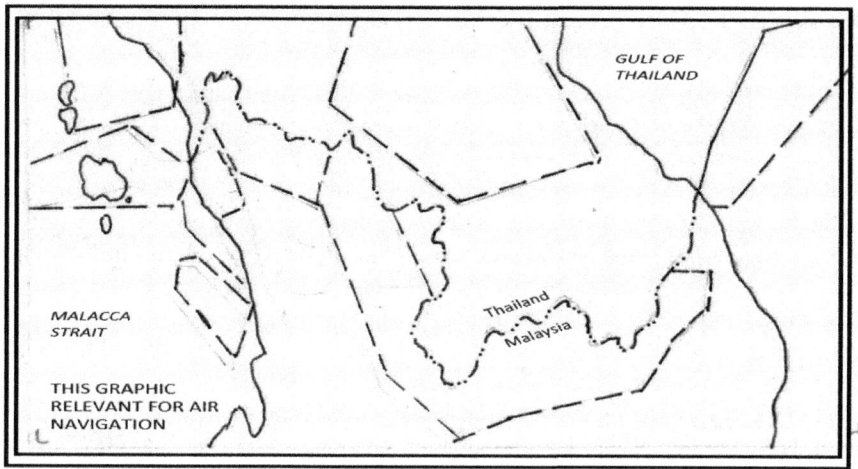

Fig. 3.6 The terrestrial political boundary between Malaysia and Thailand as portrayed on an aeronautical map shown by line symbology thus: dash and two dots

The terrestrial international political boundary as delineated on the the following graphics are shown as line symbology for depiction on the nautical chart as a series of short dashes with the words of Thailand and Malaysia on the respective sides of the boundary. The symbology utilised for the political boundary on the aeronautical map is a series of long dashes and two dots. The names of the countries are stated on the respective sides of the boundary.

In determining the Terminal point of a maritime boundary between adjacent coastal states it is vital that the location at the land/sea interface is clearly defined by description and geographical coordinates and portrayed on a graphic or map, as illustrated in Figs. 3.5 and 3.6. This fact was discussed in Chap. 1.

A MOU (*Memorandum of Understanding*) between Malaysia and Thailand was signed on 8th September 1972 for the purpose of undertaking the joint demarcation and survey of the international boundary involving about 552 km of land boundary and 95 km of riverine boundary. There were 19 priority areas for the former and four with the latter. The joint demarcation and survey of the boundary commenced on 6th July 1973 and was completed on 26th September 1985. Joint survey and delineation of the fixed and permanent international boundary along Sungei Golok (Kolok River) commenced on 1st November 2000.

The determination of a terrestrial international political boundary can be fraught with problems and the demarcation process have generally taken years to complete especially if the terrain is difficult to traverse and there are cultural, historical, and religious buildings and or heritage sites that are claimed (and counter-claimed) by the parties to the agreement. Defining and delimitating a maritime boundary, especially, the outer limits of the continental shelf can prove equally challenging as discussed below.

3.5 Natural Continental Shelf

Article 76 of the 1982 Convention defines the natural continental shelf (CS), as a submerged extension of the land modified by erosion and deposition, gives way to the continental slope, the continental rise and finally to the abyssal plain. Figure 3.7 illustrates the geomorphic margin. (Forbes, 1995: 94 and 95).

The Truman Doctrine the United States of America declared on 28th September 1945, dominion over "land beneath the sea". Other states followed suit. In 1958, the *Geneva Convention on the Continental Shelf* confirmed that the mineral resources of the seabed and subsoil and the sedentary living resources on the continental shelf were the property of the adjacent state. States that had the necessary technology were also permitted to exploit the continental slope beyond 200 m.[11]

The complexity of fixing the outer limit of the continental shelf has been a problem for coastal states since 1945. The situation was not improved by the definition incorporated in Article 1 of the Geneva Convention on the Continental Shelf 1958, which stated that:

> ... the term continental shelf is used as referring (a) to the seabed and subsoil of the submarine areas adjacent to the coast but outside the area of the territorial sea, to a depth of 200 metres, or beyond that limit, to where the depth of the superjacent waters admits to the exploitation of the said areas; (b) to the seabed and subsoil of similar areas adjacent to the coasts of islands. (Forbes, 1995:5)

The 1958 legal definition thus bore no resemblance to the geomorphic definition of the continental shelf and only a rough resemblance to the geomorphic definition of the continental margin.

Article 76:1 of the 1982 Convention redefined the area of seabed sovereignty of adjacent coastal states as the seabed and subsoil of the submarine areas that extends beyond the territorial sea throughout the natural prolongation of its land territory to the outer edge of the continental margin, or to a distance of 200 M from the baseline from which the territorial sea is measured where the outer edge of the continental margin does not extend to that distance. In paragraph 3 of the same Article, the continental margin is defined as comprising the submerged prolongation of the land

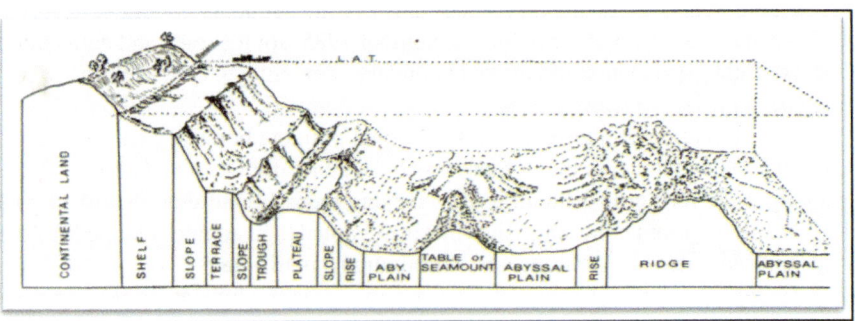

Fig. 3.7 Delineation of the geomorphic margin. *Source* Forbes (1995:8)

mass, the slope and the rise but does not include the deep ocean floor with its oceanic ridges. Its average gradient must lie between 1: 10 to 1: 50 (Article 76; Forbes, 1995:76; Forbes, 2001).[12]

According to the 1982 Convention, the regimes of the Territorial Sea and the Contiguous Zone take precedence over the regime of the Continental Shelf in that part of the seabed that underlies the water columns of the territorial sea and contiguous zone (Article 76; Forbes, 2001: 179).

The establishment of a definition of the continental margin required the development during the negotiating processes of the *Third United Nations Convention on the Law of the Sea* (the 1982 Convention) of detailed and precise formulae as to how the limits of such margins were to be determined.

One approach, which came to be known as the Irish Formula because it originated from the Irish delegation, provided alternative means of establishing the outer limit of the continental shelf in cases where the continental margin extended more than 200 M from the baseline. One alternative was to construct straight lines joining fixed points, at each of which the thickness of sedimentary rock was at least one percent of the shortest distance from such point to the foot of the continental slope. The other was to construct straight lines joining fixed points not more than 60 M from the foot of the continental slope (refer: Fig. 3.8).

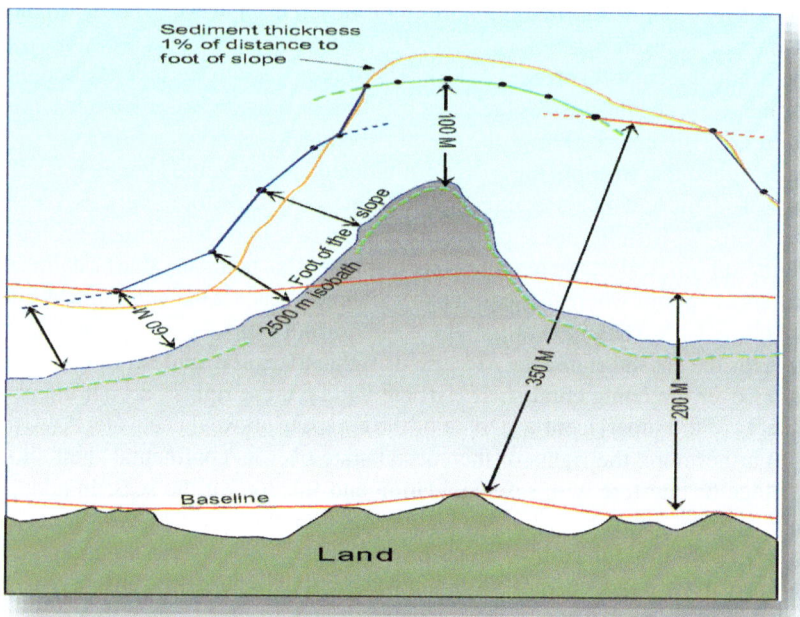

Fig. 3.8 Constraints in determining the extended limits of the continental shelf

The Soviet Union proposed that in no case could the continental shelf extend beyond 300 M from the baseline. Where the margin terminated somewhere between 200 and 300 M, the outer limit of the shelf would be the edge of the margin, which would be determined based on geological and geomorphological data and, in the absence of such data, by reference to the second variant of the Irish Formula.

In the case of narrow continental margins, the 200 M formula as put forward by a group of Arab States had been recognised and generally accepted. Several coastal states had by that time implemented an exclusive zone of a width of 20 M.[13]

Article 76 also authorised the creation of a Commission on the Limits of the Continental Shelf (CLCS), whose function was to provide competent advice to states on the establishment of the outer limits of their continental shelves. The composition, functions, and workings of the 21-member Commission are specified in Annex II of the Convention.[14]

The Commissioners were directed to the Statement of Understanding Concerning a Specific Method to be used in Establishing the Outer Edge of the Continental Margin, when making their recommendations on such matters relating to the southern region of the Bay of Bengal. This request was in direct response to the argument put forward by Sri Lanka's delegates that the Irish formula was based on a configuration common to most, but not all, continental margins. In such cases the formula resulted in a limit of coastal state jurisdiction which coincided with the limit of the margin's significant resource potential. Where the geological and geomorphological characteristics of the continental margin were quite different, however, the Sri Lankans claimed that this result was not achieved.

The continental margin off Sri Lanka has two characteristic features. Although the foot of the slope occurs close to the coast it is followed by an extensive rise beneath which lies the greater portion of the margin's valuable sedimentary rocks. The Sri Lankans argued that to apply the Irish Formula in this case would be to deprive their country of about half of its continental margin. They therefore suggested a third alternative to the Irish Formula, proposing that, in situations where such geological and geomorphological characteristics were displayed, the boundary line be delineated by reference to points where the sedimentary thickness reaches the minimum by either variant of the Irish formula in areas where that formula applies.

Whereas the coastal state may exercise its rights to explore and exploit its natural resources over the continental shelf (Article 77: 1), these rights "do not affect the legal status of the superjacent waters or of the air space above those waters" (Article 78: 1). Furthermore, the rights of the coastal state over its continental shelf should not infringe or interfere with safe navigation and freedom of the seas. In addition, within this zone, aliens have full navigation rights provided they observe the rules and safety zones designated by the coastal state. They also possess the rights to lay submarine cables; conduct research in the water column, but with consent from the coastal state if the research is of the seabed; fish in the water column, but not catch sedentary species (Article 77: 4).

The coastal state has complete authority to legislate for the protection of the seabed environment (Articles 194 and 200) provided that such regulations do not unduly interfere with the rights and duties of aliens. The Convention also permits aliens to conduct research in those areas of the shelf which are more than 200 M from the baseline of the coastal state and which have not been designated by that state as areas within which exploration "… will occur within a reasonable period of time" (Article 246: 6).

The Continental Shelf is the area of the seabed and substratum that extends beyond the territorial sea to 200 M from the territorial sea baseline, and beyond that distance to the outer edge of the continental margin, as defined in Article 76 of the 1982 Convention. The continental shelf is largely co-extensive with the EEZ, within 200 nautical miles from the territorial sea baseline system.

Coastal or island states possess sovereign rights over the continental shelf for the purposes of exploring and exploiting mineral and other non-living resources of the seabed and substratum, together with sedentary organisms.

3.5.1 Hedberg's Proposal

In 1972, Hollis D. Hedberg, an oceanographer, whilst noting that the continental slope base cannot be used directly as the national–international maritime boundary, but must act as a guide, envisaged the following steps:

> In view of the impracticability of defining accurately a geological line boundary on the ocean floor, the slope should be used to define a zone within which the precise political boundary will later be placed. To accommodate irregularities and uncertainties in the position of the slope base, the boundary zone should be at least 100 kilometres wide; and to satisfy coastal states that had already claimed territorial seas of 200M, it may be necessary to make the zone 300 kilometres wide. In any event, the precise width must be decided by international agreement and applied uniformly throughout the world.

The landward limit of the boundary zone should be the best scientifically determined position of the slope base. An International Marine Boundary Commission comprising scientists and engineers (that is, not lawyers, sociologists, politicians, or economists) should decide this position. The seaward limit of the boundary zone would, of course, be fixed by the zone width already agreed.

Once the Commission has agreed on the location of the boundary zone, the coastal state concerned should be free to decide on the precise position of the political national–international boundary within that zone. However, for practical reasons the boundary should consist of straight lines or arcs joining points fixed by geographical coordinates of latitude and longitude.

This proposal had several advantages:

- Definition of the boundary zone using the most reasonable position for the slope base as judged by a group of independent scientific experts gives the zone a scientific legitimacy untainted by political manoeuvring.
- The finite width of the boundary zone and the location of the zone seaward of the most reasonable slope base position allow political flexibility, not least in

countering arguments likely to arise from uncertainties in the precise position of the slope base.

- The four criteria originally proposed by Hedberg as being necessary to ensure international agreement are all fulfilled.
- A perception of great tactical advantage by allowing each state to delineate its own boundary within internationally agreed guidelines, rather than having a boundary imposed on the state by some outside international authority.

There were at least two objections to Hedberg's proposal. The first was recognised and answered by Hedberg himself whereby he suggested that there would be a tendency to keep the boundaries as simple as possible for ease of clarity in delineation. The second and more serious objection was that the Hedberg proposals would be too rational for lawyers and politicians to accept. The Law of the Sea negotiations constituted legal/political acts, and it was perceived that things 'natural' had little or nothing to do with this type of negotiating process.

3.5.2 Juridical Shelf Edge: Outer Limits of the Continental Margin

A summarised interpretation of Paragraphs 4, 5 and 6 of Article 76 is offered. There are seven procedural steps to determine the outer limit of the juridical shelf:

- Baseline system (the low-water line as marked on large-scale charts, bay-closing lines, and straight baselines).
- Delineate the 200 M and 350 M lines from the baseline, continental shelf.
- Delineate the 2500-m isobath.
- Delineate the 100 M arcs beyond the 2500-m isobath.
- Delineate the foot of the slope: the point of maximum change in gradient at its base.
- The arcs of 60-M radius beyond the foot of the slope.
- The points where the ratio x = Sediment thickness divided by Distance to the foot of the slope = 0.01.
- Then a coastal or island State may claim the maximum width, that is,
- 200-M line, or (ii) 60 M beyond the foot of the slope.
- Which are within:

  – 350 nautical-mile line, or
  – 100 nautical miles beyond the foot of the slope.

- Unless there is a submarine elevation that is a natural component of the continental margin, in which case, the 350-nautical mile limit does not apply.
- In Article 76, there exist three interpretations for defining the outer limits of the Continental Shelf.

3.6 The Definition of the Continental Shelf and Criteria for the Establishment of Its Outer Limit

Provisions of Article 76 of the 1982 Convention define the continental shelf and the criteria by which coastal and island states may establish the outer limits of its legal limit of continental shelf (juridical or extended continental shelf limits). The Commission on the Limits of the Continental Shelf (CLCS) was established to approve submissions made by coastal and island states.

By 25 June 2024, the CLCS had received 94 submissions and 11 revised submissions from coastal and island states. The most recent submission was from the Government of the Philippines on 14 June 2024. The CLCS has recommended 35 submissions by this date. The readers may wish to view the details of the presentations and submissions made by the claimant states and commentary on the recommendations by the CLCS which are available at the web pages of the UN CLCS.[15] The Executive Summaries of Malaysia/Vietnam Joint Submission and that of Malaysia and other regional states are available for inspection at the CLCS website.

3.7 Summary

This chapter focused on the geographical graphics employed in the processes of defining, delineating, demarcating, delimiting international political boundaries (maritime and terrestrial) and in national developmental projects on land and at sea.

The discussion in the following chapter focuses on maritime boundary delimitation and on Malaysia's maritime jurisdictional zones—actual and potential.

Notes

1. The concepts of the EEZ and legal extended continental shelf and the rights and obligations of states over these jurisdictional zones are contained in the provisions outlined in Parts V and VI of the Third UN Convention on the Law of the Sea Convention, 1982. The text, in select languages, is available at un.org/depts/los/convention_agreements....
2. A representative sample of books on Cartography, GIS and Remote Sensing is offered here if the reader wants a greater appreciation of the disciplines. For example, Gretchen N. Peterson *GIS Cartography*, CRC Press, 3rd Edition; C. Dana Tomlin *GIS and Cartographic Modelling*, 2012; J.W. Crampton *Mapping: A Critical Introduction to Cartography and GIS*, 2010. There are many other titles. Forbes (2008a); also https://gisgeography.com/cartogram-maps/.
3. ESRI 'Maps Serve as a Compass for SDGs' by Guillaume Le Sourd, Winter 2022; GFDRR, 'Open Mapping for the SDGs', 2010; Forbes (2008d: 4–13).
4. International Hydrographic Organization web page: iho.int/en/importance-of-hydrography.
5. See: The Status of Topographic Mapping in the World, A UNGGIM-ISPRS Project, 2012–2015. See also GEBCO and IHO websites for trends in bathymetric surveying and mapping.
6. Ref: Borden Dent and others, *Cartography: Thematic Map Design*, 6th Edition, 2008; Peter A Burrough and others (2015) Principles of Geographical Information Systems, Oxford, 3rd

Edition, 352 pp; D.J. Wright and D.J. Bartlett (2000) Marine and Coastal Geographical Information Systems, New York: Taylor and Francis; Forbes 2008d: 4–13; Map Scale National Geographic Education https://education.nationalgeographic.org/resource/map/scale.

7. Mark Monmonier, *How to Lie with Maps*, 3rd Edition, 2018.
8. Erik W. Grafarend and others, Map Projections Cartographic Information Systems, 2nd Edition, 2014. There are numerous publications that offer explanations on map scale.
9. Nigel Calder, *How to Read a Nautical Chart*, Revised Edition, 2012; UK Admiralty, *NP5011 Symbols and Abbreviations used on Admiralty Charts*, 8th Edition, 2020. See also: J.A Butler and others (Eds.) Marine Resource Mapping: An Introductory Manual, Rome: F.A.O. 1987; National Ocean Service (NOAA) https://oceanservice.noaa.gov/facts/chart_map.html.
10. V.L. Forbes, *The Maritime Boundaries of the Indian Ocean Region*, Singapore: SUP, 1995: 75; V.L. Forbes, *Indonesia's Delimited Maritime Boundaries*, Springer, 2014; Armstrong, P.H. and Forbes, V.L. (2005); 243–264.
11. The Law of the Sea Articles referenced here are from the 1958 Geneva Conventions and 1982 UN LOS Convention. (See Note 1, above); see also V.L. Forbes, *The Maritime Boundaries of the Indian Ocean Region*, Singapore: SUP, 1995.
12. Forbes (1995: 89–101); V.L. Forbes, Conflict and Cooperation in Managing Maritime Space in Semi-enclosed Seas, Singapore: SUP, 2001. The Legal Status of the Continental shelf is discussed at p. 178.
13. The legal limit of the outer continental shelf is discussed by V.L. Forbes, *Indonesia's Delimited Maritime Boundaries*, Springer 2014, pp. 30–31; S.V. Suarez (2008) *The Outer Limits of the Continental Shelf: Legal Aspects of their Establishments*, Springer, 294 pages; and Joanna Mossop (2016) The Continental Shelf Beyond 200 nautical Miles: Rights and Responsibilities, Oxford University Press; see also THESIS_DOCTORATE_FORBES_Vivian_Louis_2022.pdf (uwa.edu.au).
14. Commission on the Limits of the Continental Shelf. https://www.un.org/Depts/los/clcs.
15. Submissions through the UN S-G to the CLCS. https://www.un.org/....commissions.

Chapter 4
Malaysia's Maritime Jurisdictional Limits

The first situation arises where substantial activities subject to coastal state jurisdiction are being conducted or are likely to be conducted in an area of actual or potential conflict.
Bernard H. Oxman
International Maritime Boundaries: Political, Strategic, and Historical Considerations
Annex 186, The University of Miami, Inter-American Law Review, 1994–95, p. 245

Abstract Malaysia's sovereignty extends, beyond its landmass and Internal Waters, to a belt of sea termed the Territorial Sea of Malaysia. Beyond this legal concept Malaysia claims an Exclusive Economic Zone and Continental Shelf—natural and outer continental shelf, the latter requires scientific justification and the area free from any sovereignty and territorial dispute. Malaysia, in the parlance of the Law of the Sea may be termed a 'zone-locked state' as it is perceived to be restricted in its claim to full-widths of certain maritime zones in many geographical marine areas. This narrative, supported with 31 illustrations, discusses Malaysia's maritime jurisdictional limits with a focus on its international maritime boundaries.

Keywords Maritime boundary · Delimitation · Exclusive economic zone · Continental shelf · Joint development area · Agreed common area · *Letter of intent*

4.1 Introduction

The sovereignty of the coastal and island State over a territorial sea adjacent to its coast has been a principle established by customary international law. Over the centuries the width of this zone of sea has varied from three to as much as 200 nautical miles (M). Provision for the sovereignty of a coastal state to extend its outer limit of the territorial sea was given in Section I of Part I of the 1958 *Geneva Convention on the Territorial Sea and the Contiguous Zone* (the 1958 Geneva Conventions). Malaysia was party to this Convention, which it ratified on 21st December 1960.

Provision for that sovereignty has also been reiterated in Articles 2 and 3 of the 1982 *United Nations Law of the Sea Convention* (hereafter referred to as the 1982 Convention). Malaysia demonstrated its acceptance of the 1982 Convention by signing this historic legal and political document on 10 December 1982.

The 1982 Convention entered into force on 16 November 1994, on the first anniversary of the deposit of the 60th instrument of ratification. The document, as amended in 1996, is widely accepted. Malaysia was the 107th State to demonstrate its acceptance of the 1982 Convention on 14th October 1996. Some reservations on various aspects of the Convention were voiced by the Government of Malaysia, as had many other States, in documents available at the UN's website. However, it is not the intention of this narrative to discuss such matters.

The law of the sea has evolved over a period of five hundred years or more through customary and treaty law and the practice of States. The 1982 *United Nations Law of the Sea Convention* (the 1982 Convention) is a comprehensive political and legal document, which includes directives for international law, politics, and international relations. An *Agreement Relating to the Implementation of Part XI of the 1982 Convention* (the Agreement) is an essential package on international law. The 1982 Convention and the Agreement entered into force on 16th November 1994 and 28th July 1996, respectively, as a package deal.

Its status, as of 25th October 2023, suggests that the document is an international treaty which is crystallised in international law. Whereas initially there were 157 States and political entities signatories to the document in 1982, a present total of 169 have deposited their instruments of ratification. A mere 152 States have ratified the Agreement and 93 States have ratified the *Agreement for the Implementation of the provisions of the Convention Relating to the Conservation and Management of Straddling Fish Stocks and Highly Migratory Fish Stocks*. By their actions of ratification or accession to the 1982 Convention and its related documents, the Parties have obligations and duties according to the provisions of the convention and, at the same time, have the rights to extended maritime jurisdiction and resource utilisation within the prescribed zones.[1]

Malaysia's sovereignty extends, beyond the land and Malaysia's internal waters, to a belt of sea around Malaysia, of 12 M width, described as the Territorial Sea of Malaysia. This belt of sea, established through legislation enacted by the Government of Malaysia's *Ordinance* No. 7 of 2nd August 1969, is 12 M. This fact is reiterated in Act 750, the *Territorial Sea Act* of 2012. Hamzah and Forbes (2021) offered an informative study on this topic.[2]

Figure 4.1, the Index Map, portrays the extent of geographical coverage of each of the suite of maps, included in this study, that purport to show the spatial extent of Malaysia's maritime jurisdictional limits, which is the theme of this chapter and the focus of Sect. 4.7. A modicum of overlap to each adjoining map will assist the reader to use the information on each successive map and the flow of the lines that delineate the maritime boundaries—negotiated or potential—as shown on the following 21 maps. Each rectangular box, depicted by red lines represents the approximate geographical extent of each map in this series; an exception being that of the map of Brunei Bay.

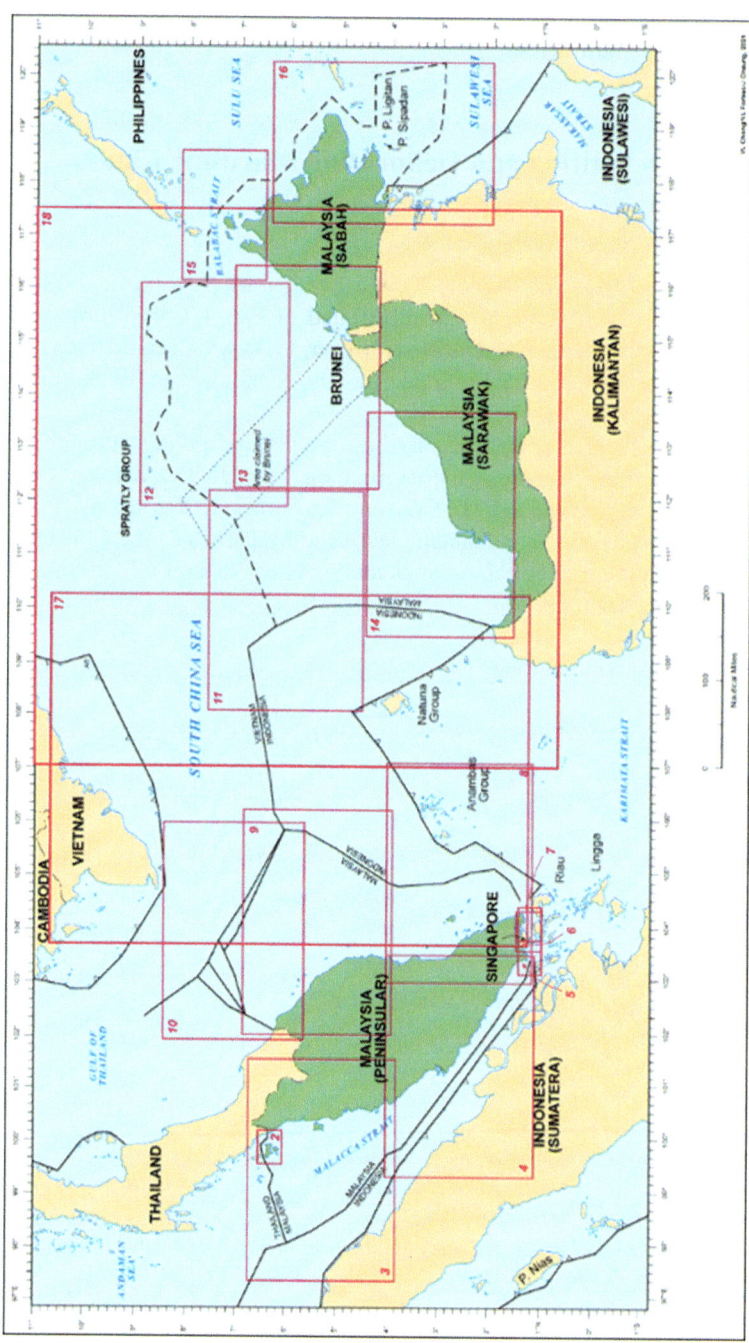

Fig. 4.1 An index map depicting arrangements of graphics for Sect. 4.7 of this Chapter

Figure 4.1 also depicts the seas that are adjacent to Peninsula Malaysia, Sabah and Sarawak and the neighbouring States. Malaysia, the nation, is presented in a green tint, its neighbours in a buff colour.

4.2 Defining, Delimiting and Delineating Maritime Limits

International Maritime Law

The limit of the *territorial sea* (Article 3), the *contiguous zone* (Article 23) the *exclusive economic zone* (Article 55) are illustrated in Fig. 4.2. The limit of the *continental shelf* is prescribed in Article 76. However, rocks which could not sustain human habitation or economic life of their own have no economic zone or continental shelf regimes (Article 121 (3)).

The State is responsible for defining its base points and giving due publicity through a public notice and depositing the information at the UN's Division of Ocean Affairs and Law of the Sea. (UN 1982 Convention; Forbes 2020a: 62–87; Forbes 2020b: 28 pages). Declarations and statements made by Parties to the Convention as well as Protests relating to certain actions and practices taken by States may be found in the UN webpages.[3]

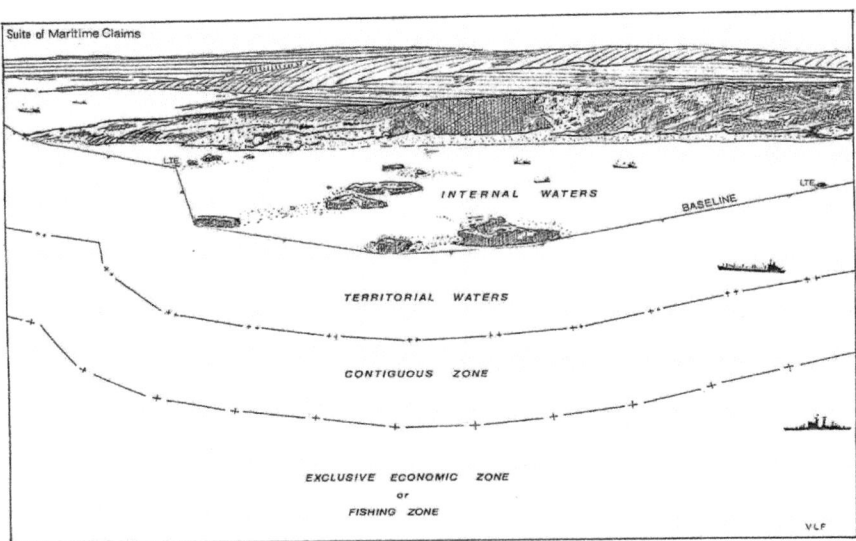

Fig. 4.2 Maritime jurisdictional zones. The inner limits of a natural continental shelf underlays the EEZ and may include the CZ. *Source* Author

4.3 Maritime Zones in Adjacent Seas

Malaysia's adjacent seas and straits not only contain valuable natural marine biotic and mineral resources but are also conduits for the sea lanes for intra-state, inter-regional and international shipping. The harvesting, exploitation, sustainable development, and indeed, protection of the natural resources and marine environment need diligent and effective management. Malaysia's maritime space, which can be 'zone-locked,' for the most part, abuts similar space of at least six other states and is traversed extensively by coastal and international shipping. Clearly defined maritime jurisdictional limits and zones are critical for decision-makers as well as for those authorities and agencies charged with implementing and enforcing the laws of the state and in meeting international obligations.

Malaysia's sovereignty over two islands in the Sulawesi Sea was acknowledged in a judgement handed down by the International Court of Justice (ICJ) in December 2003. A case concerning Pulau Batu Puteh (Pedra Branca) and associated geographical features, Middle Rocks, and South Ledge, located at the eastern approaches to the Straits of Malacca and Singapore was heard at the International Court of Justice (ICJ). A decision was handed down on 23rd May 2008.[4]

The Governments of Malaysia and its neighbour, Brunei Darussalam, discussed the issue of delimitation of their common maritime boundary in the South China Sea in August 2003. In 2009, the leaders of the two adjacent States signed a *Letter of Intent*. The precise details of the *Letter* are not in the public domain and therefore an analysis of that agreement is not possible. Suffice to say that the relationship between the two Governments is cordial and stable.

In May 2009, and again in November 2017, the Government of Malaysia took the initiative to make a submission to the Commission on the Legal Continental Shelf for its claim in the southern sector of the South China Sea. The May 2009 submission was a joint effort with Vietnam; the November 2017 lodgement was a unilateral action, which is discussed below.

The Governments of Indonesia, Malaysia and the Philippines will at some stage need to negotiate and resolve their common maritime boundaries in the north-western sector of the Sulawesi Sea. Likewise, the Governments of Indonesia, Malaysia and Singapore will require to settle the closure of two gaps that presently exist (by 15th May 2024) in the territorial sea boundaries in the eastern and western approaches to the Straits of Malacca and Singapore.

4.4 Territorial Sea Basepoint and Straight Baseline

The criteria for defining the territorial sea basepoints is presented in Article 7. The base points must be located on land territory and situated on or landward of the low-water line. No straight baseline segment may be drawn to a base point located on the land territory of another State.

Article 7:4 infers that only those low-tide elevations (LTE) which have had built on them lighthouses or similar installations may be used as base points for establishing straight baselines. Other low-tide elevations may not be used as base points unless the drawing of baselines to and from them has received general international recognition.

Article 7:6 stipulates that a State may not apply the system of straight baselines in such a manner as to cut off the territorial sea of another State from the High Seas or an EEZ. In addition, Article 8:2 provides that, where the establishment of a straight baseline has the effect of enclosing as internal waters areas which had not previously been considered as such, a right of innocent passage as provided in the Convention shall exist in those waters. Article 35(a) has the same effect with respect to the right of transit passage through straits.[5]

Article 7:2 makes provision for appropriate base points to be located along the furthest seaward extent of the low-water line in the case of a coastline that is deeply indented or is masked by islands, or is highly unstable.

The straight baseline segments drawn joining these base points remain effective, notwithstanding subsequent regression of the low-water line, until the baseline segments are changed by the coastal State in accordance with international law reflected in the Convention.

Article 9 permits the delineation of the straight baseline may be drawn across the mouth of the river between points on the low-water line of its banks.

Article 11, implies only those permanent artificial harbour works which form an integral part of a harbour system, may be used as part of the baseline for delimiting the territorial sea.

Article 13 implies that the low-water line on a low-tide elevation may be used as a basepoint if that LTE is situated wholly or partly at a distance not exceeding the breadth of the territorial sea measured from the mainland or an island. Article 14 authorises the coastal State to determine each baseline segment using any of the methods permitted by the Convention.[6]

Consider delineating straight baseline along the coastline of the river systems depicted in Fig. 4.3. A set of drawings in Fig. 4.4 illustrate various interpretations of the provisions contained in Article 10 on bay-closing lines.

Article 10:2 defines a 'juridical bay' as a well-marked indentation on the coast whose penetration is in such proportion to the width of its mouth as to contain land-locked waters and constitute more than a mere curvature of the coast. An indentation is not a juridical bay unless its area is as large as, or larger than, that of the semi-circle whose diameter is a line drawn across the mouth of that indentation.

Article 10:3 implies that the indentation is that area lying between the low-water mark around the shore of the indentation and a line joining the low-water mark of its natural entrance points.

If the distance between the low-water marks of the natural entrance points of a juridical bay of a single State does not exceed 24 nautical miles (M), the juridical bay may be defined by drawing a closing line between these two low-water marks, and the waters enclosed thereby shall be considered as internal waters (Article 10:4). Where the distance between the low-water marks exceeds 24 M, a straight baseline

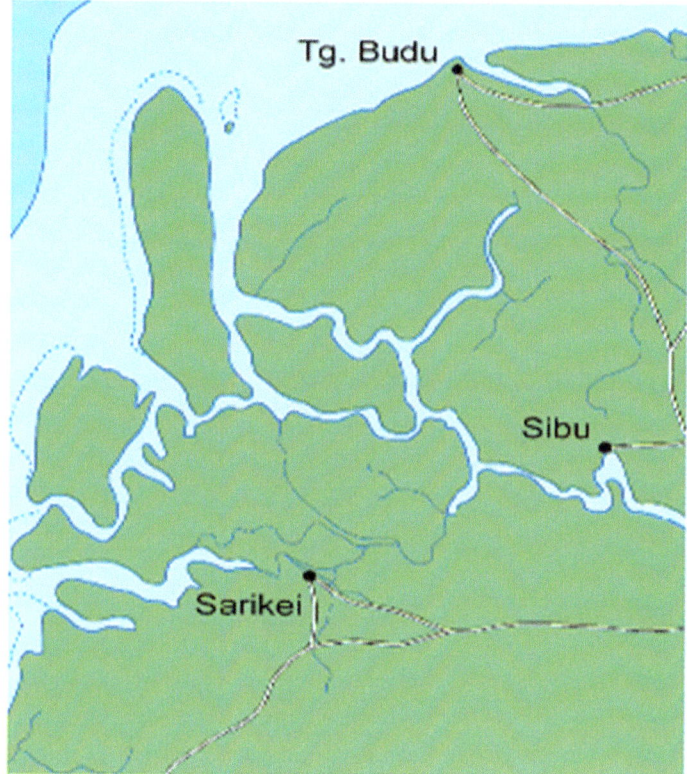

Fig. 4.3 Coastline and river mouths. *Source* Author

of 24 M shall be drawn within the juridical bay in such a manner as to enclose the maximum area of water that is possible within a line of that length.

Article 10:6 exempts so-called historical bays from the rules described above.[7]

4.5 Delineation of Baselines; Charts and Publicity

Article 16:1 requires that the normal baseline be shown on large-scale nautical charts, officially recognised by the coastal State. (Fig. 4.5) Alternatively, the coastal State must provide a list of geographic coordinates specifying the geodetic data. Drying reefs used for locating basepoints shall be shown by an internationally accepted symbol for depicting such reefs on nautical charts, pursuant to Article 6.

To comply with Article 16:2, the coastal State must give due publicity to such charts or lists of geographical coordinates and deposit a copy of each such chart or list with the Secretary-General of the United Nations. Malaysia complied with this requirement on 22nd August 2022.

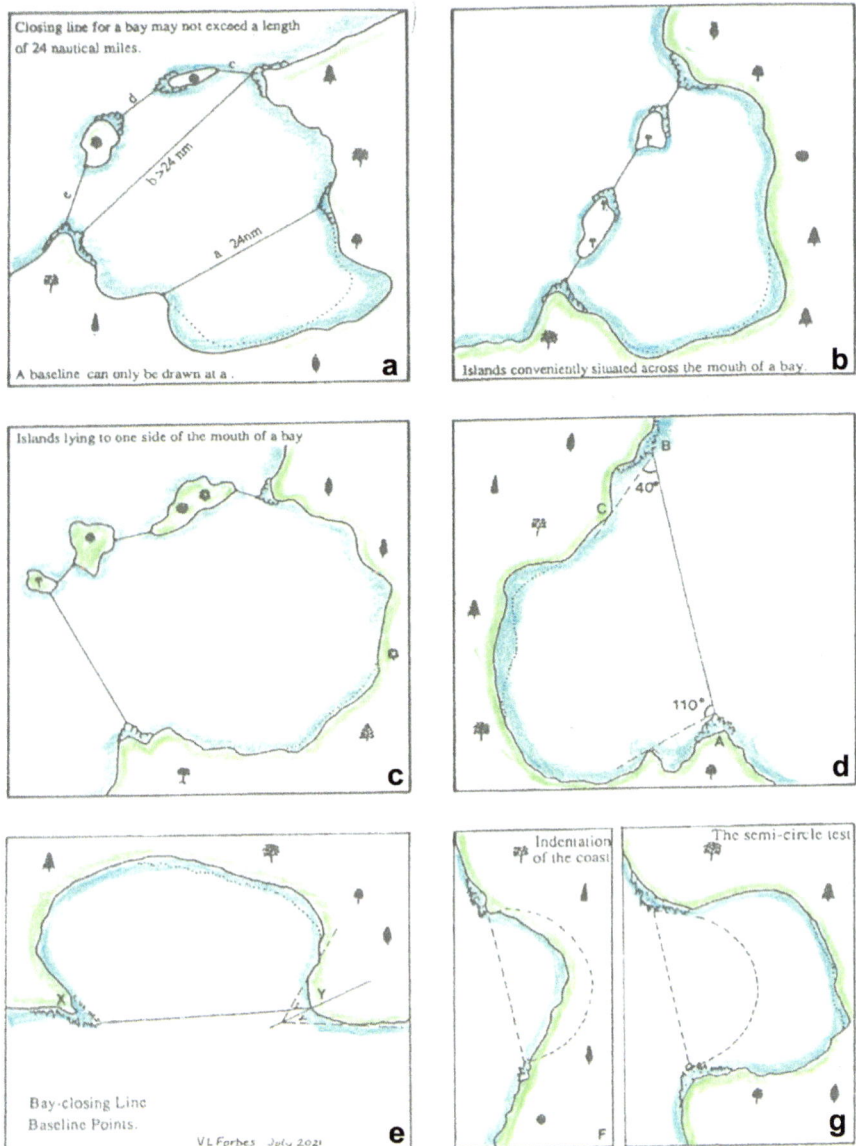

Fig. 4.4 Bay closing lines as a Territorial Sea datum. *Source* Author

All coastal and island States who desire to take advantage of claiming the suite of maritime jurisdictional zones as stated in the provisions of the 1982 Convention are obligated by Articles 16 and 84 of the said convention to deposit with the UN Secretary for the Law of the Sea a copy of the relevant charts or maps and/or a list of geographical coordinates that delineate and define the base points, the outer limits

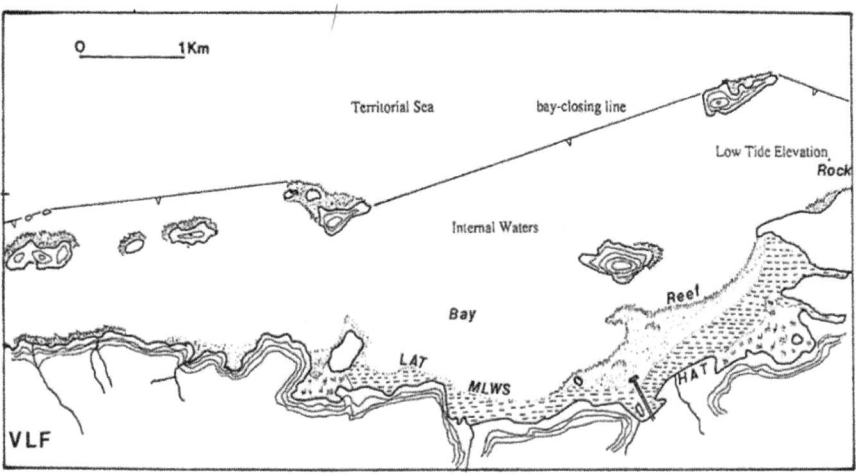

Fig. 4.5 Delineation of a state's straight baselines

of the territorial sea, the approximate limits of the Exclusive Economic Zone and the outer limits of the juridical continental shelf. (Fig. 4.6).

In the instance of the last-named concept, it is possible that any claim by a coastal and/or island State to an outer continental shelf will have to be approved by the Commission of the Legal Continental Shelf in the unlikely situation of a sovereignty and territorial dispute over the same area of the seabed and marine features on or near the waterline.

4.6 Determination and Delimitation of a Maritime Boundary

The processes of delimitating a maritime boundary would include determining an alignment of the boundary by defining reference points from arcs of equidistance would be delineated on charts of appropriate map-scales. The first step will be to ensure that parties to the agreement are satisfied with the datum employed at the land/sea interface—the terrestrial border post on or close to the low-water mark, for example, at the Eastern Border Post of the Indonesia/Malaysia boundary on Pulau Sebatik (Refer to Figs. 1.8, 2.22 and 2.23). Each negotiating team will wish to ensure that the other's territorial sea basepoints are in accord with the provisions of the 1982 Convention.

The alignment of the Territorial Sea boundary, according to the 1982 Convention must be ranged on an equidistance principle in the situation of states with adjacent coasts, as depicted in the hypothetical example illustrated in Fig. 4.7, or on a median-line basis for states whose coasts are opposite to each other. Beyond the 12-M limit

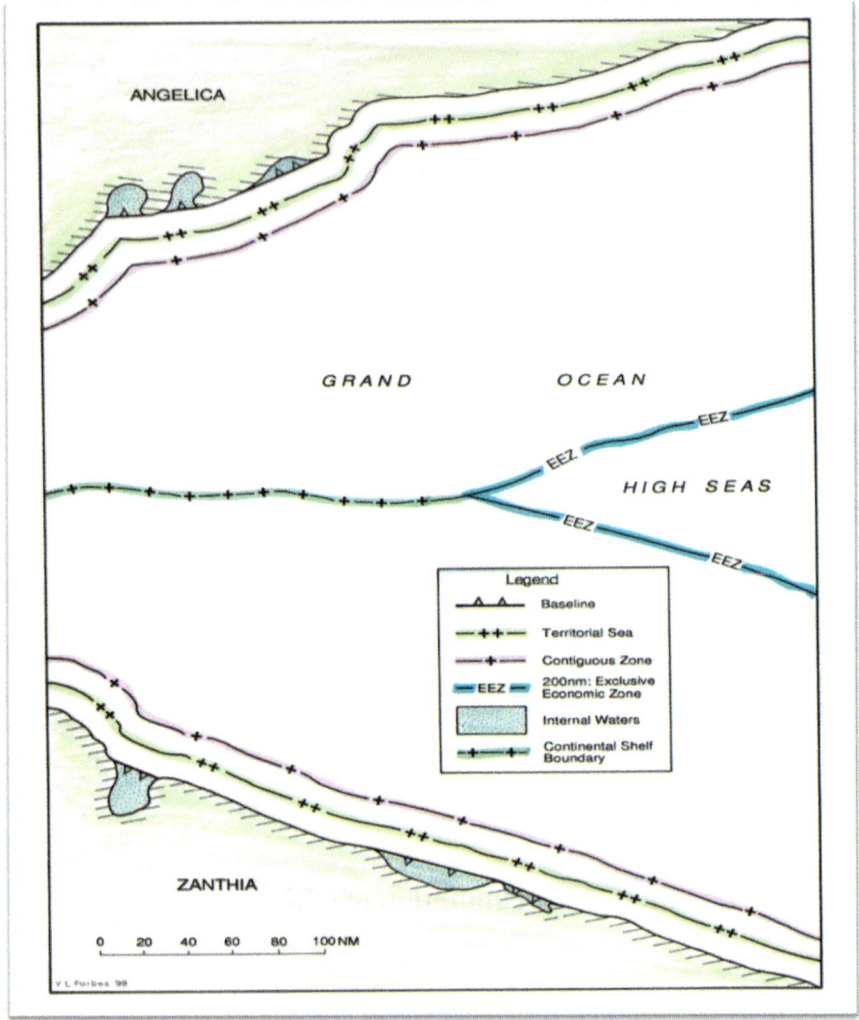

Fig. 4.6 Hypothetical maritime boundaries between states with opposing coasts

the boundary will be projected to cover the concepts of the Contiguous Zone, the natural continental shelf, and the Exclusive Economic Zone if the distances permit and provided a third or more States are not closer than 400 M away.

The narrative and map of the delimitated maritime boundary will be documented on an Agreement with the listing of the geographical coordinates of the terminal and turning points. Examples of maritime boundary agreements appear in the Annexe of this present volume and the boundary alignments and arrangements of Malaysia and its maritime neighbours.

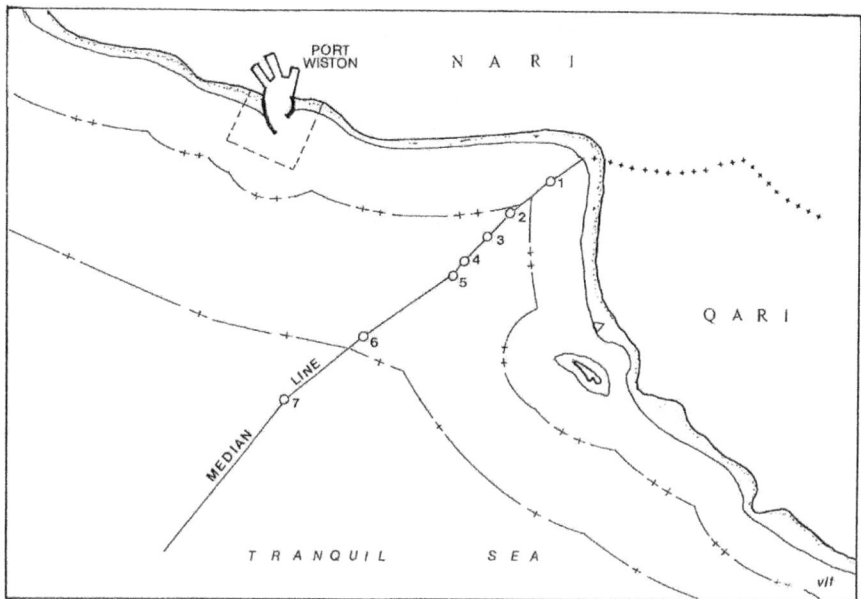

Fig. 4.7 Hypothetical maritime boundary between adjacent states. *Source* Author

4.7 Malaysia's Actual and Potential Maritime Limits

The semi-enclosed seas of Southeast Asia are partially or wholly politically determined through negotiations between the respective littoral States and by unilateral declaration and are delineated on this map. The latter are legally and naturally subject to negotiations when the parties to the disputes are ready to undertake the initiative.[8]

Figure 4.8 shows information relevant to Malaysia's actual and potential maritime jurisdictional claims. Portion of Indonesia's archipelagic baseline system is delineated as defined in the revised Indonesian Proclamation of 2002, which is discussed in detail in Chapter Seven, below. The Territorial Sea baseline systems (straight baselines) for Cambodia, Thailand, and Vietnam in the southern portion of the Gulf of Thailand are also portrayed on relevant maps.

Malaysia's continental shelf boundary with Indonesia is in three sections. One is within the Malacca Strait. It is a series of ten straight lines, which commences at the Common Point and is then numbered from one to ten. The geographical coordinates of the points are defined (See Annex II). Points 11–20 and 21–25 (upright style) are located as two segments in the South China Sea on either side of the Anambas and Natuna archipelagos which belong to Indonesia.

The Malaysia/Thailand maritime boundary towards the northern limits of the Malacca Strait is delineated, and it connects to a Common Point that links to a portion of the Indonesia/Thailand maritime boundary which is projected north-westerly into the Andaman Sea.

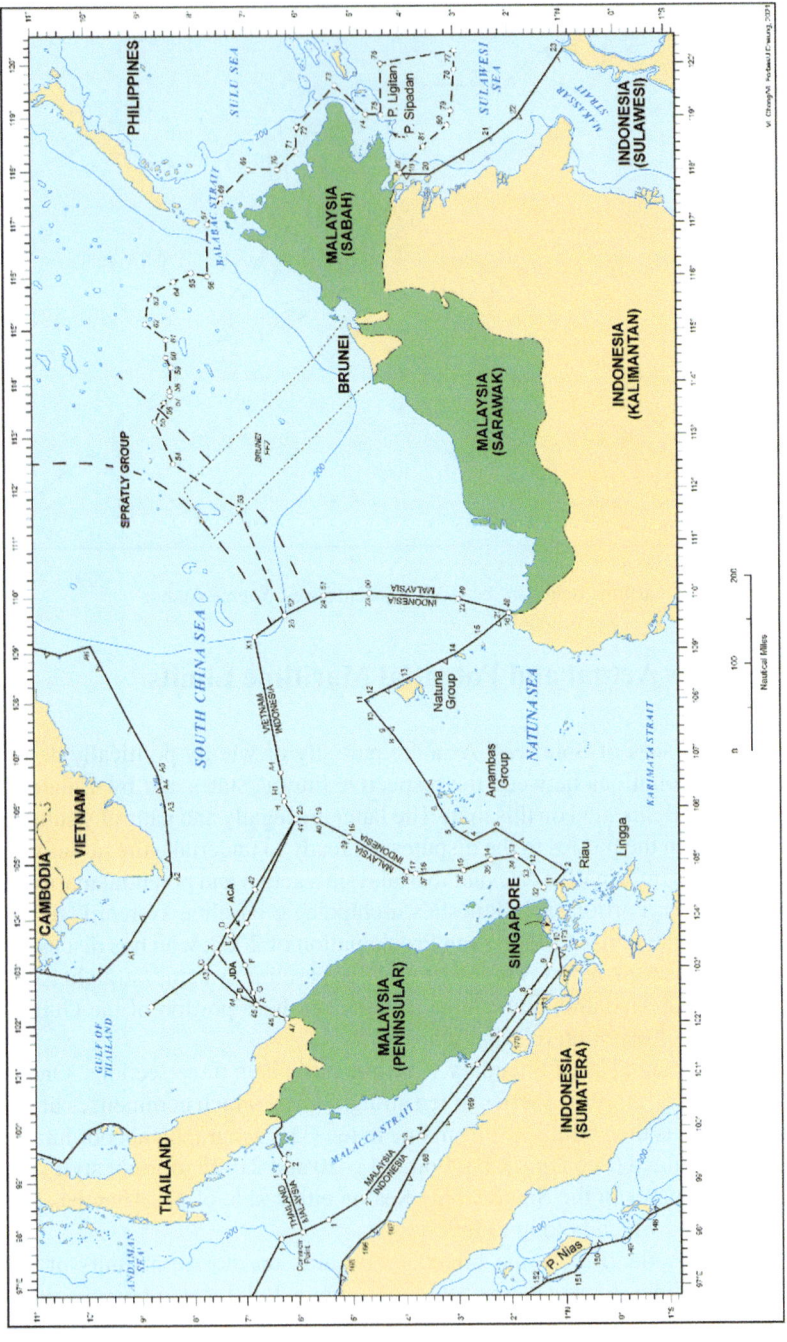

Fig. 4.8 Malaysia's actual maritime boundaries and a unilateral claim

To the east of Peninsula Malaysia, in the western sector of the South China Sea, the extent of the Joint Development Area (JDA) between Malaysia and Thailand and the Agreed Common Area (ACA) delimited by Malaysia and Vietnam are delineated on this map. The two polygons have been defined for the express purpose of exploration and exploitation of hydrocarbon reserves in the substratum of that portion of the South China Sea.[9]

Malaysia's unilateral claim to a continental shelf, beyond the 200-m isobath, numbered Points 32–47 (*italic* style) and shown as solid black lines; from 48 to 84, depicted as a pecked line in Fig. 4.8. Brunei claims a portion of this area. The overlapping claim is depicted on the map.[10]

The other maritime boundaries are those between Indonesia and Vietnam, comprising five-line segments and labelled (H, H1, A4, X1 and 25) and a north-westerly projection from Point 'C' of the JDA representing the Thailand and Vietnam maritime boundary. The long-pecked lines (200 M arcs) represent the Joint-submission claim by Malaysia and Vietnam.[11]

The discussion on the maritime limits of Malaysia that follows is geographically grouped. The first Sect. 4.8 relates to the maritime limits within the Malacca Strait; the second Sect. 4.9, discusses the maritime limits within Singapore Strait; the third Sect. 4.10, covers the area of the southern sector of the South China Sea abutting the Malaysian landmasses; and the fourth Sect. 4.11, that off the north and east coast of Sabah, Malaysia.

4.8 Malaysia's Maritime Limits Within the Malacca Strait

A maritime boundary and the limits to the Port of Langkawi are portrayed on this map, Fig. 4.9. The boundary, which runs through Chinchin Strait separates Malaysian territory from Thailand's islands in the vicinity of the Strait (*Selat*). The boundary commences at the western terminal point of the Malaysia/Thailand terrestrial boundary (Point 'A') which is northwest of Kuala Perlis. The maritime boundary, read Territorial Sea boundary, comprises a series of straight-line segments linking points A, B, C, D and E.

Langkawi is an archipelago of 104 islands lying off the north-western coast of Peninsular Malaysia. The largest of these islands is called Langkawi Island and is bigger than Pulau Pinang (Penang). The other islands in this group are Pulau Dayang Bunting (Island of Pregnant Maiden), Pulau Singa Besar (Lion Island) and Beras Basah Island. The smallest of these islands are atolls about 20 square metres or even smaller. Langkawi's major ports are Jeti Teluk Ewa, Langkawi and Kuah, Langkawi which fall under the administration of the Marine Department of Malaysia. The island is a major resort centre and has a duty-free status which has brought economic benefits to the locals on the island and as well as to the adjacent mainland communities of Malaysia and Thailand.

The extent of Malaysian coastline covered in Fig. 4.10 is from Kuala Perlis, near the Malaysia/Thailand boundary to Pulau Jarak which lies about 40 nautical miles

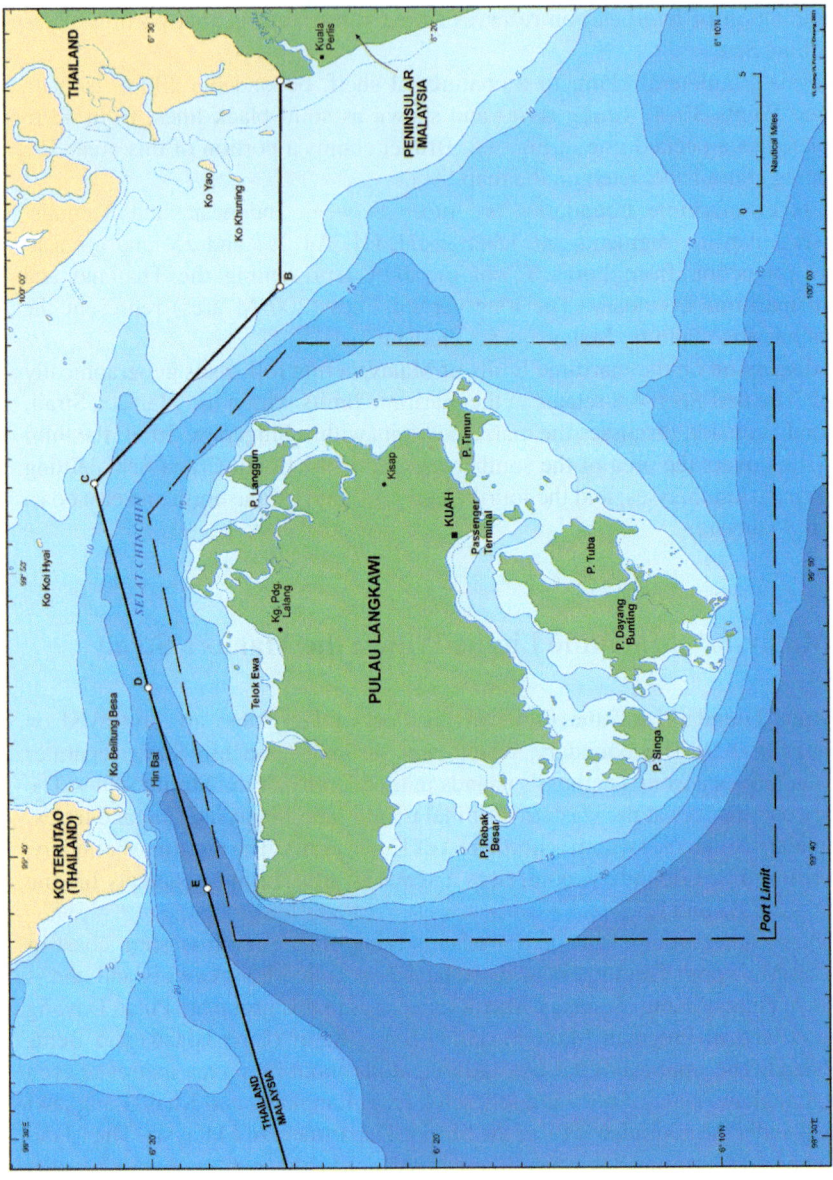

Fig. 4.9 Partial maritime limits between Malaysia and Thailand

(M) southwest of Pulau Pangkor and the Port of Lumut. The map encompasses the northern half of the Malacca Strait.

Indonesia's Archipelagic Waters base points are depicted as 165, 166 and 167. The lines linking these points are the straight baselines, the datum for measuring the breadth of that country's Territorial Sea. These lines are also the outer limits of Indonesia's Archipelagic Waters.

Arcs of a radius of 12 nautical miles (M), representing the limit of a State's Territorial Sea is shown in red around the islands and off the coastline. In the case of Indonesia, the arcs are measured from that country's straight baseline system. This map also offers an indication of the approximate alignment of the submarine cables (the *wavy* lines) for telecommunication purposes that transcend international political boundaries. Isobaths for the 20-, 50- and 200-m are depicted.

On this map, three sets of continental shelf boundaries are illustrated: one in its entirety, namely, the Malaysia and Thailand maritime boundary; and the other two as portions of a greater length, which is the Malaysia/Indonesia continental shelf boundary and a northerly extension from the Common Point of an Indonesia/Thailand boundary.

The Malaysia/Indonesia continental shelf boundary was delimited as a series of straight lines linking Points (1, 2 and 3) whose geographical coordinates were defined in a bilateral agreement. The location of two islands, Pulau Jarak and Pulau Perak, some appreciable distance off the Malaysian coastline advantages Malaysia in its maritime claim in the strait.

The three major ports shown on this map are Butterworth, George Town (commercial) and Lumut (Naval). There are numerous smaller ports and jetties that are used for intra-strait trade and commercial fisheries.

The southern half of the Malacca Strait and adjacent landmasses are depicted in Fig. 4.11. The maritime jurisdictional limits portrayed are portion of Indonesia's archipelagic baseline system, arcs of 12-M Territorial Sea, the southern sector of the Malaysia/Indonesia continental shelf boundary and an agreed Territorial Sea boundary defined between Indonesia and Malaysia by straight lines linking labelled Points 1–8 as shown in italic style.

Points 2–4 of the TS boundary are coincident with Points 5–7 (inclusive) of the continental shelf boundary. From Point 4 the alignment of the TS deviates in a southwesterly aspect to link to Points 5–8. Point 8 of the TS is identical to Point 10 of the continental shelf boundary which is less than five miles southwest of Pulau Kukup.

Tanjong Tuan and the anchorage at Malacca are the major commercial ports on this map. Other coastal towns are havens for boats engaged in fishery operations. A marina is established at Port Dickson.

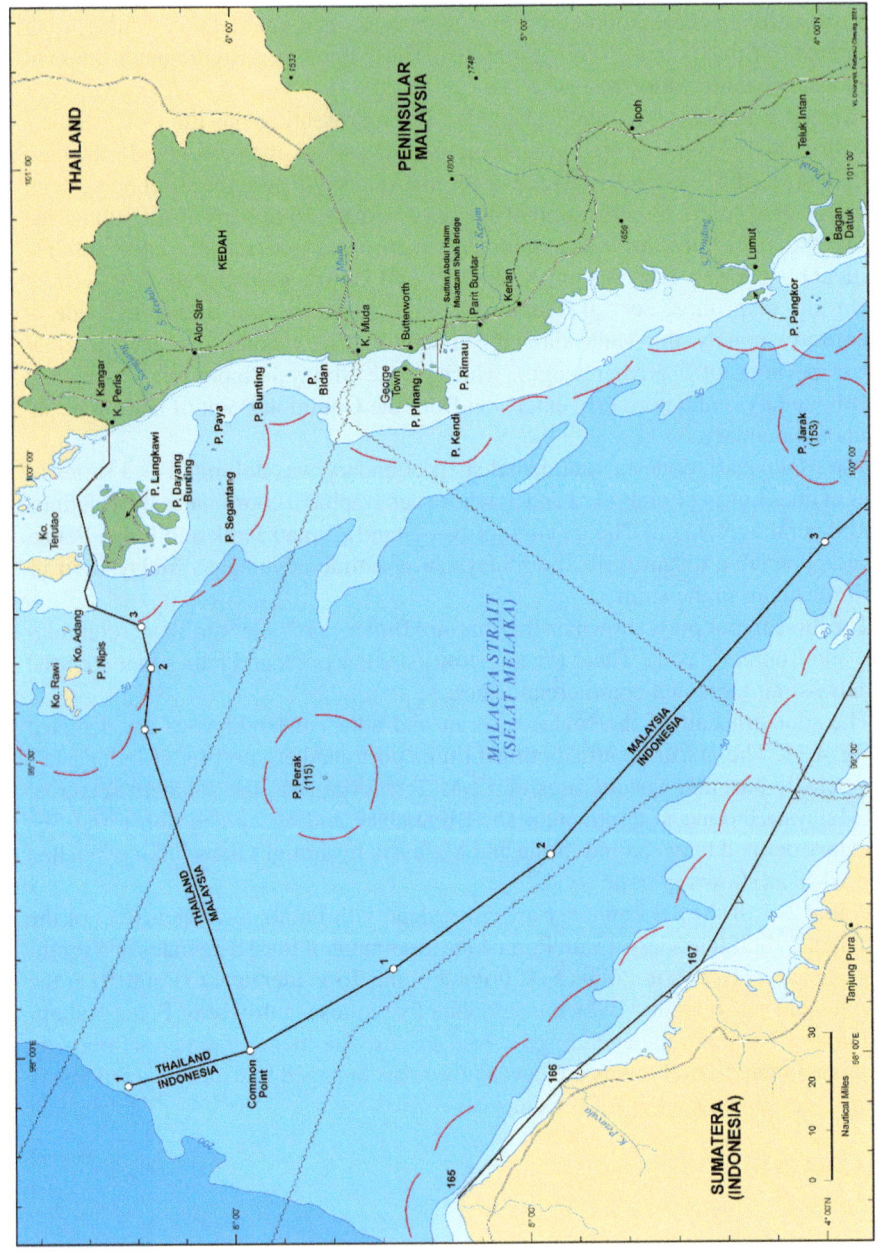

Fig. 4.10 Malaysia's maritime limits in the vicinity of Perak and Jarak Islands

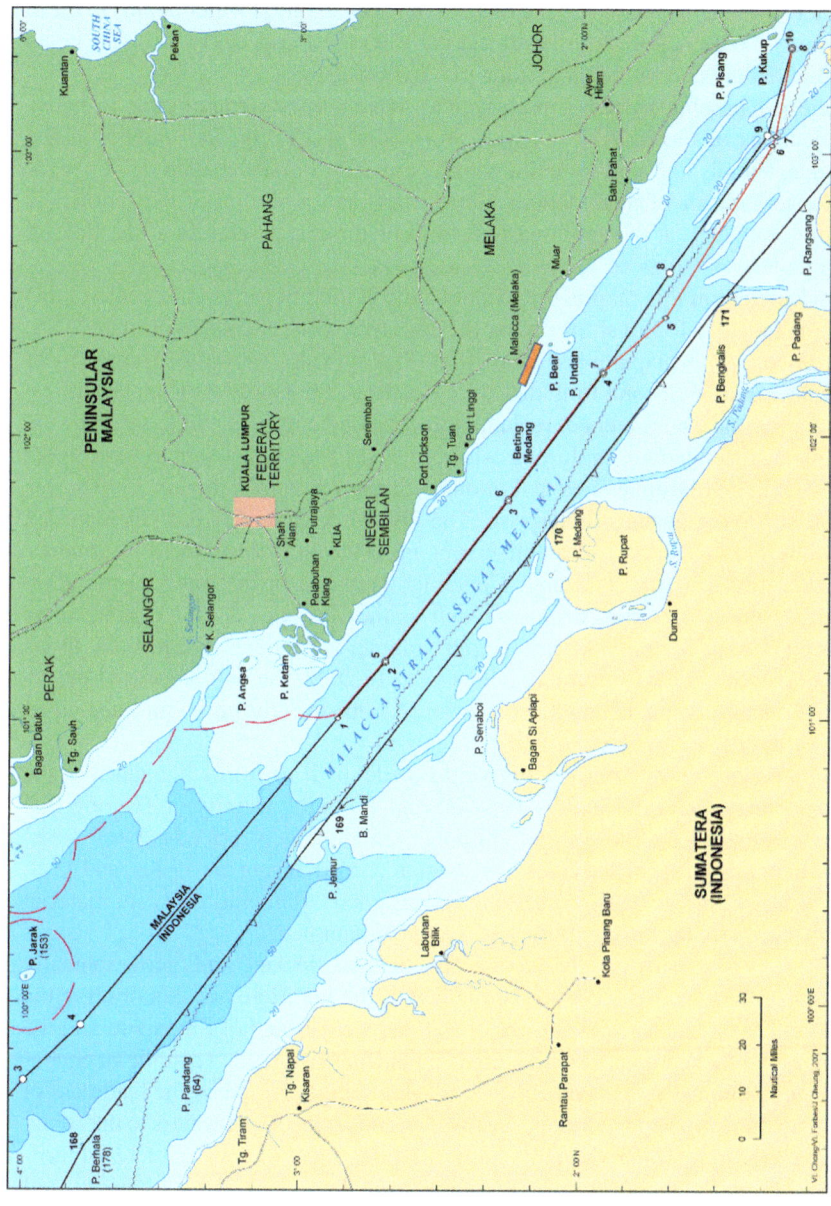

Fig. 4.11 Malaysia's maritime limits in the southern sector of Malacca Strait

4.9 Malaysia's Maritime Limits Within Singapore Strait

The western approaches to the Singapore Strait are shown in Fig. 4.12. Numerous straits, not all of them labelled on the map, exist in the vicinity. Water depths in these straits rarely exceed 20 m. Marine traffic daily is dense especially as small boats engaged in fishing and inter-island trade ply these waters.

Indonesia's baseline system is portrayed as straight lines linking Points 173–177 and projections of the lines to the west and east respectively. Two territorial sea boundaries, namely, the Malaysia/Singapore boundary, Points labelled W1 to W25, in the western sector of Selat Johor (Johor Strait) and a portion of the Indonesia/Singapore boundary, at the south-eastern limit of the map, which is a line linking Points 1 and 2.

Also depicted on this map is the extent of the Traffic Separation Scheme (TSS) which is mandatory for all ships plying these waters. Water depths within the TSS, encompassed by this map are less than 20 m. The map also shows the Causeway at Johor Bahru and the Second Link that connects Tuas, Singapore with Malaysia.

The approximate alignment of reclamation developments being undertaken by the Government of Singapore for Jurong Island and the Tuas Extension are portrayed. A major container terminal sited at Tanjong Pelapas was established in 1999 and is presently a thriving business enterprise. The Port of Tanjung Pelepas has recorded tremendous growth since its establishment in 1999 and was named 'Container Terminal of the Year' in 2004 and 'Container Terminal of the Year' in the 2006 *Lloyd's List* Asia Maritime Awards as well as the Logistics Award organized by Lloyd's FTB Asia. In terms of throughput volume, it handled 4.02 million TEUs in 2006 a rise of 15.2% from the previous year. In 2022, the port surpassed the 11 million TEUs throughput handled in one year. Expansion of the port with a dedicated storage yard will be completed by 2026.

Of significance on this map is the 'yet to be defined' maritime boundary of the three States that would link Points 10, W25 and 1C in the vicinity of Tanjong Piai and the Indonesian base line system. The area is affectionately termed a 'Grey Zone' within which many illegal activities have been recorded during the 1990s, and later, that included people smuggling, trade in illegal goods, dumping of waste material and acts of armed robbery and piracy, actual and attempted.

Figure 4.13 illustrates vividly the concepts of 'zone-locked' and 'geographically disadvantaged' states in the context of the law of the sea. Here one can observe the masking effect of Malaysia's southern coastline by the island of Singapore and the resultant 'zone-locked nature' whilst the islands comprising the Republic of Singapore—a geographically-disadvantaged state—are semi-enclosed by the landmass of Malaysia to the north and the Indonesian islands and archipelagic waters to the south. Thus, all three littoral States cannot claim full entitlement of a width of 12-M territorial sea within Singapore Strait. The map shows the territorial sea boundaries of Malaysia and Singapore in Selat Johor (East and West streams) and Indonesia and Singapore in the main Channel of Singapore Strait, in their entirety. A portion of

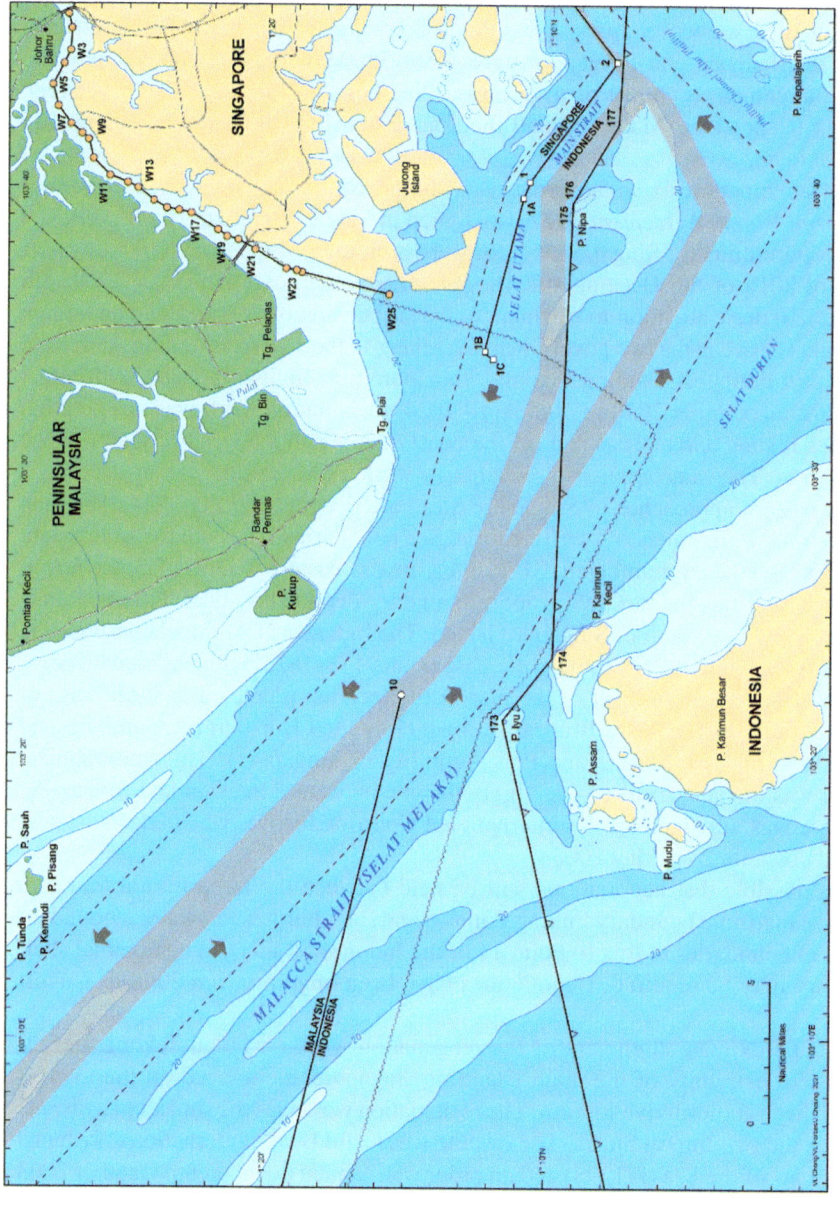

Fig. 4.12 Malaysia's maritime limits at the western approaches to Singapore Strait

Indonesia's extensive archipelagic baseline system is shown here as lines linking base points 175–181.

Extensive reclamation has been undertaken, for many decades, by Malaysia and Singapore within the Johore Straits and in the Singapore Strait by Singapore, particularly in Tuas, western Singapore and Pulau Tekong and Changi Area. The Governments of Malaysia and Singapore have yet (15th May 2024) to agree upon the precise alignment (delimit) of a Territorial Sea boundary in the vicinity between Tanjung (Cape) Piai (Malaysia) and the Tuas Extension (Singapore's reclamation undertakings).

The alignment was in accordance with the *Straits Settlements and Johore Territorial Waters Agreement* 1927 was signed in Singapore on 7th August 1995.

New port limits for Singapore's MPA were promulgated in *Port Marine Circular* No. 9 of 2018 of 6th December 2018. *Port Marine Circular* No 06 of 2019 (8th April 2019) drew attention to mariners of the suspension of Port Limits pending an agreement between the two governments. It relates to the area surrounding the Johore Bahru port limits off Tanjung Piai and Singapore port limits off Tuas. With effect from 29th September 2020 to 30th April 2024, reclamation work was in operation at Pulau Tekong and at Tuas and off the coast of Johor in the western sector of the Johor Strait. The same applies to reclamation work by Malaysia.

The eastern approaches to Singapore Strait are portrayed in Fig. 4.14. The map also shows the alignment of two Territorial Sea boundaries. They are the Malaysian/ Singapore boundary numbered E1 to E47, in the eastern sector of Selat Johor, inclusive, and the Indonesia/Singapore boundary depicted as lines linking Points 3–6, inclusive just south of Singapore Island. The Indonesian Archipelagic Baseline system is depicted as lines joining the base points 178–182, inclusive.

Towards the eastern limit of the map is the location (depicted as a red star) of Pulau Batu Puteh (or Pedra Branca upon which is sited Horsburgh Lighthouse and administered by the Marine and Port Authority of Singapore). This geographical feature was the subject of a sovereignty dispute between Malaysia and Singapore. The case was adjudged by the International Court of Justice, whose decision was handed down on 23rd May 2008.

It is possible that negotiations will be held between the three littoral States to finalise a maritime boundary linking Points 8 and E47. Further eastward, a boundary or sets of boundaries to link to Point 11 of the Indonesia/Malaysia continental shelf boundary may be required. Negotiations have been ongoing from 2008 to August 2023.

The approximate alignment of the reclamation work on Pulau Tekong and off the eastern extremity of the main island of Singapore, as inferred on the map, is derived from authoritative sources. The reclamation work was the subject of a dispute between the two littoral States. The dispute was heard by the International Tribunal for the Law of the Sea (ITLOS) and an order was handed down on 8th October 2003 which unanimously:

> **Directed** *Singapore not to conduct its land reclamation in ways that might cause irreparable prejudice to the rights of Malaysia or serious harm to the marine environment, taking especially into account the reports of the group of independent experts.*

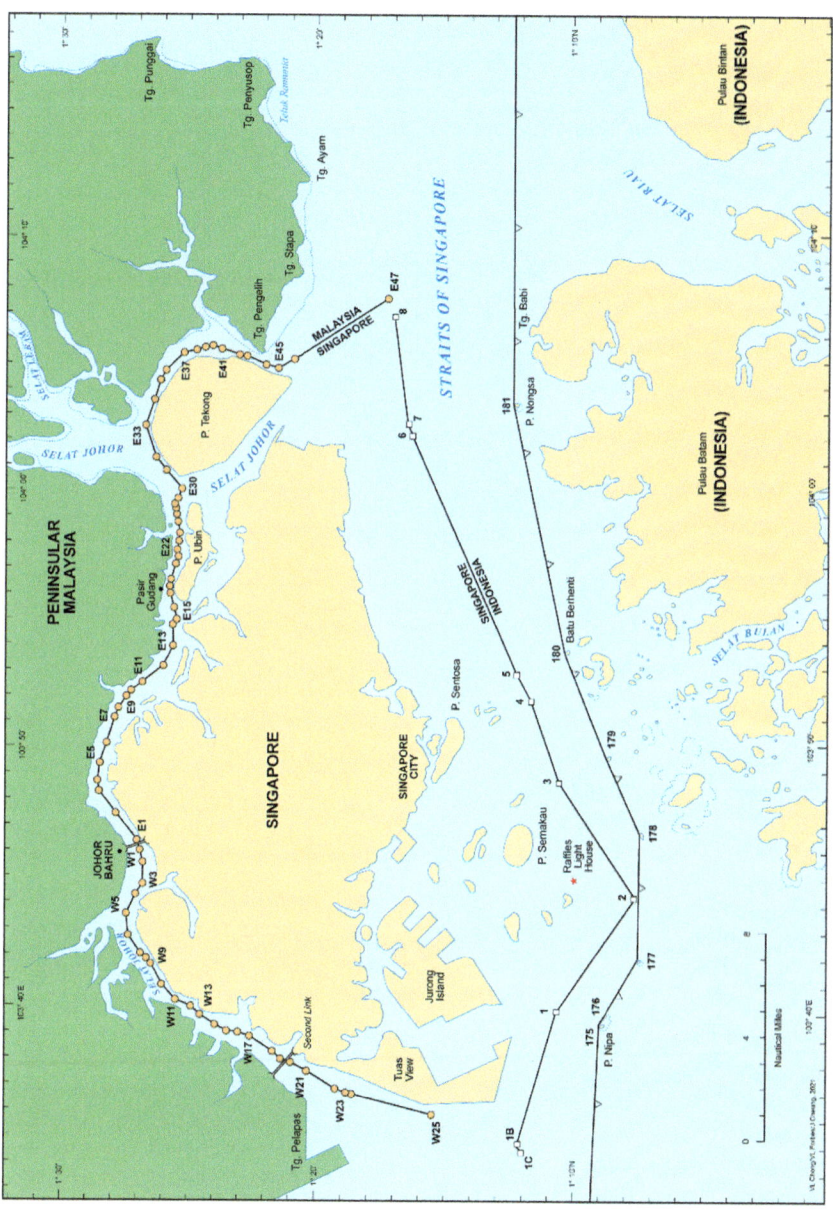

Fig. 4.13 Partial delineation of maritime limits within Singapore Strait

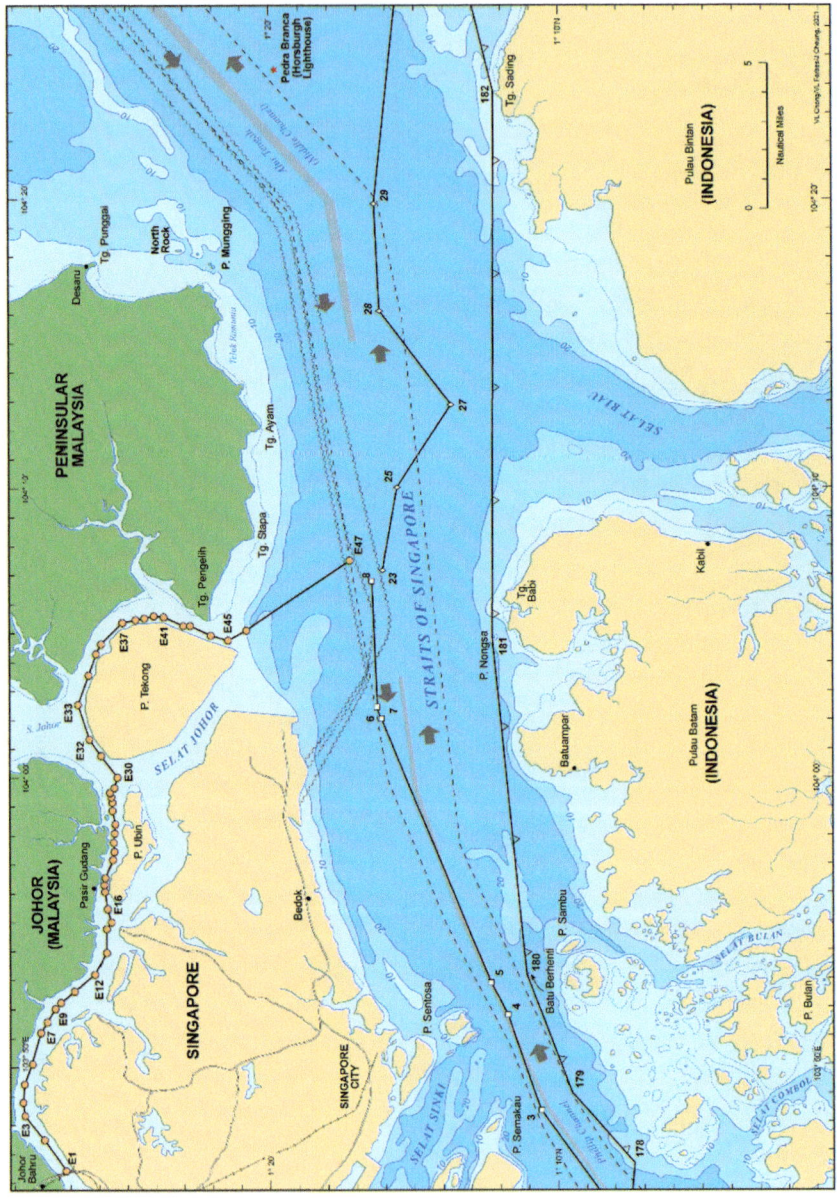

Fig. 4.14 Partial maritime limits at the eastern approaches to Singapore Strait

Decided *that Malaysia and Singapore shall each submit the initial report referred to in article 95, paragraph 1, of the Rules, not later than 9th January 2004 to this Tribunal and to the Annex VII arbitral tribunal, unless the arbitral tribunal decides otherwise; and,*

Decided that each party shall bear its own costs.

Both Governments accepted the decisions of the ITLOS and agreed to work towards resolving their concerns and differences relating to the reclamation within the Straits of Johor.

Further developments along the foreshores of both States and hinterland of Malaysia continue at a rapid rate that includes port infrastructures and the construction of a bridge across the mouth of Sungai Johor.

One of many recent cooperative developments undertaken by Malaysia and Singapore within Johore Strait is the Johor Bahru-Singapore Rapid Transit System (RTS) Link. It will be a four-kilometre shuttle service between the Malaysia terminus at Bukit Chagar station in Johor Bahru and the Singapore at Woodlands North Station. The cross-border RTS Link will stretch 2.7 km in Malaysia and 1.3 km in Singapore. The predicted (January 2024) peak capacity of up to 10,000 passengers per hour in each direction. A journey time of about five minutes is envisaged.

Facilities for Customs, Immigration and Quarantine (CIQ) will be co-located at the Bukit Chagar (Malaysia) and Woodlands North, Singapore stations. Planned completion of the facilities and the RTS is set for December 2026. The new link will significantly improve connectivity between Johor Bahru and Singapore and assist in easing congestion along the Causeway and the Second Link to the southwest.

The RTS Link will cross the border to the east of the Causeway somewhere between E1 and E2 (top left-hand corner of the map) that delimits the Territorial Sea boundary between the two neighbours which are depicted on Fig. 4.14.

The three marine features, as depicted in Fig. 4.15, located at the eastern approaches to the Singapore Strait, are Pedra Branca (a name given by Portuguese navigators for White Rock) or Pulau Batu Puteh, by its Malayan name, and the others are Middle Rocks and South Ledge.

Pedra Branca is a granite island, measuring 137 m long, with an average width of 60 m and covering an area of about 8560 m^2 at low tide. It is located at Lat. 1° 19′48″ N and Lon. 104° 24′ 27″ E. It lies approximately 24 M to the east of Singapore, 7.7 nautical miles to the south of the Malaysian state of Johor and 7.6 M to the north of the Indonesian island of Bintan. On the island stands Horsburgh Lighthouse, which was erected between March 1850 and October 1851 when it commenced operations.

Middle Rocks and South Ledge are the two marine features closest to Pedra Branca. Middle Rocks is located 0.6 M to the south and consists of two clusters of small rocks about 250 m apart that are permanently above water. The eastern rock in the group, located at Lat. 1° 19′ 15″ N., Lon. 104° 24′ 36″ E., has a charted elevation of 2.2 m; its western counterpart, sited at Lat. 1° 19′ 15″ N., Lon. 104° 24′ 24″ E has an elevation of 1.6 m.

South Ledge, whose geographical coordinates are Lat. 1° 17′ 51″ N. and Lon. 104° 23′ 33″ E. is about 1.6 M to the south-south-west of Middle Rocks and 2.2 M to the south-south-west of Pedra Branca. It is a rock formation only visible at low-tide

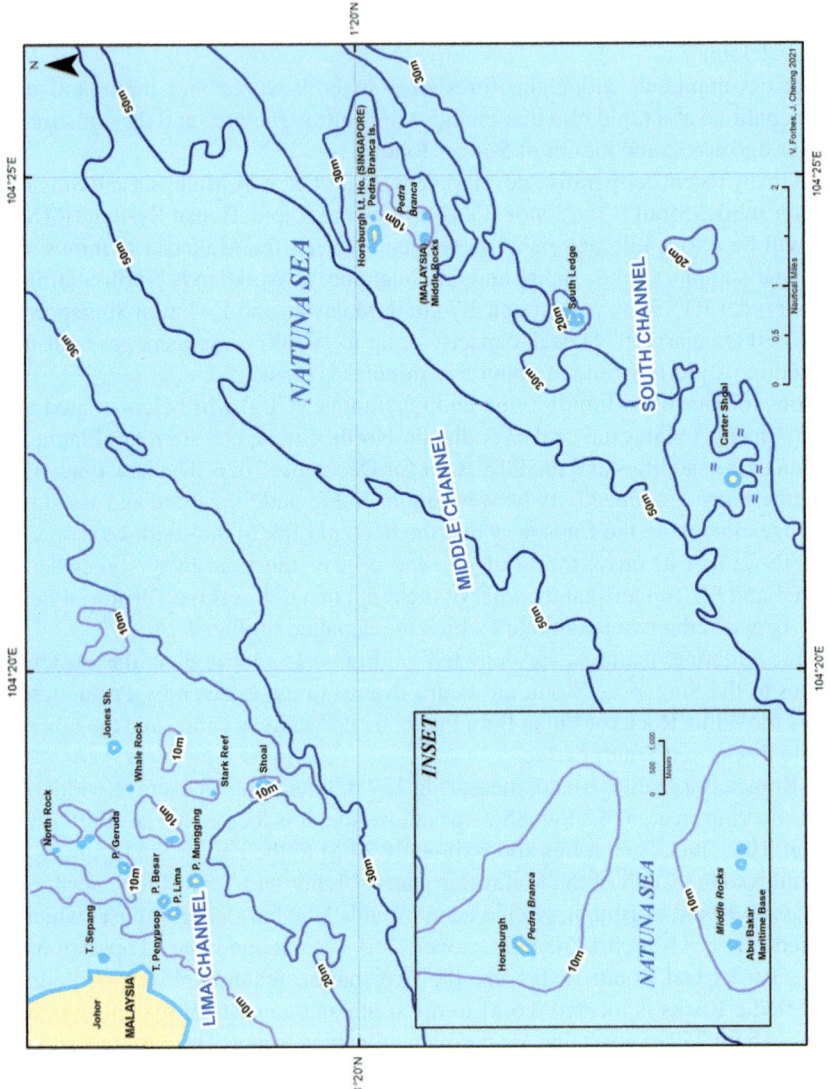

Fig. 4.15 Middle rocks and adjacent marine features

Fig. 4.16 Middle rocks the site of Abu Bakar Marine Base. It was inaugurated on 1st August 2017. The rocks are about one kilometre south of Pedra Branca. *Source* Author

event. This feature lies about 5.5 M north of Indonesia's Archipelagic base point No. 182 which is located on a small island off Tanjong Sading. Within the meaning of the 1982 Law of the Sea Treaty it is termed a Low-Tide Elevation (LTE). Each of the three above-named features may be used as basepoints by the sovereign State in establishing a datum to measure the width of its Territorial Sea and other maritime jurisdictional zones where it is deemed necessary.

A structure called Abu Bakar Marine Base has been established which is attached to Middle Rocks. (Fig. 4.16) It comprises a 300 m-long jetty, a helipad and lighthouse designed to safeguard Malaysia's sovereignty and utilised to conduct marine scientific research.

Sovereignty over Pedra Branca (Fig. 4.17) was awarded to Singapore by the International Court of Justice on 23rd May 2008; whilst Malaysia is recognised as the sovereign holder of Middle Rock. Sovereignty over the third disputed rock, South Ledge, is to be determined later by the Parties to the dispute (Malaysia and Singapore) when the Parties delineate the limits to their respective territorial seas. The court noted that sovereignty over South Ledge would belong to the "state in the territorial waters of which it is located." The Judgement of the Case is described in the Annexe, below.

In 2017, the then Government of Malaysia applied for a review of the 2008 ICJ Ruling. However, the Pakatan Harapan Government that took office in 2018, withdrew the application. In January 2023, the incumbent Government of Malaysia, whilst respecting the 2008 ICJ Ruling, considered that withdrawing the 2017 application to review the ruling was "improper".

In January 2024, a Royal Commission of Inquiry (RCI) was established to examine the reasons for the withdrawal of the review application. It is opined that the inquiry was an internal matter and to better prepare the authorities to cope with similar sovereignty issues in the future. Malaysia and Singapore have set up a Joint Technical

Fig. 4.17 Pedra Branca and Horsburgh Lighthouse, Natuna Sea. *Source* B. D. Proudfoot and present author

Committee to implement the 2008 ICJ Ruling on Pedra Branca, Middle Rocks and South Ledge. This Committee will discuss the sovereignty over South Ledge.

4.10 Malaysia's Maritime Limits: Southern Sector South China Sea

The southern half of the eastern seaboard of Peninsular Malaysia, from Tanjong Sepang to a few miles north of Tanjong Gelang is portrayed in Fig. 4.18 together with the offshore islands, the continental shelf boundary between Malaysia and Indonesia, numbered as Points 11–17 and a portion of Indonesia's archipelagic baseline system in the vicinity of the Anambas Group of islands, identified as Points 3–7. The projected lines south-westward from Point 3 and north-eastward from Point 7 illustrate the alignment of the baselines.

The outer limits of the Territorial Sea of Malaysia and Indonesia are depicted as arcs of circle of 12-M radius.

The wavy lines, coloured grey, show the approximate alignment of the submarine cables and the pecked lines are the recommended routes that Masters of ships are advised to take in the vicinity during adverse weather conditions during the Monsoons. The major ports of Mersing and Kauntan are shown on this map.

The northern half of the eastern seaboard of Peninsular Malaysia, from Tanjong Gelang to the Malaysian/Thai terrestrial (in reality, riverine) boundary, just northwest of Tumpat is featured in Fig. 4.19. In addition, the continental shelf boundary between Malaysia and Indonesia is shown as a series of straight lines linking Points 17–20, and a portion of the Indonesia and Vietnam continental shelf boundary

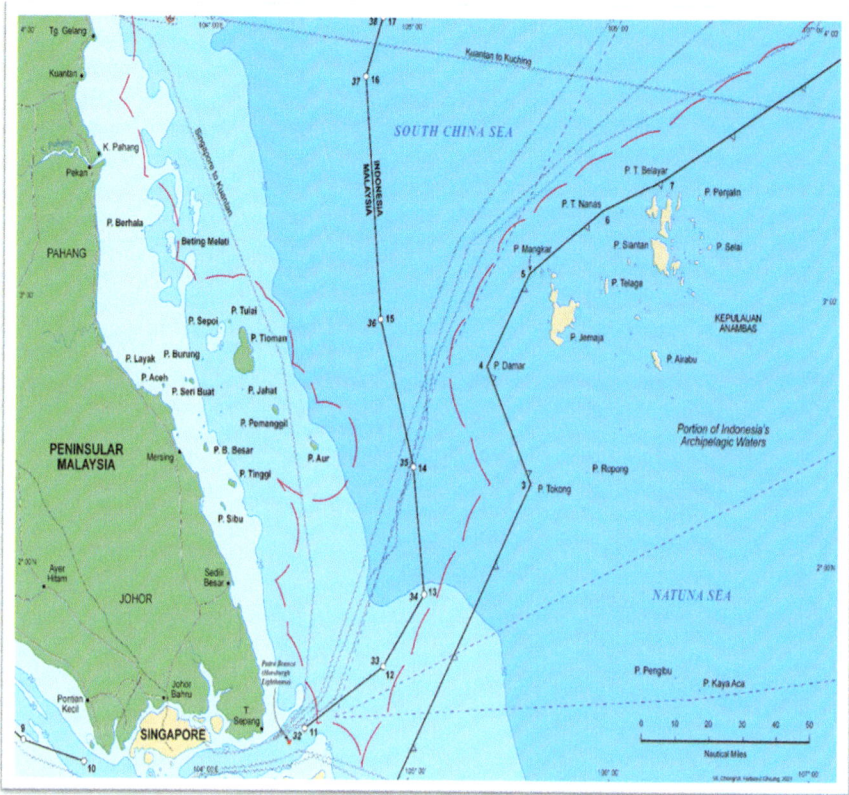

Fig. 4.18 Malaysia's maritime limits off the southern Malay Peninsula

commencing at Point A which is coincident to Point 20. This infers that this anchor point of the seabed and water column boundaries are recognised by the three littoral States, namely, Indonesia, Malaysia, and Vietnam.

Pipelines from the *Tapis* and associated offshore hydrocarbon reserves are shown as approximate directional alignments towards the coast in the vicinity of the processing facilities at Kertih. Also illustrated are the alignments of the submarine telecommunication cables in the vicinity of the coast.

The northern sector of the Malaysia/Indonesia continental shelf boundary, numbered as Points 23, 24 and 25; the eastern half of the Indonesia/Vietnam maritime boundary, and portions of the lines defined in unilateral claims by Brunei and Malaysia (Points 52 and 53), respectively, are the main features in Fig. 4.20 which is devoid of land but for a small island, with its associated baseline and Territorial Sea, that forms part of Indonesia's Natuna Group.

The whole map portrays the southern sector of the South China Sea where water depths vary between 100 and 1000 m and greater as shown by the isobaths. The edge of the natural continental shelf is the 200-m isobath. This map does not depict the

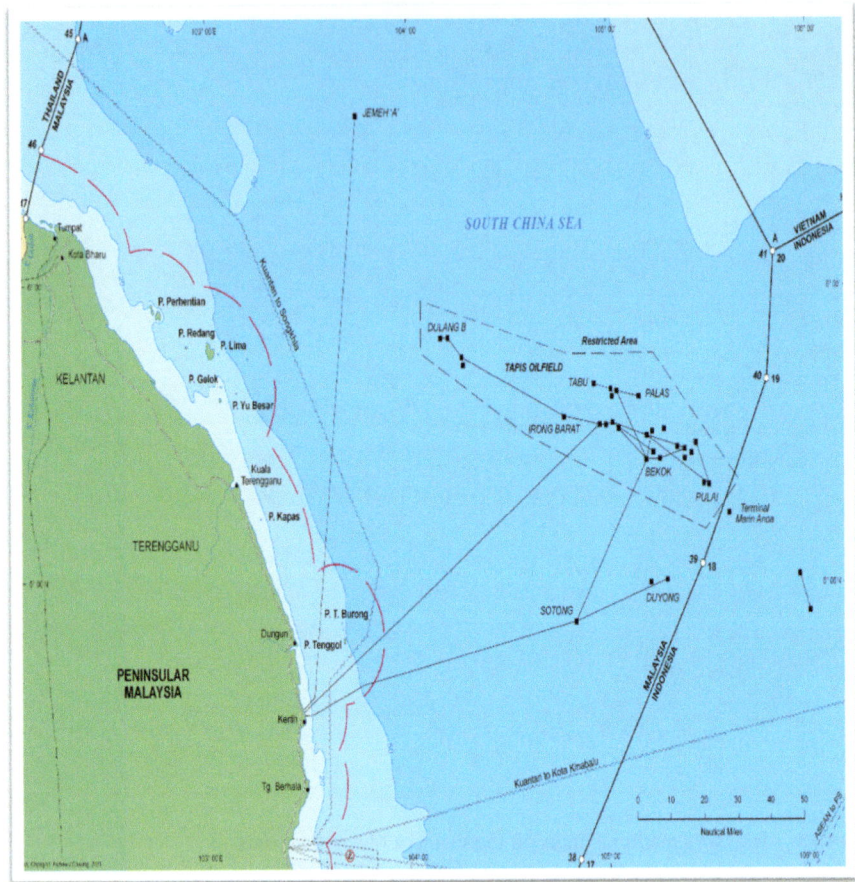

Fig. 4.19 Malaysia's maritime limits off the northern Malay Peninsular

slope of the continental shelf and other underwater features as it is not the intention to do so.

The approximate alignments of the submarine cables laid on the seabed of the South China Sea in the vicinity are also portrayed.

Figure 4.21 shows the eastward extension and northern limit of Malaysia's unilateral claim to a continental shelf, defined by geographical coordinates, and numbered 54–66 inclusive in the vicinity of the southern Spratly Archipelago. The map also portrays the marine features of the Spratly Archipelago claimed as sovereign territory by Malaysia.

The pecked lines numbered from A(1) to 237 (and beyond) indicate the 200 M arcs, the limit of the Malaysian claim in the Joint Submission with Vietnam to the CLCS in 2009.

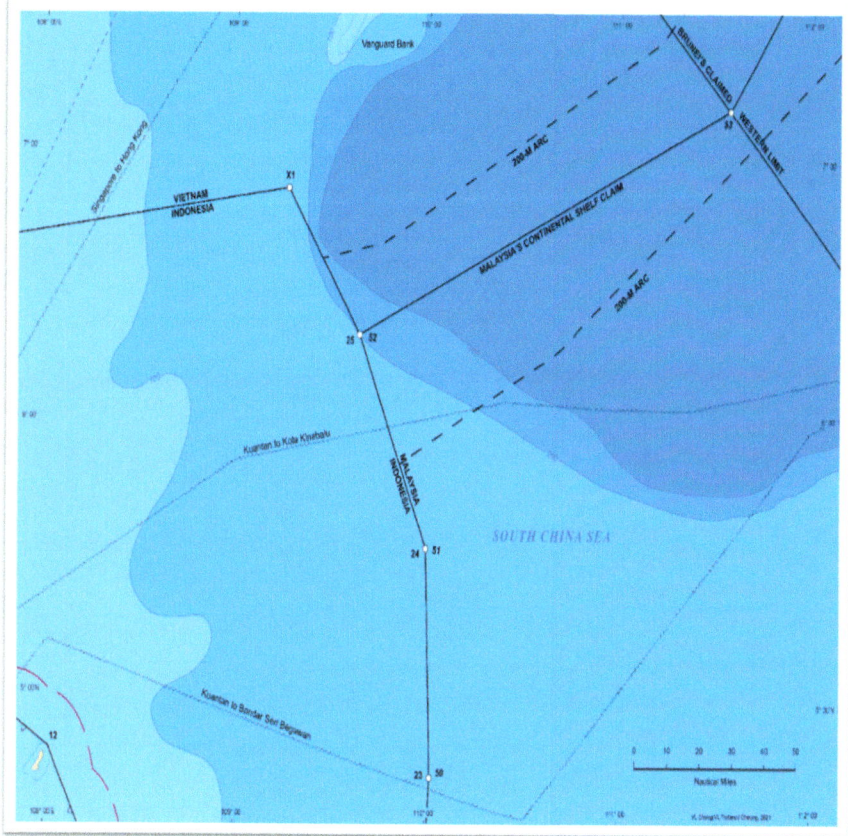

Fig. 4.20 Malaysia's maritime limits: vicinity of Vanguard Bank, South China Sea

Overlapping a portion of this claim is that of Brunei's EEZ depicted as lines delineating the eastern and western boundaries. The lines are an extension from the lateral limits of that country's Territorial Sea claim. Negotiations between the Governments of the two nations will be necessary to resolve any disputes relating the alignment of maritime boundaries especially regarding exploration and exploitation of potential hydrocarbon fields.

The map also illustrates the approximate alignment of the submarine cables in the vicinity as well as some pipelines.

The land mass portrayed in Fig. 4.22 comprises portions of Malaysia's Sabah, and Sarawak States, and the Federal Territory of Labuan; Brunei (has two parcels of land which is separated by Sarawak) and a modicum of Indonesia's Kalimantan Province. No negotiated maritime boundaries are depicted on this map; however, it does portray the eastern and western limits of Brunei's claimed EEZ limits and Points 1, 2 and 5–11 inclusive that relate to a Territorial Sea zone.

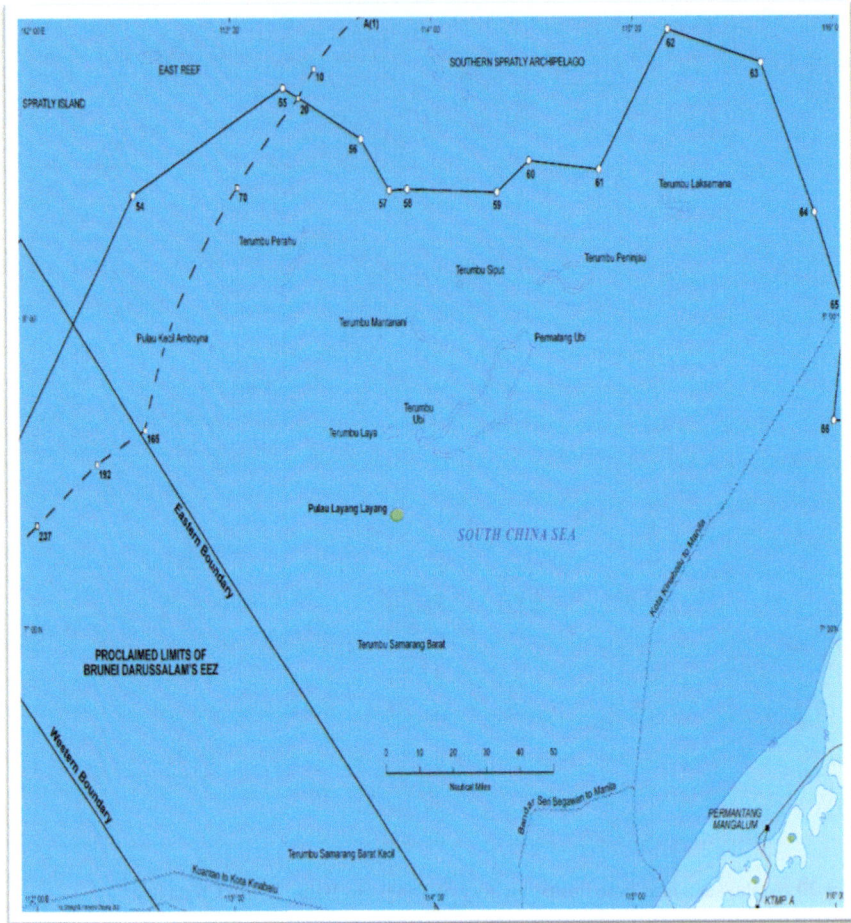

Fig. 4.21 Malaysia's maritime limits: vicinity of the Spratly Archipelago

The line segments joining Points 5–11 indicate the seaward limit of Brunei's Territorial Sea. Malaysia's Territorial Sea limits are depicted as arcs of 12 M from the coasts of the offshore islands and the mainland.

The map also shows the approximate positions of offshore oil wells and submarine pipelines, the alignment of submarine cables and the recommended close-inshore route (depicted as a pecked line in black) for safe navigation during the northeast Monsoon.

Figure 4.23 presents the coastline of Sarawak from Tanjong Datu, a geographical feature, on which is located the western terminal point of a terrestrial international political boundary between Indonesia's Province of Kalimantan and Malaysia's Sarawak State, to a point on the coast some 50 M northeast of Bintulu.

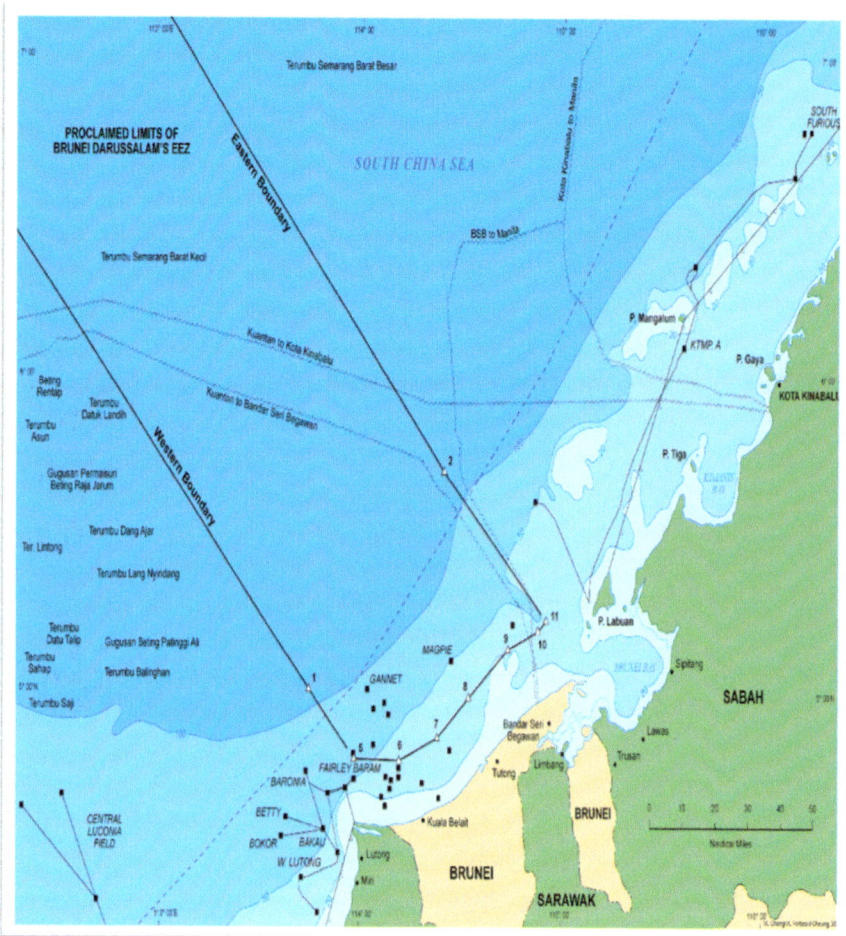

Fig. 4.22   Malaysia's maritime limits off Sabah and Sarawak

A portion of the continental shelf boundary between Indonesia and Malaysia is shown on the map which is indicated by the lines linking Points 21 and 22 and the northerly projection from the latter point. The 12-M arcs, shown in red, infer the seaward limit of the Territorial Sea.

Projecting seaward from Bintulu are submarine pipelines that link the port to the offshore hydrocarbon fields of *Central Luconia* and *Bayan*. The pecked lines in black are the recommended shipping routes for commercial ships during adverse weather conditions during the North-east Monsoon.

The jurisdictional maritime limits in Brunei Bay are portrayed in Fig. 4.24. This map demonstrates the complexity of determining the limits to maritime jurisdiction within a bay whose coastline is shared by two or more political entities. Furthermore,

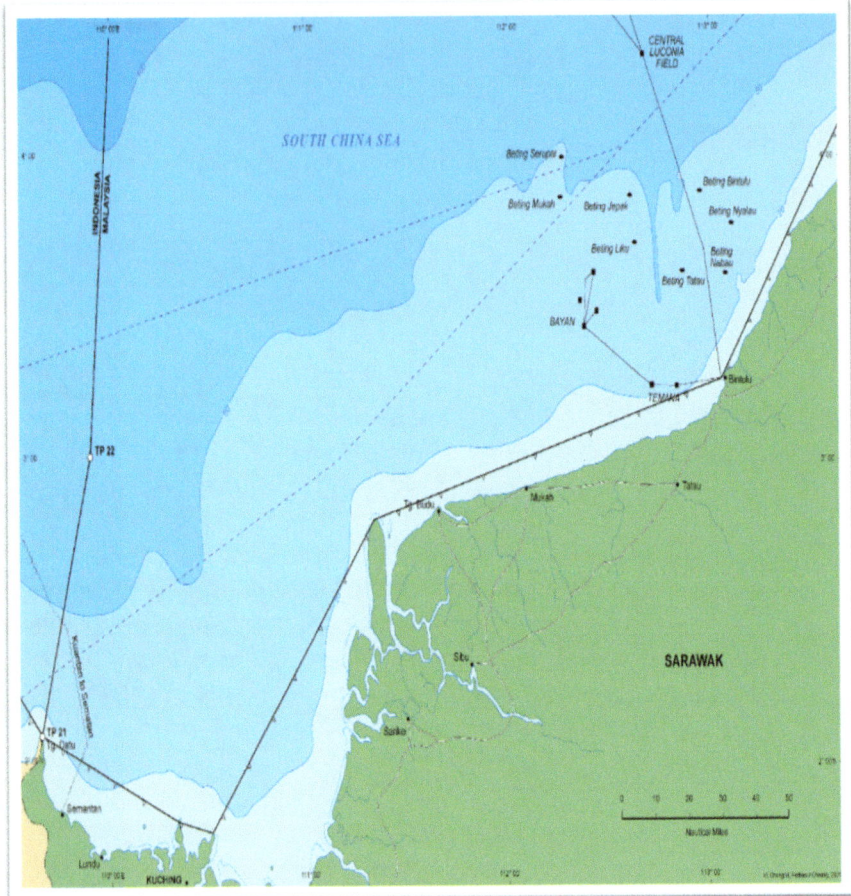

Fig. 4.23 Malaysia's maritime limits: southwestern sector of South China Sea

the limits were defined at a moment in time when colonial administrators were negotiating with the local rulers of the States. Three *Order in Council* documents, which
are reproduced in Annexe II, below, proclaimed sets of geographical coordinates for
each boundary.

The limits around Pulau Labuan and adjacent islands which collectively is known
as Federal Territory of Labuan, is portrayed on the map. Of significant interest on
this map are the delineation of Condominium Corridors and the four reference points
that were used in 1958 to determine the limiting lines.

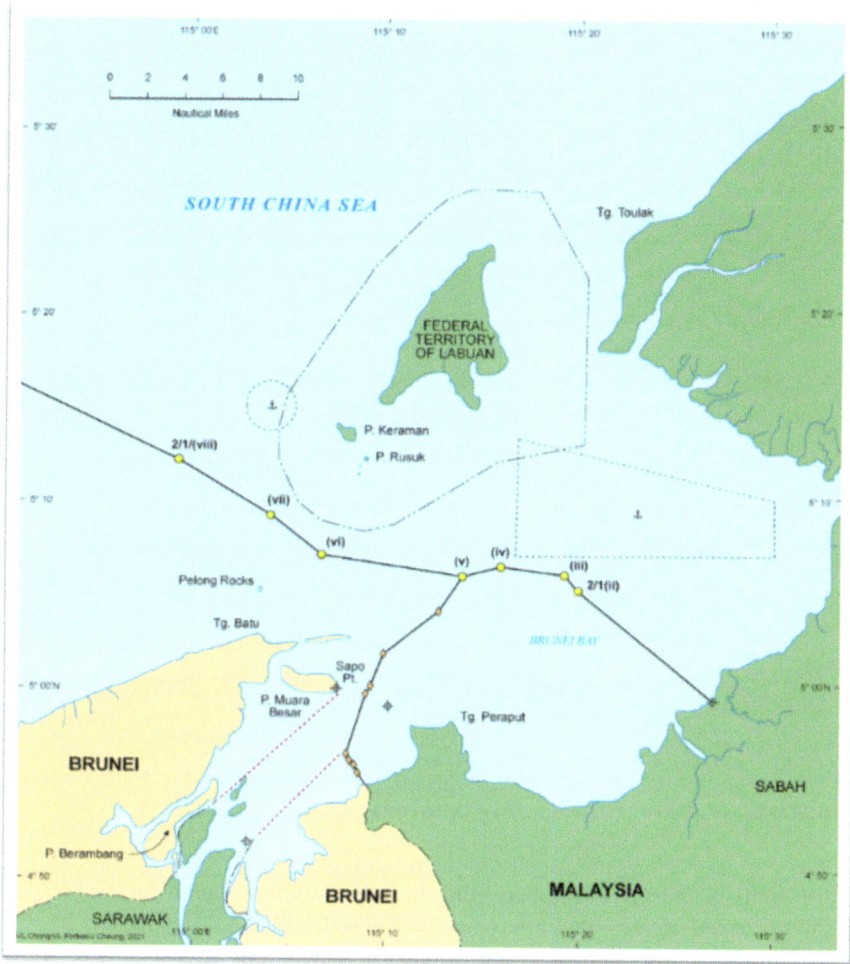

Fig. 4.24 Malaysia's maritime limits off Brunei Bay

4.11 Malaysia's Maritime Limits off Sabah

Balabac Strait and its adjacent seas—South China and Sulu—are portrayed in Fig. 4.25. Malaysia's unilateral claim to a continental shelf (north-eastern limit) is illustrated by the lines joining Points 67–69 and the respective extensions—westward and south-eastward directions. This portion of the claim is identical to an International Treaty Limit of The Philippines of an agreement between Spain and the United States of America which was signed on 10th December 1898 as described in Article III of the said Treaty.

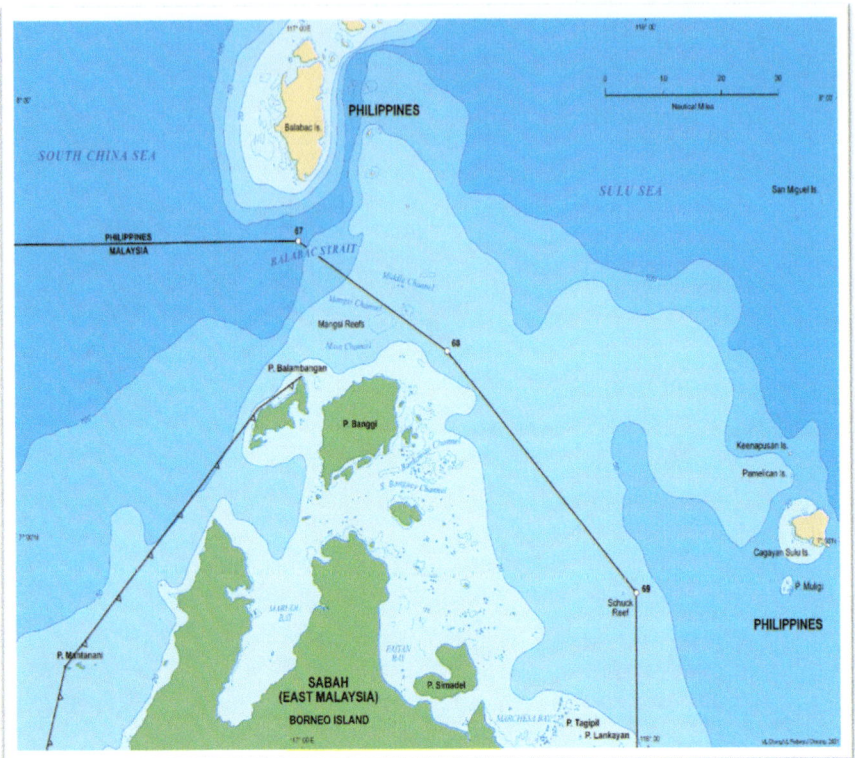

Fig. 4.25   Malaysia's maritime limits off northern coast of Sabah

The limits of the Territorial Sea from the coastline of the mainland and islands are shown, in red, as arcs of circles of radius 12 M. The bathymetry for the 20-, 50- and 100-m isobaths are depicted on this map as blue contour lines.

Figure 4.26 portrays the eastern seaboard of Malaysia's Sabah State and a portion of Indonesia's Kalimantan coast. It also shows the eastern limits of the continental shelf claimed by Malaysia indicated by Turning Points (TP) 70–84 inclusive. From TP 70–76 inclusive, the line segments are identical to the Philippine territorial limits alluded to in an agreement between Spain and the United States of America.

The map also depicts the Indonesian archipelagic waters baseline system off the coast of Kalimantan which in the context of this graphic commences at the land boundary terminal between Indonesia and Malaysia on the eastern coast of Pulau Sebatik—an island the northern half of which is administered by Malaysia and whose southern sector is governed by Indonesia. TP 84 is coincident with the land boundary terminal demarcation at the intersection of Lat. 4° 10′ N on the east coast of Pulau Sebatik, in the vicinity of East Pillar, alluded to in Chap. 1.

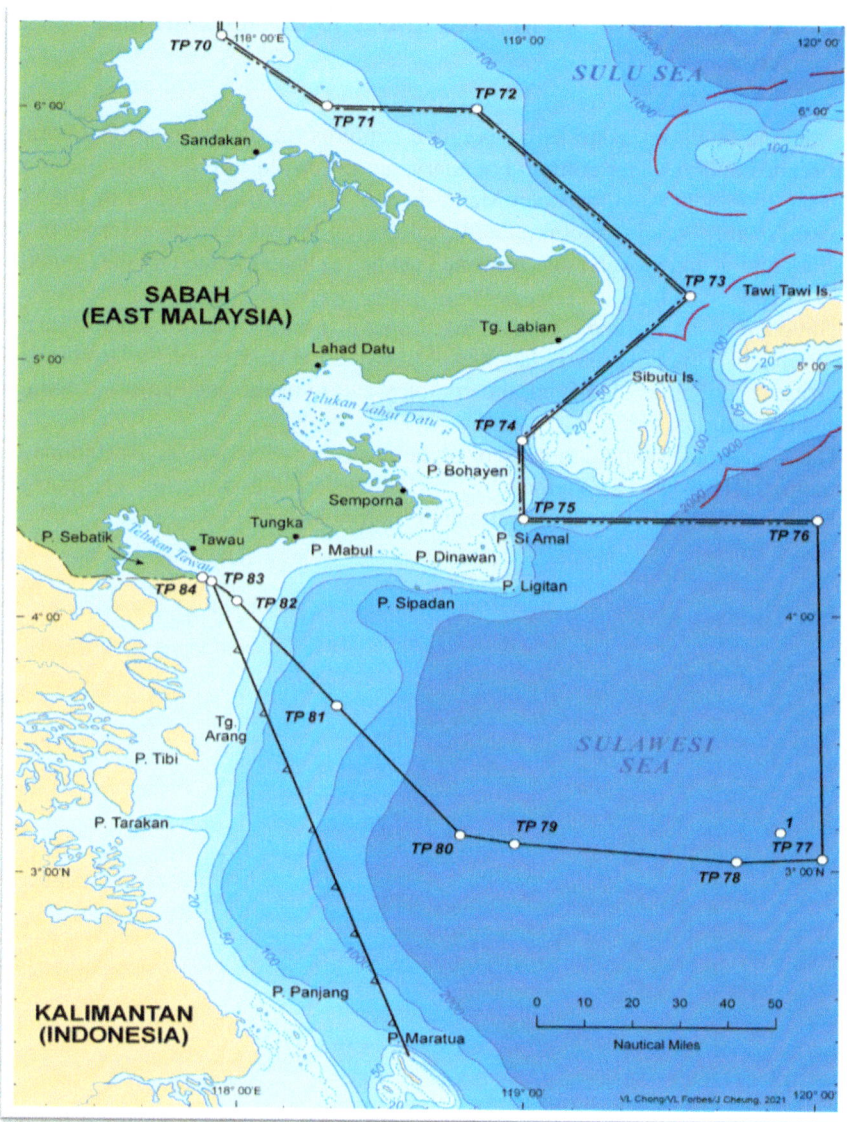

Fig. 4.26   Malaysia's maritime limits off the east coast of Sabah

Of significance on this map are the locations of Pulau Ligitan and Pulau Sipadan and their proximity to the coastline of Sabah. Sovereignty issues over these geographical features became the centre of focus in a case heard by the International Court of Justice. After that judgment, a dispute reared its head when competing exploration companies were awarded exploration rights (often overlapping blocks) offshore from Pulau Sebatik by authorities in Indonesia and Malaysia.

4.12   An Agreed Common Area and a Joint Development Area

Two marine areas set aside as cooperative ventures, in the absence of a permanent seabed and/or water column boundary, are portrayed in Fig. 4.27. One is the Joint Development Area, between Malaysia and Thailand, established in 1989 for the express purpose of exploration and exploitation of hydrocarbon reserves in the offshore region and in water depths more than 50 m. It is identified by the Points A, B, C, X, D, E, F and G.

The other, is an Agreed Common Area, negotiated in 2002, between Malaysia and Vietnam. This polygon is labelled A1, B1, C1, D1, E1 and F1. Both zones are irregular polygons signifying the complex nature of the negotiation to achieve equitable solutions.

The Common Point of three sets of boundaries is labelled as A (Indonesia/Vietnam), E1 (Malaysia/Vietnam), 20 (Indonesia/Malaysia) and 41 (part of Malaysia's unilateral claim to a continental shelf boundary).

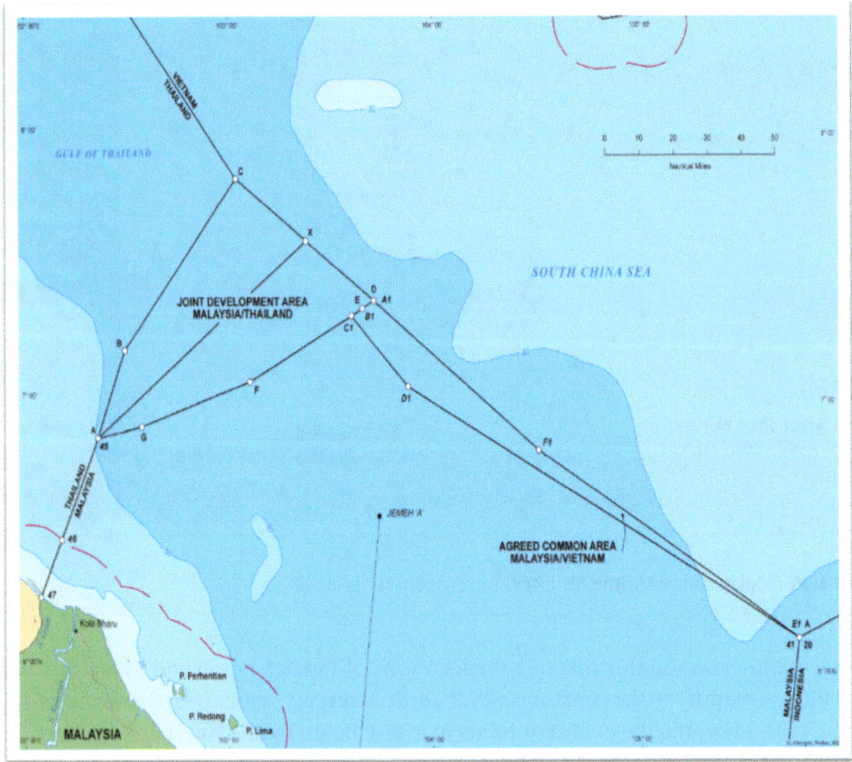

Fig. 4.27   Cooperative approaches to sustainable development of resources

The map also portrays the 12-M arcs signifying the limits of the Territorial Sea for Malaysia, Thailand and Vietnam and the political divisions of the portion of the South China Sea and the southern sector of the Gulf of Thailand.

A portion of the submarine pipeline from '*Jemeh A*' hydrocarbon field towards the processing plant on the adjacent coast is depicted on the map.

4.13 Malaysia's Outer (Extended) Continental Shelf Claim

On 6th May 2009 Malaysia and Vietnam lodged a Joint Submission to the CLCS relating to their continental shelves in the southern sector of the South China Sea. On 12th December 2019, Malaysia submitted a claim for an Outer Continental Shelf (OCS) extension in the South China Sea.[12] Protests relating to these claims were made by the Governments of China (PRC) and the Philippines. The PRC attached a copy of the 'Nine-Dash Line' map, which purports to refute any existence of a continental shelf claim beyond 200 M in the South China Sea. The PRC, and the Government of Taiwan (Republic of China, ROC, or Chinese Taipei) claim a continental shelf for all the features in the middle portion of the South China Sea in accordance with the original 1947 map that depicted the '11-Dashed Lines'.

Malaysia and Vietnam informed the Commission that there are unresolved territorial disputes in the Defined Area of the Joint Submission. Further, *they stated that the Joint Submission will not prejudice matters relating to the delimitation of boundaries between the States with adjacent or opposite coasts.*

The limits, consisting of 810 Fixed Points whose geographical coordinates, based on the World Geodetic System 1984 (WGS84) are published in the submitted document.[13]

The limits are generated and bound by the intersection point of the envelope of arcs of 200 nautical miles (M) of Malaysia and the Philippines in the east (Point A), the intersection of two converging envelope of arcs of Malaysia's 200 M limits towards the south west from A (Points B and C), the intersection point of Malaysia's 200 M limit and the boundary line under the Agreement between the Governments of Indonesia and Malaysia relating to delimitation of the Continental Shelves between the two countries of 1969 towards the south west (Point D). Point 25 under the Agreement towards the north-west (Point E), Point 25 under the Agreement between Indonesia and Vietnam on the delimitation of their Continental Shelf Limit of 2003 towards the north-west (Point F), and the intersection point under the aforementioned Agreement towards the north-west (Point G) and the envelope of arcs of Vietnam's 200 M limits towards the north-east (Points H and I).[14] See Fig. 4.28.

Submission for Extended Continental Shelf: November 2017

A document prepared, an Executive Summary of 28th November 2017,[15] became a partial Submission to the CLCS pursuant to Article 76 of the 1982 Convention for the delineation of the outer limits of the continental shelf in the South China Sea which was lodged on 12th December 2019.

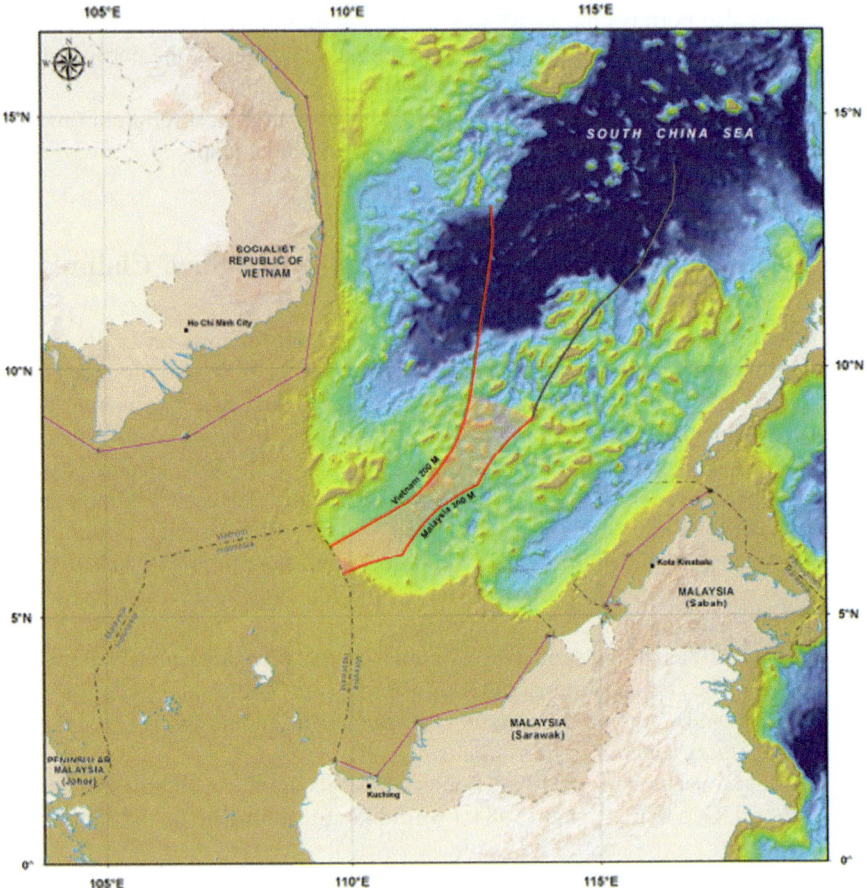

Fig. 4.28   The 2009 joint submission made by Malaysia and Vietnam. *Source* Executive summary of joint submission, 2008; see Note 12

   The outer limits of the continental shelf beyond 200 M were defined by 96 fixed points determined by 60 M from the foot of the continental slope (FoS), in accordance with Article 76, Paragraph 4(a)(ii) of the 1982 Convention. The fixed points, whose geographical coordinates in WGS84, are connected by straight lines not exceeding 60 M in length, in accordance with Article 76, Paragraph 7 of the 1982 Convention. In accordance with Article 76, Paragraphs 5 and 6 of the 1982 Convention, all fixed points comprising the line of the outer limits of the continental shelf are not more than 350 M from the baselines from which the breadth of the territorial sea is measured, or do not exceed 100 M from the 2500-m isobath constraint line.[16]

Figure 4.28 illustrates the outer limits of the continental shelf established for this joint Partial Submission. The geographical coordinates are in WGS84. The details for each of the 96 Points include the Latitude and Longitude in degrees, minutes and seconds of arc value, the method used to define the individual points and the distance between the points in units of kilometres and nautical miles. Figures 4.29 and 4.30 portray Malaysia's extended (outer) continental shelf claim (Table 4.1).

The Governments of the PRC and the ROC claim over the marine features are identical in accordance with the 'Nine-Dash Line,' map as first published in 1946 and subsequently modified between 1947 and 2009. The legality and validity of the 1946 Map and later versions was rejected in the Hague Tribunal South China Sea Ruling of 12th July 2016.[17]

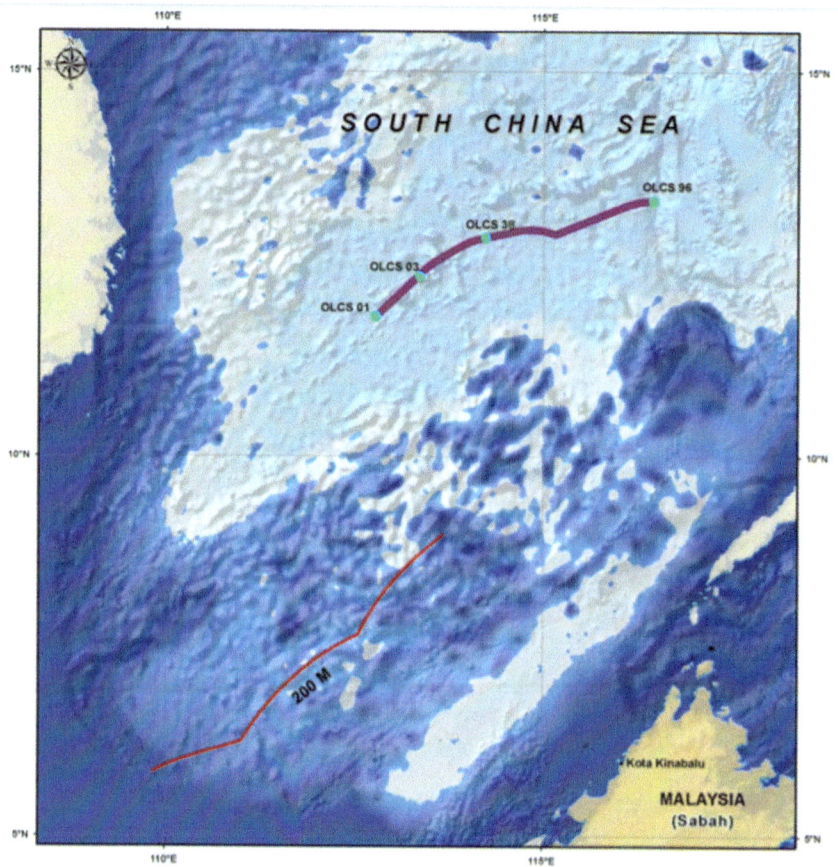

Fig. 4.29 The outer limit of the continental shelf claimed by Malaysia. *Source* Executive summary of Malaysia's claim, 2017. Available UN CLCS

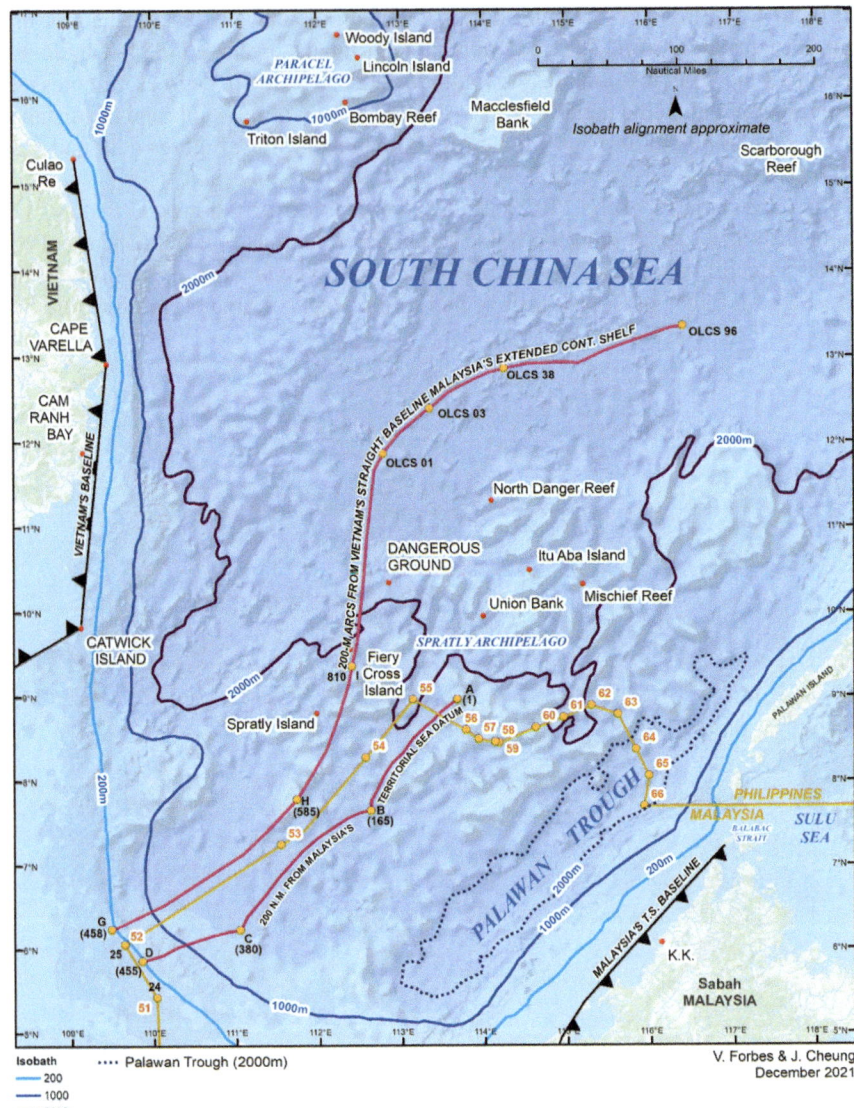

Fig. 4.30 Malaysia's claimed extended continental shelf in the South China Sea. *Source* Author;
MYS_EN_DOC-01_281117 available at UN website

Point	Latitude N (° ′ ″)	Longitude E (° ′ ″)	Description
A is 1	08 59 04.1	113 40 37.6	Intersection of envelope of arcs of 200 M limits from Malaysia and the Philippines

(continued)

(continued)

Point	Latitude N (° ′ ″)	Longitude E (° ′ ″)	Description
B is 165	07 39 42.8	112 33 43.1	Intersection of two converging envelopes of arcs of Malaysia's 200 M limits
C is 380	06 14 59.8	111 01 27.3	Intersection of two converging envelopes of arcs of Malaysia's 200 M limits
D is 455	05 51 09.7	109 50 29.2	Intersection of Mal. 200 M and the boundary line of the agreement: Mal/Indo of 27 October 1969
E is 456	06 18 11.0	109 38 45.0	Point 25 of the agreement of 27 Oct. 1969 between Malaysia and Indonesia re Cont. Shelf boundary
F is 457	06 18 12.0	109 38 36.0	Point 25 of the agreement between Vietnam and Indonesia of 26 June 2003, re: Cont. Shelf boundary
G is 458	06 24 55.7	109 34 06.7	Intersection of Vietnam's 200 M limit and the boundary of the agreement Vietnam and Indonesia of 2003
H is 585	07 41 59.6	111 33 37.3	The point of the envelope of arcs of Vietnam's 200 M limits
I is 810	09 30 15.4	112 25 40.3	The point of the envelope of arcs of Vietnam's 200 M limits

Select List of Geographical coordinates of the Outer Limits of the Continental Shelf (OLCS) in the South China Sea. All coordinates are in WGS84

Point	Latitude N (° ′ ″)	Longitude E (° ′ ″)	Distance in M	Method
OLCS 01	11 5016.2	112 47 16.8		Fixed point 60 M from FoS 01
OLSC 02	12 2526.5	113 24 12.8	50.35 M from 01	Fixed point 60 M from FoS 07
OLCS 16	12 4352.5	113 52 17.5	20.11 M from 15	Fixed point 60 M from FoS 08
OLCS 36	12 5618.4	114 36 16.5	25.94 M from 35	Fixed point 60 M from FoS 09
OLCS 69	12 5423.4	115 09 34.4	1.00 M from 68	Fixed point 60 M from FoS 11
OLCS 95	13 2012.9	116 31 08.7	1.00 M from 94	Fixed point 60 M from FoS 11
OLCS 96	13 2012.3	116 31 20.3	0.19 M from 95	Fixed point on 350 M

Apart from the few points mentioned above, all the defined points are 1852 m (one nautical mile) apart

Source MYS_ES_DOC-01_281117 Executive Summary of Malaysia's Partial Submission

Table 4.1 Geographical coordinates of the outer limit of continental shelf as claimed

Point identification 1969 MB 1979CS	Latitude N (° ′ ″)	Longitude E (° ′ ″)	Comment
21 and 48	02 05 00	109 38 48	Southern terminal point
22 and 49	03 00 00	109 54 30	
23 and 50	04 40 00	110 02 00	
24 and 51	05 31 12	109 59 00	
25 and 52	06 18 12	109 38 36	Mal/Indo and Indo/Viet TPs
53	07 07 45	111 34 00	
54 (equidistant)	08 23 45	112 30 45	Amboyna Cay and Spratly Is
55 (from named features)	08 44 24	113 16 15	Barque Canada R/ Cuareteron
56	08 33 55	113 39 00	Barque Canada R/ Cuareteron
57	08 24 24	113 47 45	Barque Canada/Cornwallis S
58	08 24 26	113 52 24	Mariveles R/Cornwallis S. R
59	08 23 45	113 52 24	Erica Reef/Cornwallis Reef
60	08 30 15	114 29 10	Investigator R/Tennent Reef
61	08 28 10	114 50 07	Investigator R/Tennent Reef
62	08 55 00	115 10 35	Commodore R/Alicia Annie R
63	08 49 05	115 38 45	Commodore R/1st Thomas Sh.
64 (equidistant)	08 19 55	115 54 05	Commodore Reef and ?
65	08 01 30	116 03 30	Boundary Point defined in the 1898 Treaty and 1930 Treaty
66	07 40 00	116 00 00	Western terminus 1930 Treaty

4.14 The Exclusive Economic Zone in the Celebes Sea

A Delimited Boundary Between Indonesia and the Philippines

A study by Forbes (2019: 6–25)[18] provides a narrative on the delimited maritime boundary between Indonesia and the Philippines in the Sulawesi Sea. The 'Agreement between the Republic of the Philippines and the Republic of Indonesia Concerning the Delimitation of the Exclusive Economic Zone Boundary' was signed on 23rd May 2014.

The EEZ boundary, as negotiated, appears to be based on an equidistant principle. The geographical coordinates of two terminal and six turning points were defined and are shown in Table 4.2. The azimuths and distances between the points are illustrated in Table 4.3. The EEZ boundary between the Contracting Parties is defined

Table 4.2 Turning Points of EEZ boundary in Sulawesi Sea

Point	Latitude (N)	Longitude (E)
1	3° 06′ 41″	119° 55′ 34″
2	3° 25′ 38″	121° 21′ 31″
3	3° 48′ 58″	122° 56′ 03″
4	4° 57′ 42″	124° 51′ 17″
5	5° 02′48″	125° 28′ 20″
6	6° 25′ 21″	127° 11′ 42″
7	6° 24′ 25″	128° 39′ 02″
8	6° 24′ 20″	129° 31′ 31″

Source Forbes 2019:17; article 1 of agreement, 2014

by geodetic lines connecting Points 1–8 expressed in geographical coordinates based on the World Geodetic System of 1984 (WGS84) Datum, and in the sequence given below.

About 180 nautical miles of the boundary is perceived to fall within Malaysia's unilateral continental shelf claim in the northwest sector of the Sulu Sea (refer Fig. 4.31; and commentary at Sect. 1.10, above).

This Agreement specified that it does not prejudice any rights or positions of the Contracting Parties about the delimitation of the Continental Shelf boundary. The actual location on the sea of the points and geodetic lines referred to in Paragraph (1) of Article I shall be determined by methods to be mutually agreed upon by the competent authorities of the Contracting Parties. Any differences in the interpretation, application or implementation of this Agreement shall be resolved amicably by consultation or negotiation, through diplomatic channels.

The value of 646 M differs from the official statement which infers the overall length of the boundary is 657.21 miles or 1162 km. The official value is more accurate especially if it has been computed employing geographical coordinates based on the geodetic datum of WGS-84. Points 1–8, inclusive appear to be located equidistant from the relevant points on the coastlines of the opposites States. If a Common

Table 4.3 Azimuth and distances between points

Points	Azimuth	Distance (M)
1–2	075°	90
2–3	077°	98
3–4	055°	136
4–5	087°	36
5–6	051°	133
6–7	093°	88
7–8	093°	65

Total distance is 646 M (author's plotting on a small-scale chart)
Source Forbes 2019:18

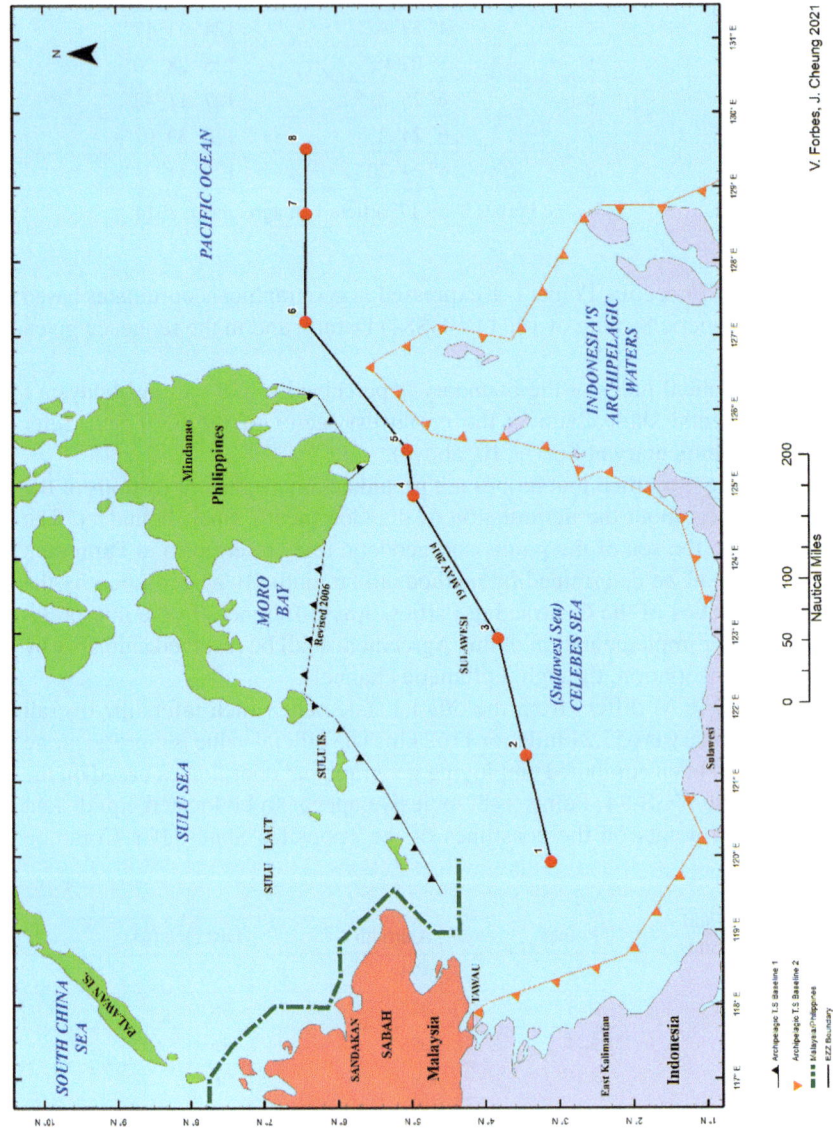

Fig. 4.31 Defined EEZ limit in the Celebes Sea between Indonesia and the Philippines

Point (CP) or tri-junction is to be negotiated between Malaysia, Indonesia, and the Philippines it is possible that it will be about 44 M westward of Point 1.

Determination of a maritime boundary, whatever name, and purpose it is given, is required for the western sector of this semi-enclosed sea between Malaysia and Indonesia and a tripoint will need to be established which will involve the Philippines to enter the negotiations.

4.15 Summary

A suite of maps offered in this chapter portray Malaysia's maritime jurisdiction in detail, as the map scale permits. The depiction of the exact location of Territorial Sea basepoints as announced in August 2022 has not been possible because of the scale of maps employed and the lateness of the announcement in the context of the compilation of the maps. To depict each point would require maps at a much larger scale.

Negotiations between Malaysia and its neighbours are in progress on resolving the maritime boundary closures at the time of publication of this volume. Malaysia and its South China Sea littoral neighbours have sovereignty and jurisdictional issues to be resolved.

The following chapter discusses the issues of safety of mariners and other users of ocean space, above and below the sea surface which is always paramount. It is vital that administration of this spatial extent of national and international jurisdiction is complementary, and legislation is implemented through rules and regulations that are enforced and implemented by the authorities and agencies that are tasked with such duties.

Notes

1. Refer to UN pages: <legal.un.org/avl/ha/gclos/gclos.html> for Status of the Conventions. See also J. A. Roach and R. W. Smith Excessive Maritime Claims, 3rd Ed. Vol 73, Brill, 2012 <https://doi.org/10.1163/9789004217720> .
2. B. A. Hamzah and V.L. Forbes (Editors) (2019) *Maritime Security and the Sulu Zone: Readings on History, Peace Making and Terrorism*, Publisher: UPNM, National Defence University of Malaysia, Kuala Lumpur, 537 pp.; and Forbes, V. L. 'Celebes and Sulu Seas: Maritime Jurisdictional Limits', Ch. 2 in the book with the above title, pp. 6–25. V. L. Forbes, *Indonesia's Delimited Maritime Boundaries*, Springer, 2014.
3. The Declarations and Statements made by Parties to the 1982 Convention may be found at UN https://www.un.org/depts/los/convention_agreements/convention_declarations.htm.
4. Refer: The International Court of Justice for details of the case <icj-cij.org/case 130> The Judgement of the Case is described in the Annex of the present volume. See also Jayakumar and Koh (2009) and Kadir Mohamad (2009).
5. Message from the President of USA, Senate Treaty Doc. 103–39, 104D Congress 2nd Session, 1994 Washington, DC US Gov Printing Office]; see also J. Ashley Roach 'The Maritime

Claims Reference Manual and Law of Baselines International Law Studies, Vol 72, pp.181–196 <https://digital-commons.usnwc.edu/ils/vol72/iss1/10/>.
6.   Letters of Transmittal and Submittal and Commentary regarding the Law of the Sea Convention] Bernard H. Oxman The 1994 Agreement and the Convention *The Am Jour of Int Law*, Vol. 88, No 4 (October 1994, pp.687–686).
7.   Part II, Section 2 Articles 5–14 inclusive; see also Note 6 (above).
8.   Forbes, 2019: 13–27.
9.   Forbes, 2019: 13–27.
10. Forbes, 2020: 62–87.
11. Forbes, 2014.
12. Forbes, 2020: 81–83; see also CLCS/64 for Presentation and Executive Summary and CLCS/55/2 < both available at the UN CLCS website >.
13. As noted in Note 12, above; see also Joint Submission Part 1: Executive Summary, dated May 2009; Malaysia Partial Submission Part I Executive Summary, dated November 2017.
14. Forbes, 2020: 81–87.
15. Refer: MYS_EN_DOC-01_281117 available at UN website.
16. See Notes 10 and 15, above.
17. The Hague Tribunal's South China Sea Ruling https://docs.pca-cpa.org/2016/07… See also https://www.cfr.org/councilofcouncil/global-memos/hague-tribunal-south-china.
18. Forbes V. L., 'Celebes and Sulu Seas: Maritime Jurisdictional Limits', in B. A. Hamzah (see Note 2 above) Ch.: pp 6–25.

Chapter 5
Traffic Monitoring and Maritime Safety

The complexity of processes in safety-critical domains, such as maritime traffic management, is increasing due to continuing technical, organisational, and environmental developments. The VTS is currently undergoing drastic changes, primarily driven by strategies and projects focusing on increasing the overall efficiency of the maritime transportation system through advanced technology.
Gesa Praetorius, 2014 (unpublished PhD thesis)
Vessel Traffic Service (VTS); A Maritime Information Service or Traffic Control System
Chalmers University of Technology, Malmo, Sweden, pg. iii.

Abstract The security of marine craft and the safety of mariners and other users of ocean space, above and below the sea surface are paramount. Administrators and stakeholders are obligated to ensure that the sea lanes that are adjacent to the coastlines of maritime states are equipped to handle marine traffic are effectively managed. In this chapter a suite of maps and other graphics portrays marine and aeronautical traffic within and over Malaysia's maritime jurisdictional limits. The narrative alludes to the fact that the Straits of Malacca and Singapore are the vital and most utilised seaways, daily, connecting the ports of the Indian Ocean and Pacific Oceans, especially those of East Asia.

Keywords Malaysian air space · Adjacent sea · Air traffic system · Marine electronic highway · Vessel traffic information system · Vessel traffic service (system) · Landfall lighthouse

5.1 Introduction

The safety of mariners and other users of ocean space, above and below the sea surface is always paramount. For this reason alone, it is vital that administration of this spatial extent of national and international jurisdiction is complementary, and legislation is implemented through rules and regulations that are enforced and implemented by the authorities and agencies that are tasked with such duties. Furthermore, there are obligations on all stakeholders to ensure that all systems operate in an efficient and sound manner. In this chapter a suite of maps portrays marine and air traffic within and over Malaysia's maritime jurisdiction limits.

The Straits of Malacca and Singapore are the vital and most utilised seaways connecting the Indian Ocean with the South China Sea and Pacific Ocean and together offer the shortest route for tankers trading between the Middle East and Far East Asian countries. Consequently, traffic transiting the region is considerably heavy, conservatively estimated at around 100,000 vessels annually. In addition, there are many local vessels engaged in trade across the straits; numerous fishing vessels can be encountered in most areas; naval ships—regional and extra regional, the latter causing some consternation; and many pleasure craft and work boats. An average of 66 containerships per day transit the Malacca Strait (or about 24,000 annually) and approximately 80% of oil by volume is carried on tankers bound for ports in Northeast Asia.[1]

For a near flawless flow of traffic there must be clear and precise communication between agencies at local, national, and international levels; however, the onus always rests with the users—mariners, port operators and government appointed agencies and authorities. A typical example of excellent communications between agencies and stakeholders would be the regular announcements of the location of offshore drilling activities in the hydrocarbon exploration and exploitation industries to ensure safety to navigation and air traffic controllers and pilots.

Offshore Drilling

A network of more than 10,000 km of pipeline (generally submersible) links more than 380 offshore platforms with 19 floaters and about 14 onshore terminals, within the range of supply and helicopter bases to support offshore operations. Oil drilling rigs and production platforms and buoys associated with the drilling operations are frequently moored in the vicinity of these structures may be encountered off the coasts of Malaysia and in open waters. The positions of these rigs and buoys are frequently changed and are generally promulgated by radio navigational warnings. Major oil and gas fields lie off the NW coast of Sarawak between Tanjung Datu and Tanjung Baram. West Lutong Oil Field, centred at position is the largest.

Other major oil and gas fields lie off the NW coast of Sabah between Tanjong Toulak and Pulau Kalampunian. Major oil fields, gas fields, and terminals are located off the E coast of Peninsular Malaysia.[2]

In the following Chapter a suite of maps depict the offshore oil and gas activities in waters under Malaysian jurisdiction. In the immediate section a brief narrative is offered relating to Malaysian air and maritime space in the adjacent seas.

5.2 Air Traffic Systems in Malaysian Airspace and Adjacent Seas

Limits of Flight Information Reporting (FIR) Sectors for commercial and military aircraft transiting over the semi-enclosed seas abutting Malaysia's coastline are delineated in Fig. 5.1. Also shown are the mandatory points for Reporting and locations from which requests may be initiated as well as information sought relating to meteorological and weather data over the region.

The efficient flow of air traffic across FIR boundaries is achieved by ensuring that flight plans, and associated messages are transmitted, processed, and transferred between FIRs in a continuous and competent manner. The methods and procedures used to file and/or originate flight plans impact on the quality of the air traffic services rendered. Shoddy flight planning has been reported as a contributor to increased workload for air traffic controllers due to the increased time required to interact with aspects of the flight plan.

Shortly after the disappearance of Malaysia Airlines flight *MH370* on 5th March 2014, a special Multidisciplinary Meeting on Global Flight Tracking (MMGFT) was convened at the ICAO Headquarters in Montréal, Canada, to propose recommendations for future actions. The Global Aeronautical Distress and Safety System (GADSS) concept of operations was initiated at this meeting. The GADSS concept describes in an evolutionary manner the execution of actions in the short, medium, and long terms with each action resulting in benefits. One of the main decisions taken was the need for operators to pursue global tracking of airline flights at a faster pace.

During the first week of December 2021, Richard Godfrey, a British aeronautical engineer believed that the aircraft crashed into the Indian Ocean about 2000 km west of Perth, Western Australia. The point of impact with the ocean's surface deduced by data calculations locates the wreck at about Latitude 33° S., and Longitude 95° E.[3]

With reference to Fig. 5.1, the geographical coordinates of an area of the intersections of the Flight Information Reporting Centres (FIRC), given as Latitude (N) and Longitude (E), in degrees and minutes of arc value. Points 'A' to 'U' with their respective geographical coordinates are listed as follows:

Point A: 6° 27′ N 99° 36′E; Point B: 1° 38′N 102° 20′E; Point C: 1° 13′ N 103° 30′ E; Point D: 1° 13′ N 103° 30′ E; Point E: (not listed as it is within the Strait of Johor—eastern sector); Point F: 1° 20′ N 104° 20′ E; Point G: 2° 36′ N 104° 45′ E; Point H: 3° 40′ N 103° 40′E; Point I: 4° 50′ N 103° 44′ E; Point J: 6° 00′ N 103° 08′ E; Point K: 6° 45′N 103° 40′ E; Point L: 7° 00′ N 103° 00′ E; Point M: 7° 00′ N 103° 40′ E; Point N: 10° 30′ N 114° 00′ E; Point 0: 8° 25′ N 116° 30′ E; Point P: 7° 30′ N 117° 30′ E; Point Q: 6° 00 113° 15′ E; Point R: 2° 00′ N 108° 30′ E; Point S: 2° 00′ N 108° 30″ E; Point T: 1° 00′N 108° 30′ E; Point U: 1° 00′ N 108° 58′ E.

A generalised portrayal of the intense network of commercial aeronautical routes—intra-State, inter-State and international—originating from, and bound for, the major commercial airports of Malaysia and to other regional cities and the location of select places—cities and towns—are shown in Fig. 5.2.

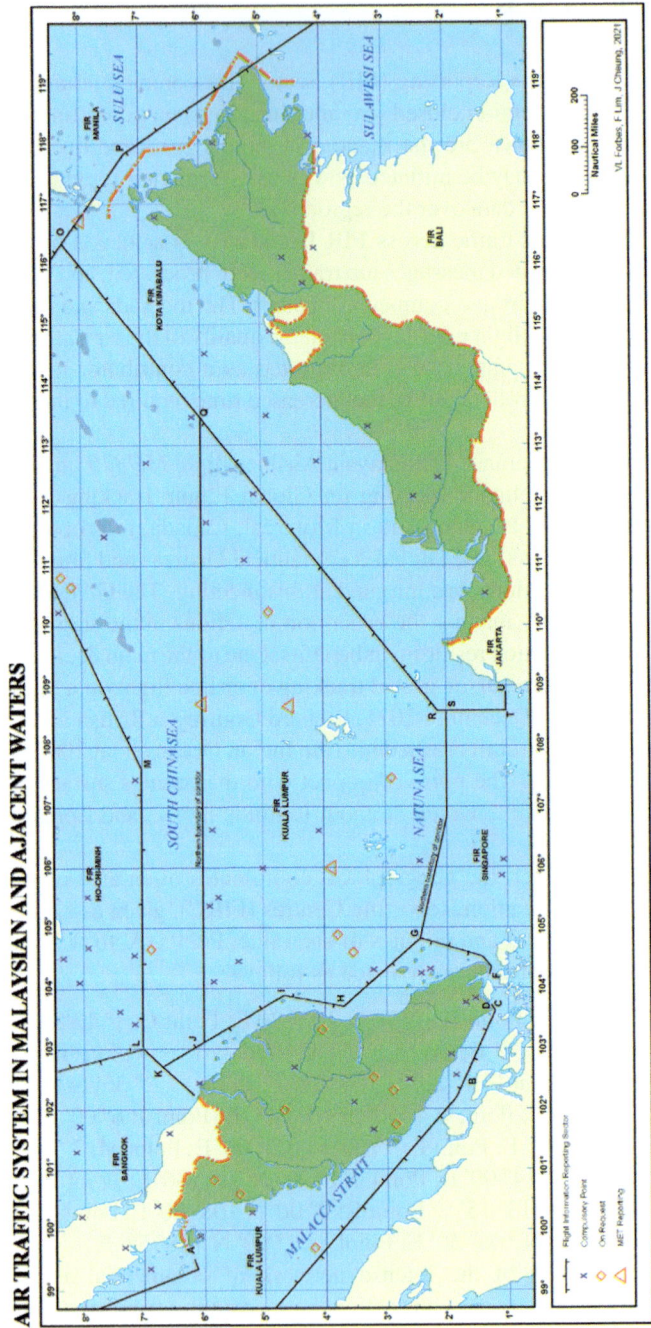

Fig. 5.1 Air traffic systems over Malaysian airspace and adjacent seas

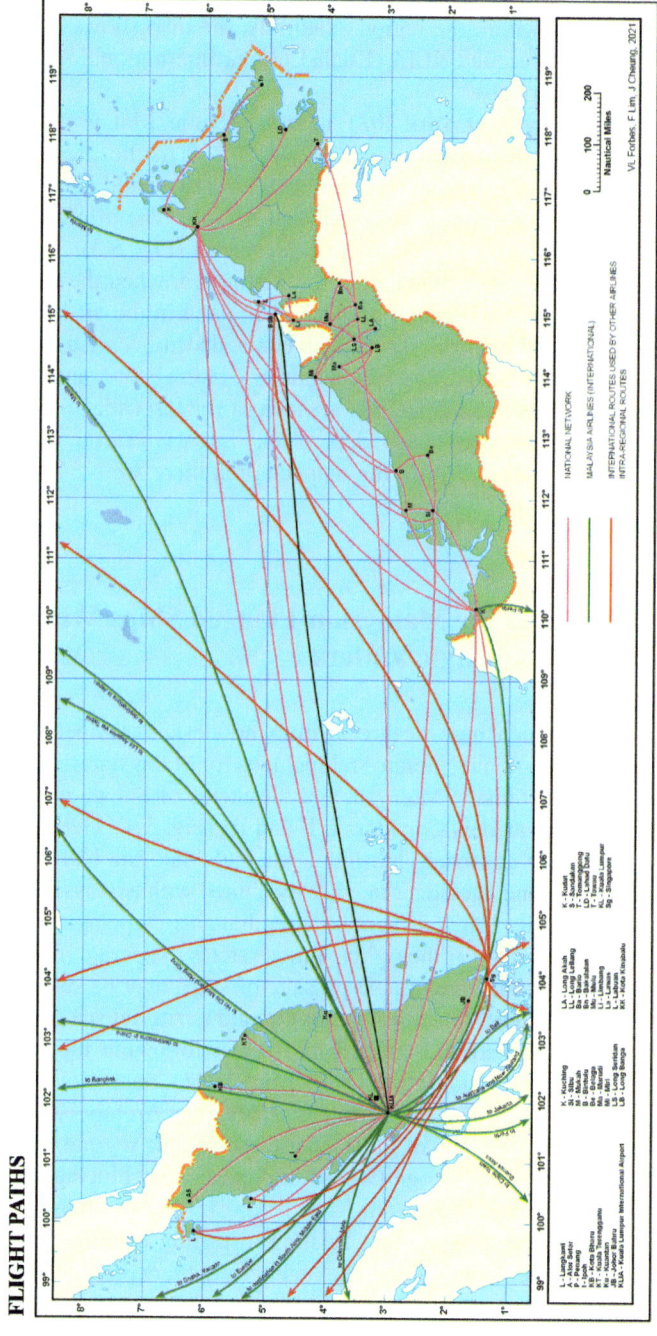

Fig. 5.2 Selected flight paths over Malaysian airspace

The lines depicted in pink form the national network; those in green showing the Malaysia Airlines international routes; and the ones in red the international airline routes that operate within the region. Naturally, the map shows the convergence of aircraft movement around KLIA (Kuala Lumpur International Airport) and Singapore's Changi Airport hubs.

As a matter of interest, on 21st April 2021, at around Noon, Malaysian Standard Time (MyST), (GMT + 8 h or more correctly, UTC + 8 h) the present author had a hunch to enquire as to how many commercial aircraft would be flying over Malaysian airspace, however defined, at that instant in time. The information was derived from data recorded utilising the flight information (Fig. 5.3). The reader, will recall, this was the moment, at the height for this region, when the COVID-19 Pandemic, severely restricted movement on land, at sea and in the air in Malaysia and elsewhere throughout the world. There was controlled movement of citizens within their local area and the airline and tourist industries were economically stifled. There were far fewer flights—the permitted ones were for 'essential services and emergencies' only. Hence the map, below, depicts only a few aircraft in flight at the instant the image was captured. Within the mapped area there were 54 aircraft in flight at the instant of viewing.

5.3 Defined Corridors for Malaysian-Flagged Ships Within Indonesia's Archipelagic Waters

The defined corridors, of which there are three, within Indonesia's archipelagic waters in the vicinity of the Anambas and Natuna Archipelagos for Malaysian-flagged ships plying between ports of Peninsula Malaysia and the States of Sabah and Sarawak and ports beyond are delineated as Corridors I, II and III (Fig. 5.4). Here is another example of the cooperative approach that Indonesia and Malaysia had adopted in managing maritime space and is evident in a Treaty signed with Malaysia in 1983.[4]

Articles 47:6 and 51 of the 1982 Convention infer that an archipelagic state make provision for neighbouring states if the rights of those states to operate in 'traditionally recognised' fishing grounds and transit passage of ships may be suspended by the adoption of the archipelagic concept of baselines and territorial sea.

The Treaty, which is ratified and entered into force, following the enactment of Law No. 1 of 1983, accommodates Malaysia's pre-1982 Convention rights and legitimate interests in the vicinity of the Anambas and Natuna Archipelagos, and by inference, that country's recognition of Indonesia's archipelagic regime.

One aspect of international recognition of Indonesia's archipelagic concept requires that neighbouring states and distant-water fishing (D-WF) nations be conversant with Indonesia's laws and regulations relating to fishing activities within the IEEZ (Indonesia's Exclusive Economic Zone). Such activities are problems at global, regional, and local.

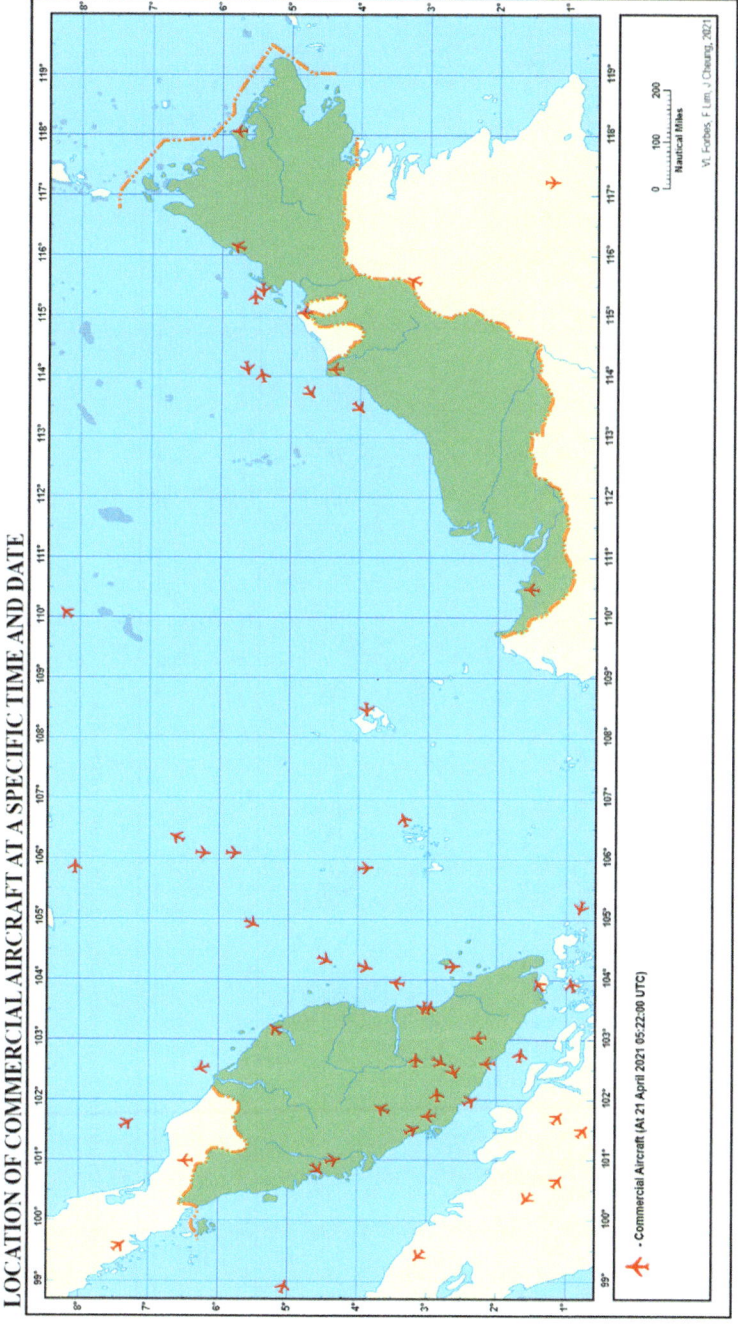

Fig. 5.3 Location of commercial aircraft at a specific date and time in 2021

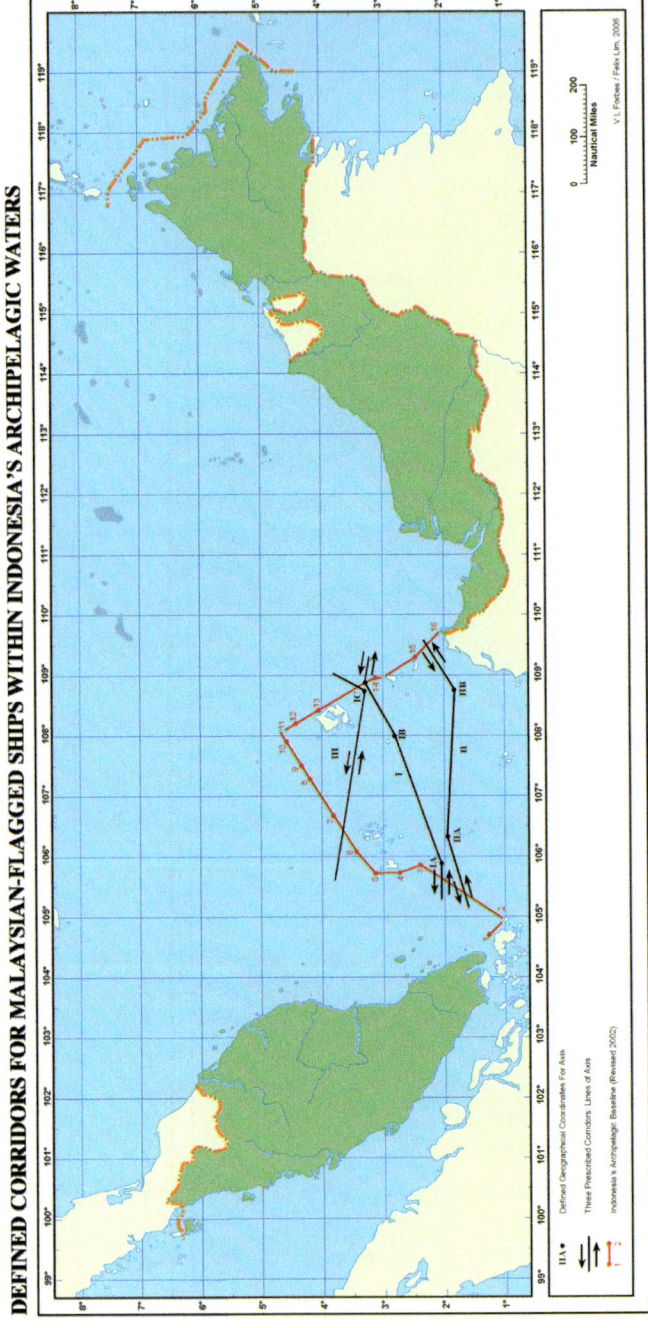

Fig. 5.4 Defined corridors for Malaysian-flagged ships in the Natuna Sea

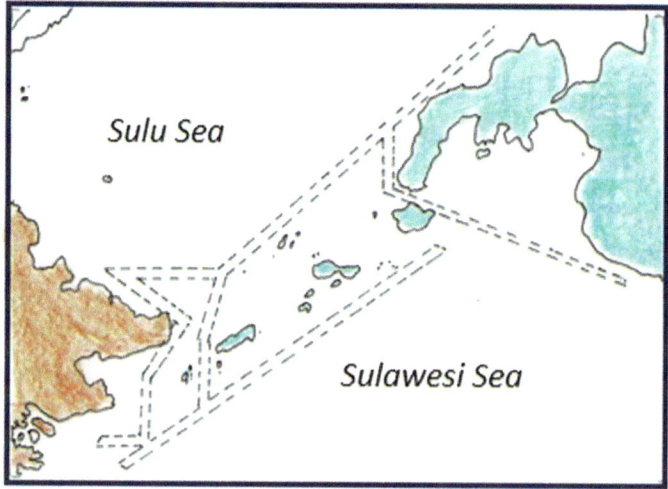

Fig. 5.5 Defined transit corridors: Sulawesi and Sulu Seas. Malaysia's landmass is depicted in brown, and The Philippines coloured as blue. *Source* Author's sketch map, 2023

Pilotage is compulsory for all major ports and offshore terminals in Peninsula Malaysia, Sarawak, and Sabah. Pilots for minor ports can be arranged through Kuching, provided sufficient notice is given. When a pilot is available and a vessel requires the services of a pilot, the standard flag and flashing light signals for requesting a pilot should be made.

Defined transit corridors have been defined by authorities of Indonesia, Malaysia and the Philippines in the Sulawesi and Sulu Seas in the vicinity of the Jolo Archipelago, Eastern Sabah, and Mindanao Island (Fig. 5.5).

Submarine Cables for Domestic and International Transmissions

A representative sample of an array of submarine cables laid on the sea floor for the express purpose of carrying telecommunications between continents, and in this instance, particularly within the geographical framework that is the focus of this volume as portrayed in Fig. 5.6.

Along the Peninsula Malaysia coast there are nine subsea cable landings. On the west coast: they are at Penang, Kuala Kurau, Bandar Bukit Tinggi, Morib, Melaka, Rengit; on the east coast they are at Mersing, Kuantan and Cherating. On the Sarawak and Sabah coasts the subsea cable landings are at Kuching, Bintulu, Miri, Kiamoan and Kota Kinabalu.

Malaysia has seen large economic impacts from improvements in connectivity, especially from submarine fibre optic cables (subsea cables) that connects this country to other countries around the world. Subsea cables are the global backbone of the internet, connecting people, businesses, and economies around the world. The importance of high-quality, reliable internet connectivity and infrastructure is more

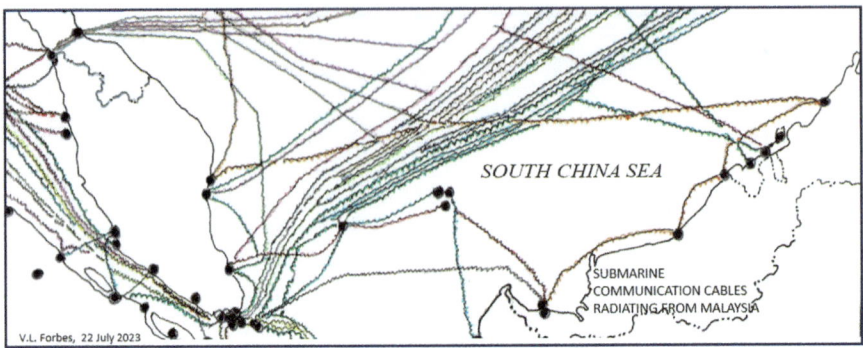

Fig. 5.6 Approximate routes of submarine communication cables. *Source* Sketch map by Leeanna B. Clothier and V. L. Forbes, 23 May 2023

apparent now than ever, as the heavier reliance on remote work and online communication during the COVID-19 pandemic drives a dramatic surge in global internet usage.

Internet connectivity in Sabah, it was announced in September 2020, was given a huge boost as the government planned to triple the bandwidth capacity of the *Sistem Kabel Rakyat Malaysia* (SKRM) submarine cable linking West and East Malaysia. This was revealed during the launch of Gerbang Sabah, an undersea fibre optic landing station at Tanjung Aru. The network spans over 3800 km.

The economic impact of the internet connectivity delivered by submarine fibre optic cables (subsea cables) is evident in the increase in GDP and increasing job opportunities in services from the most recent cable landings. These subsea cables are the global mainstay of the internet, connecting people, businesses, and economies around the world. In the case of Malaysia, the demand since 2009 has been remarkable.

Submarine cable connectivity in South-East Asian seas is vulnerable to challenges such as damage caused to the cables and inter-connectors by human activities at sea—intentional and un-intentional—and especially in sections of the South China Sea due to geopolitical differences relating to sovereignty over maritime space and marine features. These issues have implications for regional states and extra-regional nations. Submarine cables are critical infrastructure. The International Cable Protection Committee (ICPC) recommends that states set aside overlapping maritime boundary claims and remain unprejudiced and neutral to facilitate submarine cable installation and maintenance in contested maritime zones.[5]

Malaysia's subsea cable landing stations serve as access points for many of the subsea cables that connect Africa to Asia, Europe to Asia, and the Pacific Ring. (Fig. 5.7) The evidence of impact on overall growth in Malaysia from subsea cables has been overwhelmingly positive. However, it was opined that Malaysia had lost out when two major social media groups decided to open data centres in Indonesia and Singapore, with submarine cables connecting the USA to the two countries. The rationale was on the revocation of a cabotage exemption in November 2020.

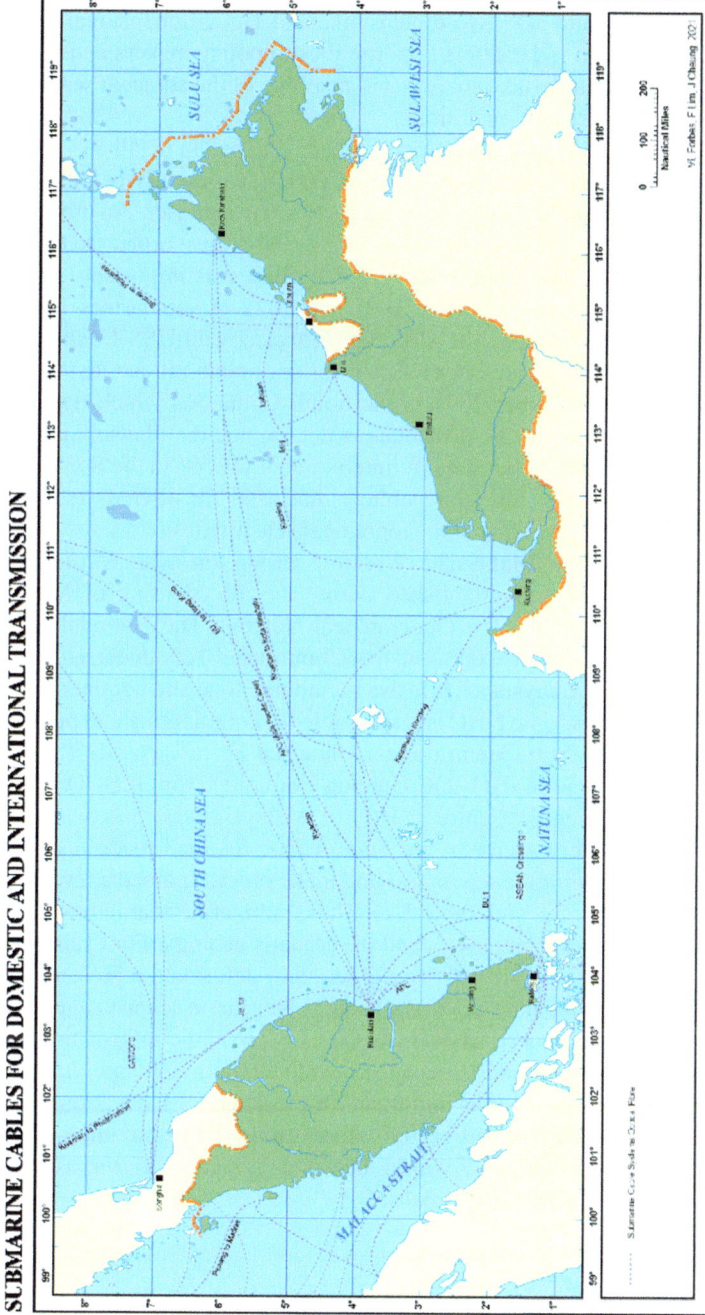

Fig. 5.7 Submarine cables for domestic and international transmissions

Laying of submarine communication cables require knowledge of the topography of the seabed, the location of ridges, seamounts, trenches, and troughs. Modern technology, for example, 2D and 3D (Two-dimensional and Three-dimensional, respectively) and sonar side-scan radar offers opportunities to policy makers and national development agencies to undertake surveys and mapping of the adjacent seas within their national maritime jurisdictional limits.

Only around five per cent of the sea floor has been mapped in detail according to General Bathymetric Chart of the Oceans (GEBCO).[6] It is presently being mapped with the aid of sonar and satellite-sourced imagery by many agencies. An initiative of GEBCO and the Nippon Foundation of Japan has a collaborative project to facilitate the complete mapping of the ocean floor by 2030. Thus, it is impossible to know how many seamounts exist. Incidents in 2005 and 2021 of submarines from the United States' Navy (for example, the USS *San Francisco* and USS *Connecticut*,[7] respectively) colliding with underwater volcano (or seamounts) illustrate the point. The latter occurred on 2 October 2021 in the South China Sea which resulted in damage to the vessel's sonar array. Investigators determined that the submarine hit an uncharted seamount whilst operating in 'international waters' in the South China Sea. The specific transit has not been publicly disclosed for obvious reasons of security, national interests, and regional geopolitical sensitivities.

In December 2006, the Malaysian Working Group on Names of Islands and Geographical Entities was established during the 4th Meeting of MNCGN (Malaysian National Committee on Geographical Names). The National Hydrographic Centre (NHC) was entrusted to lead that Committee.[8] Any undersea features which are located within Malaysia's Exclusive Economic Zone and continental shelf (natural and legal extension), the NHC is responsible for coordinating and proposing the name to the GEBCO Sub-Committee on Undersea Feature Names (SCUFN) which is appointed by the joint IHO-IOC Guiding Committee for GEBCO.

The National Hydrographic Centre (NHC) of Malaysia was tasked to undertake surveys in the southern sector of the South China Sea and made fascinating discoveries relating to submarine seamounts and in the process named a few submarine seamounts (refer Fig. 5.8 and Table 5.1). These submarine seamounts provide valuable insights into oceanic processes and the planet's geology. For example, the monitoring of real-time of seismic activity and sea-floor deformation presents geographers, oceanographers and other scientists about submarine volcanism and caldera dynamics. Importantly, such derived data is used to improve the information about the topography of the seabed and the mapped evidence on maps and charts. The updated nautical chart information is important for the safety of navigation.[9]

Just as importantly, the enhanced spatial information is of prime importance to marine scientific research especially within the South China Sea basin, as indeed, within other semi-enclosed seas and oceans.

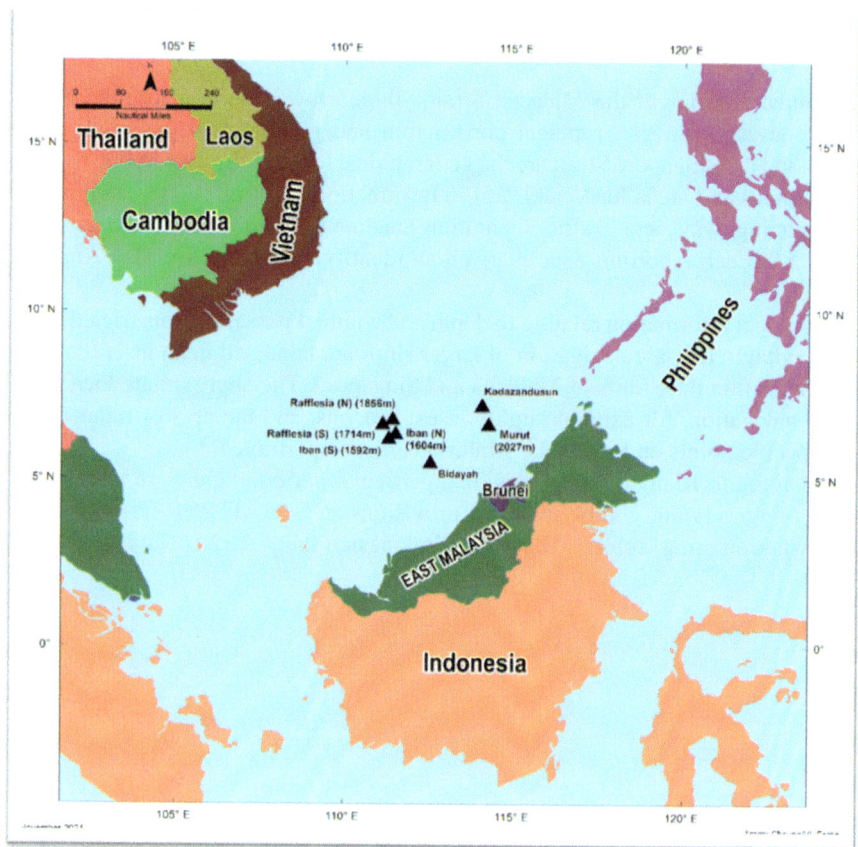

Fig. 5.8 Named submarine seamounts within Malaysia's maritime limits

Table 5.1 Discovery of submarine seamounts within Malaysia's Maritime limits

Rafflesia (N)	Hill	6 48.8	111 25.4	1855.8 m	Malaysia
Rafflesia (S)	Hill	6 40.3	111 07.7	1713.7 m	RMN
Kadazandusun	Hills	7 13.1	114 04.1	1313.0 m	"
Murut	Hill	6 38.9	114 15.5	2027.4 m	"
Bidayah	Hills	5 31.4	112 32.8	1587.0 m	"
Iban (N)	Ridge	6 23.2	111 30.7	1604.03 m	"
Iban (S)	Ridge	6 14.4	111 18.5	1591.8 m	"

Column 1: Name of feature; Column 2: Geographical feature; Column 3: Latitude in degrees and minutes, North; Longitude in degrees and minutes, East; Column 5: Depth; Column 6: Authority -RMN (NHC)

5.4   Marine Electronic Highway and VTIS

The southern sector of the Malacca Strait—the narrow portion where the marine
traffic converges—a geographical constriction and geo-political potential chokepoint—and the Singapore Strait are the geographical focus of the map in Fig. 5.9. The
limiting lines and areas for Vessel Traffic Information Systems (VTIS), the 'marine
electronic highway' and Traffic Separation Scheme (TSS) are delineated as shown
in Fig. 5.9. Each reporting zone is given an identifying name. (See also Annex V,
below).[10]

Additional information relating to depth and width of passages (constricted areas)
that are vital to the safe navigation of larger ships are tabulated in an inset for select
locations within the Straits of Malacca and Singapore. The approximate location of
aids to navigation, for example lighthouses, beacons, and the arcs of radar ranges
from specific points on Peninsular Malaysia are also portrayed.

According to Klang VTS records, in a particular period, there were three occurrences of vessels not complying with provisions of STRAITREP. They failed to
report when entering sector 6. They were: Indonesian-flagged 3207 GT Tanker *Pelita*

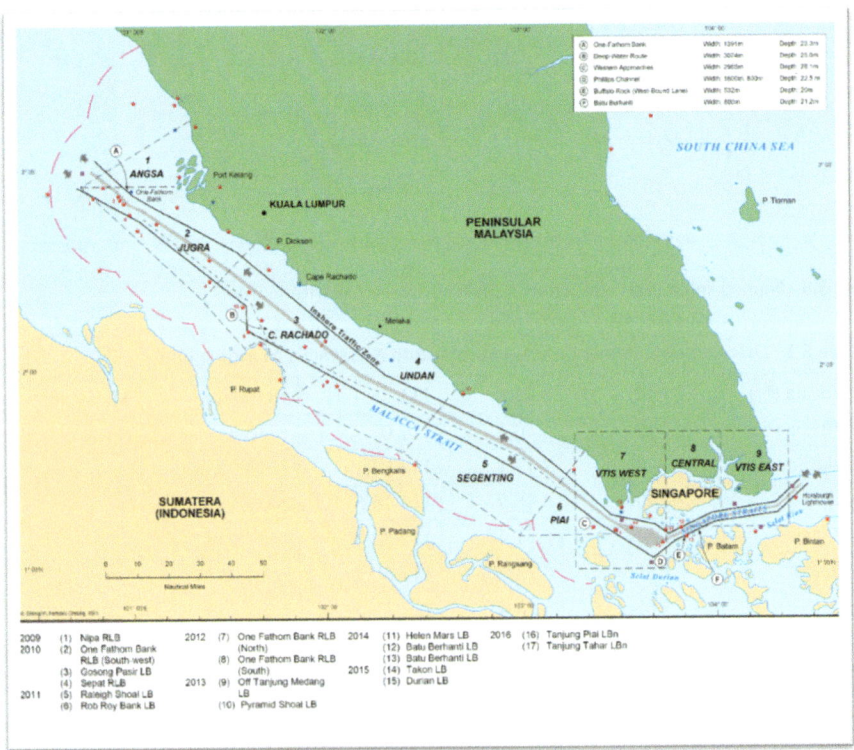

Fig. 5.9   Vessel traffic system within the straits of Malacca and Singapore

Energy on 1 November 2003; the Greek-flagged 327.923 GT Bulk Carrier *Sea Pride* on 16 November 2003; and the Panama-flagged 8061 GT General Cargo *Genius Star* on 12 February 2004.

A Vessel Traffic Service is in operation in Bintulu Port (3° 16′N., 113° 03′E.), on Sarawak.

Safety to Navigation developments by the Malacca Strait Council were undertaken in:

2009 (1) Nipa RLB 1° 11′ N; 103° 33′ E
2010 (2) 2° 50′ N 100° 50′ E; (3) 2° 42′ N 101° 02′ E; (4) 2° 30′ N 101° 22′ E
2011 (5) 2° 10′ N 102° 00′ E; (6) 1° 58′ N 102° 07′ E
2012 (7) 2° 58′N 100° 56′ E; (8) 2° 50′ N 100° 52′ E
2013 (9) 2° 16′ N 101° 42′ E; (10) 2° 22′ 101° 32′ E
2014 (11) 1° 09′ N 103° 50′ E; (12) 1° 11′ N 103° 51′ E; (13) 1° 10′ N 103° 53′ E
2015 (14) 1° 10′ N 103° 50′ E; (15) 1° 04′N 103° 44′ E
2016 (16) 1° 19′ N 103° 38′ E; (17) 1° 57′N 102° 42′ E

Figure 5.10 may be an 'intentionally busy' map and is unusual in that it shows the complex maritime issues that have, and potentially continue to exist, because of the very importance of the Malacca Strait as an international conduit for shipping travelling between the ports of the Indian Ocean and the Pacific Ocean.

In 1994, there was consensus that at least 260 commercial ships, mainly large in dimensions, transit the Malacca Strait daily. This figure does not include the plethora

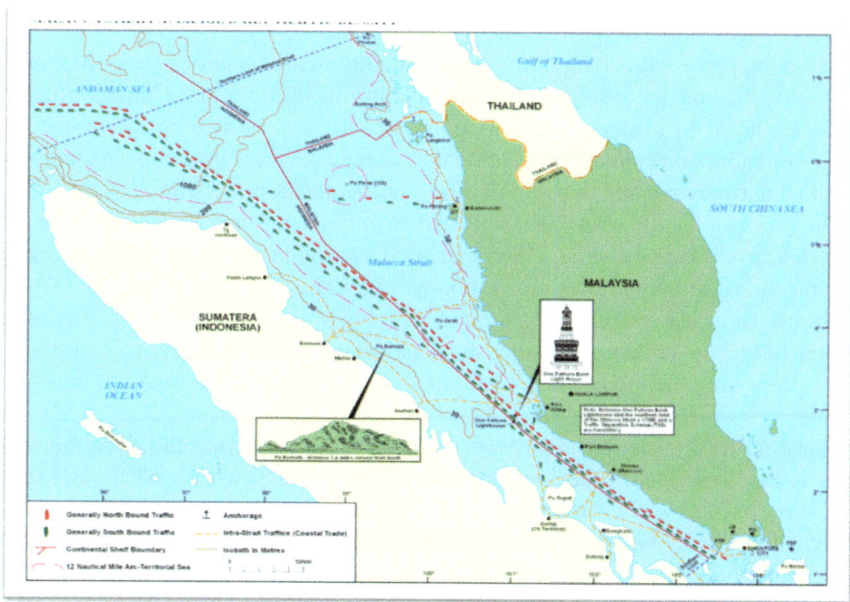

Fig. 5.10 A daily average flow of marine transportation in the Malacca Strait

of smaller ships and boats conducting intra-Strait trade as well as the numerous fishing boats and pleasure craft operating within this 500-M long waterway daily.

The traffic density in the Straits of Malacca and Singapore was projected to increase to 141,000 ships in 2020, and in dead-weight tonnage, from four billion to 6.5 billion tons. This estimate was not achieved in part due to the disruption caused by the Covid-19 Pandemic implications throughout the maritime industry and transportation. However, given the phenomenal increase in traffic to about 90,000 ships in 2021, the quality of navigational aids in the straits will have to be enhanced for a smooth flow of traffic and to prevent any accident. Under Article 43 of the 1982 Convention, it is the responsibility of the littoral states to maintain navigational aids in the straits, as it is to prevent pollution of the marine environment.

That said, for much of the northern length of the Strait the main traffic flow tends to be closer to the Indonesian coast and southward of Pulau Berhala (Indonesian territory) and Pulau Jarak (Malaysian sovereignty) the flow falls within Malaysia's territorial sea. The ships on a 'northbound voyage' are depicted as a 'red ship icon' whilst those heading for the Straits of Singapore and points further east and south are shown as 'green ship outlines' (Table 5.2).

Automatic Identification System (AIS) and Maritime Tracking

The AIS is a maritime communications device which uses extremely high frequency (VHF) radio broadcasting system to transfer data. In 2004, it was decided that all vessels over 300 gross tonnages (GT) on international voyages must have AIS equipment on board—part of the requirements set out by the International Maritime Organisation (IMO) Convention for the Safety of Life at Sea (SOLAS) Regulation V/19.2.4.

AIS-equipped vessels (shipborne AIS) and shore-based stations (non-shipborne AIS) can use the device to send and receive identifying information. This information can be displayed on an electronic chart, computer display, chart plotter or compatible navigation radar and can aid in situational awareness; and provide a means to assist in collision avoidance and assist in marine environmental protection.

The AIS can handle over 2000 reports per minute and may update information as often as every two seconds. AIS can be used as an aid to navigation by providing location and additional information on buoys and lights.[11]

Marine traffic is essentially all ships at sea plying sea lanes (*sealines*) of communications (SLOCs)—tankers, container ships, passenger ships, fishing vessels—collectively. Ships are classified into several types based on their purpose, size, and type of cargo.

Maritime tracking is an ideal way to observe (and analyse) the movement of ships as they traverse the oceans—knowing exactly what, where, when, and how (fast, slow, or otherwise) ships are travelling. It enables real-time information on the movement of ships (dynamic) and their current locations in harbours and ports to be deduced provided the mechanism is 'switched on' by the user (vessel).

Malaysia, as indeed many other countries of South and Southeast Asia have many busy ports. Cargo weighing hundreds of millions of tonnes pass through each of these ports annually. Satellite-derived imagery and the information gleaned from

Table 5.2 Statistics maintained by KLANG VTS, 2010–2020

Type	2010	2011	2012	2013	2014	2015	2016	2017	2018	2019	2020
VLCC/deep-draft CC	4333	4539	4732	4825	4993	5324	5973	6711	7517	8093	8282
Tanker vessel	16,247	16,223	17,345	18,296	18,765	18,470	19,466	20,629	20,610	20,207	20,550
LNG/LPG carrier	3579	3830	4014	4248	4173	3936	4057	4137	4547	4180	3735
Cargo carrier	8445	7996	7950	7613	6989	7146	7225	7090	6409	6273	6195
Container vessel	24,806	25,552	24,639	24,658	25,071	25,393	25,768	24,446	24,578	23,620	21,908
Bulk carrier	11,642	10,851	11,678	12,658	13,454	15,169	15,547	15,411	15,390	15,656	16,314
Ro-Ro/car carrier	2624	2545	2980	2998	3146	3111	2873	2414	2437	2433	1930
Passenger vessel	1071	877	861	1063	1041	925	1294	1776	1969	1593	548
Livestock carrier	45	47	38	55	59	75	75	50	45	36	27
Tug/tow vessel	545	414	529	563	676	469	580	533	601	670	563
Gov/naval vessel	37	57	50	58	96	86	53	54	66	102	66
Fishing vessel	20	20	52	27	51	53	25	28	36	62	128
Others	739	577	609	911	830	803	807	825	825	794	1134
Totals	74,133	73,528	75,477	77,793	70,344	80,960	83,743	85,030	85,030	83,724	81,380

them can be useful to port administrators and financial markets because they give early warning of volume changes in global trade. Based on a casual observation and deduction, by the present author over many years, and specifically on 7th July 2021, at about 1330 h Malaysian Standard Time (GMT + 8 h) there were over 400 ships and boats in the vicinity of Peninsula Malaysia at that instant in time.[12]

A *2020 Review of Maritime Transport* (UNCTAD) estimated that there were 98,140 active commercial vessels of 100 gross tons and above around the world—including those at sea and in port. The global commercial shipping fleet grew by 4.1% during 2019. There are many methods to keep track and observe maritime tracking as indeed commercial aviation globally.

VesselFinder

VesselFinder is a **free AIS tracking service** showing the ship's position and maritime transport **via a global AIS network in real-time.** A selection of graphics, courtesy of *VesselFinder*, is used as illustrations in this volume.

VesselFinder processes the ship's signals received and displays them in the form of a radar chart. It is used to display live diverse types of ships and boats around the world. If one wants to track or watch the shipping traffic in real time, the service may suit the needs. With the help of state-of-the-art technology, locating and tracking ships is no longer a problem.

There are a variety of ships such as general and specific cargo ships, cruise ships, tankers, or sailboats. For the distinction of the ship or object types these are indicated on the map in assorted colours. In addition, a distinction is made as to whether a ship is currently traveling (dynamic) or has stopped (static). The nature of a ship is also listed in the detailed information, where many additional particulars can be viewed. This can be identified by clicking on a ship or object. The size of a ship or boat is not considered in the overview or not shown accordingly. All types of ships and their associated colour are listed in the legend, below.

Due to European Union laws, the Company is no longer allowed to offer map services for the representation of ship positions. This was regretted by the company; however, the rationale employed is understandable.

Figure 5.11, is a screen capture of the instant when the ship *Buana Prosperity* 1, was underway in the Malacca Strait, about 35 M due West of Port Dickson, Malaysia. The moment, a random time, of screen capture was 1421 h (or 2.21 pm) Malaysian Standard Time (MyST) or (GMT + 8 h) on 7th July 2021. The geographical coordinates of the ship at that instant were Lat. 2° 30.733′ N, Lon. 101° 10.824′ E. The scale bar for this graphic represents 10 Nautical Miles (M). The only reason that this object was chosen was because it was relatively distant from the main traffic flow at that instant in time and place and it is a non-commercial vessel.

Within the geographical frame of this image, it may be deduced that the vessel traffic comprises mainly tanker vessels underway, many heading in a south-easterly direction and the others northbound. The pattern of the tanker movements is in accord with the Traffic Separation Scheme that is in-force in the Malacca Strait. Some vessels, portrayed as orange circle clusters, are anchored off Port Dickson and

Fig. 5.11 Location of Buana Prosperity 1, west of Port Dickson, Malaysia. *Source VesselFinder*, at 1421 h, MyST, 7th July 2021 < vesselfinder.com >

Port Linggi. These vessels may be engaged in the transhipment (transfer of cargo from one laden tanker to another) of liquid cargo.

The following suite of six screen-captured images that were made between 1:20 and 2:30 pm., Malaysian Standard Time (GMT + 8 h) on 7th July 2021 are shown in Figs. 5.12, 5.13, 5.14, 5.15, 5.16, 5.17 and 5.18 inclusive. The aim was two-fold. First, to demonstrate the traffic density of maritime transportation on a specific date/day within a few minutes of observation for this research project. Second, to illustrate the concept of generalisation in thematic cartography. A variety of map scales (small-, medium-, and large-scale maps) depict the amount of information that is portrayed on each map and clustering effect that may ensue within the geographical range of each image. The scale bar and the relevant distance are stated on each graphic. The suite of images was accessed from the webpages of *VesselFinder* on the day as mentioned above between the times of 1.30–2.21 pm Malaysian Standard Time (MyST) on 7th July 2021.

Figure 5.12 portrays the density of marine traffic within the seas of South and Southeast Asia and as far south to the north coast of Australia that includes the Arafura and Timor Sea. Of interest, is the line of traffic heading northwards from the iron-ore ports of Western Australia heading to ports of East Asia via the Lombok Strait, the Makassar Strait and through the Sulawesi (Celebes) and Sulu Seas and thence transit through the South China Sea or the western Pacific Ocean.

Figure 5.13 depicts the clustering of ships either at anchor or transiting the Straits of Malacca and Singapore. Four major ports (maritime hubs) within the mapped area attract international, and inter-regional shipping to this busy waterway. In addition, the sea lanes are abuzz with small vessels, tugs towing barges (fully loaded or partially

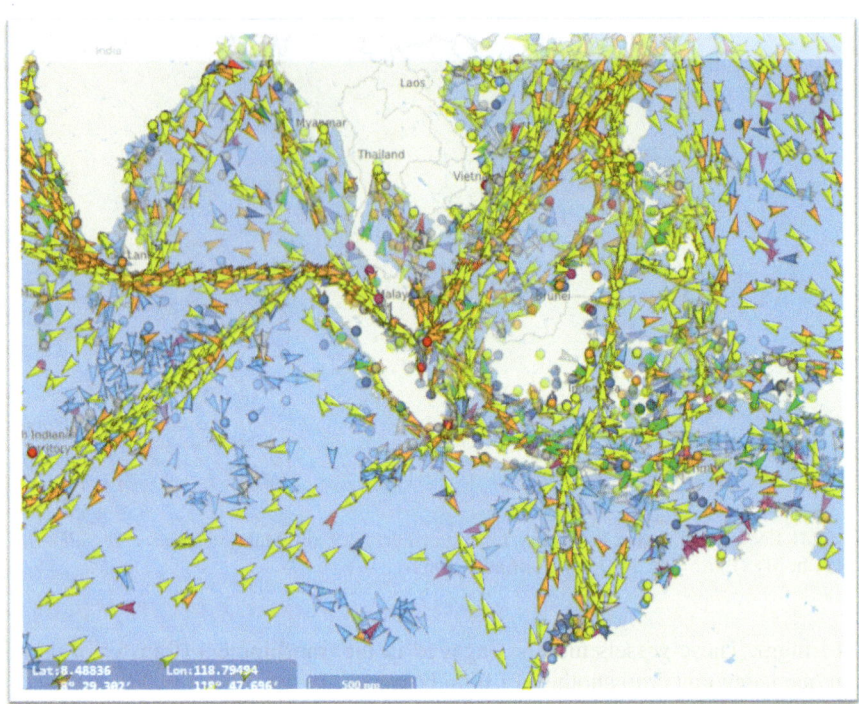

Fig. 5.12 Density of shipping within the seas of South and Southeast Asia. *Source VesselFinder*, at 1337 h, MyST, 7th July 2021

Fig. 5.13 Shipping within the Straits of Johor and Singapore. *Source VesselFinder*, at 1325 h, MyST, 7th July 2021

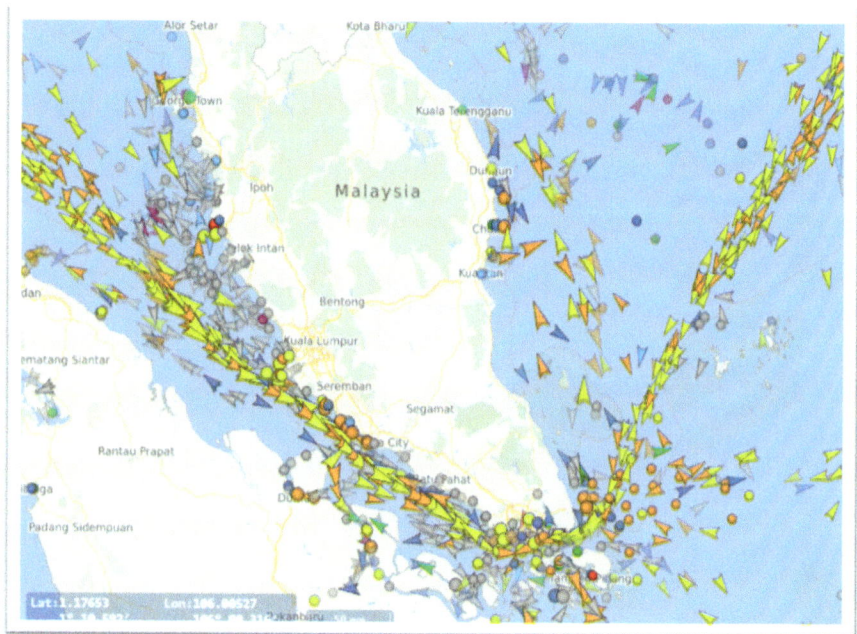

Fig. 5.14 Marine traffic density in the vicinity of Peninsula Malaysia. *Source VesselFinder*, at 1330 h, MyST, 7th July 2021

empty and workboats engaged in surveying, reclamation works and laying submarine communication cables.

Figure 5.14 displays the volume of traffic on 7th July 2021 in the vicinity of Peninsula Malaysia. As stated earlier this was a day during the height of the COVID-19 Pandemic when port authorities dictated the movement of marine traffic; airport regulators restricted flights in the region; and government health agencies imposed strict adherence to the regulations over citizens within the local areas.

As noted elsewhere in this volume, the IMO adopted the Mandatory Ship Reporting System in the Straits of Malacca and Singapore known as "STRAITREP" as proposed by Indonesia, Malaysia, and Singapore (IMO SN/Circ.201). STRAITREP came into force on 0000 h UTC on 1st December 1998 (Malaysia Standard Time: 0800 h on 1st December 1998). Masters of vessels, which STRAITREP is applicable, are advised to comply with the requirements of the adopted ship reporting system, in accordance with regulation V/8-1(h) of the *International Convention of the Safety of Life at Sea*, 1974, as amended in 1994.

Sections 5.1–5.5 of the Scheme is administered under the Authority known as KLANG VTS. Since its inception, the Authority has maintained statistics of the type and number of vessels transiting the Sections of the VTS. The reporting content in STRAITREP includes: the ship's name and call sign; geographical position and with reference to true bearing and distance from a clearly identifiable landmark; true

Fig. 5.15 Vessels in the vicinity of Port Klang, Malaysia. *Source VesselFinder*, at 1332 h, MyST, 7th July 2021

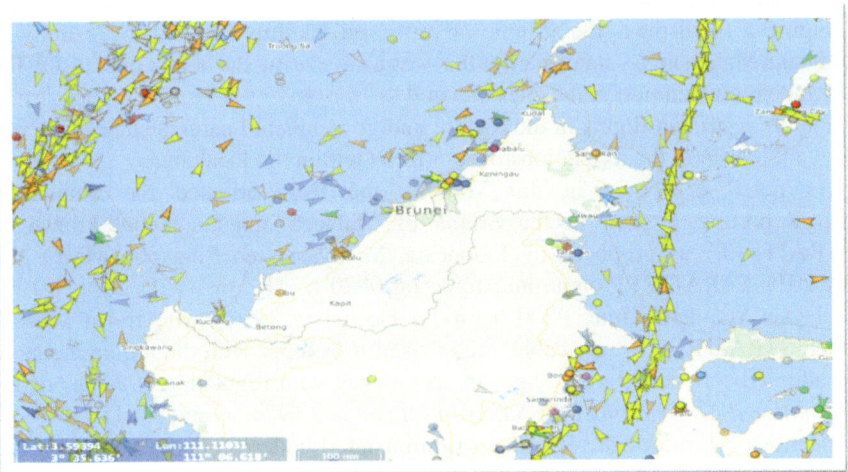

Fig. 5.16 Vessels in the seas adjacent to Sabah and Sarawak, Malaysia *Source VesselFinder*, 1334 h, MyST, 7th July 2021

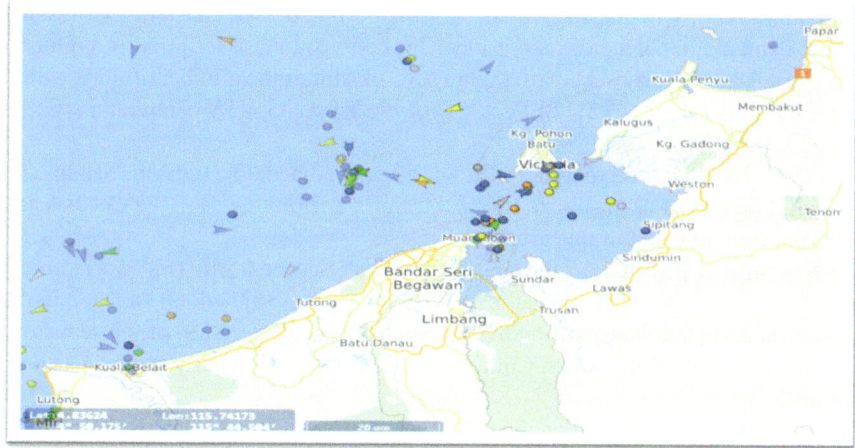

Fig. 5.17 Vessels in the vicinity of Brunei Bay and Labuan, Malaysia. *Source VesselFinder*, 1335 h, MyST, 7th July 2021

Fig. 5.18 Colour code legend for the above illustrations (*On the way equates to 'Underway'- moving through the water*)

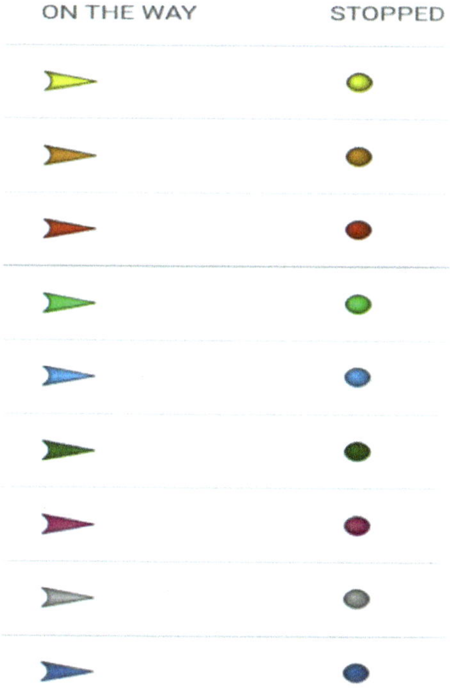

course and speed; whether hazardous cargo is on board; whether the ship has any defects, damage, deficiencies and/or other limitations; and description of pollution or dangerous goods lost overboard. Figure 5.15 illustrates the number of ships within a certain radius (about 26 M) of Sect. 5.1 of the KLANG VTS centre at a specific time of the nominated date. (Fig. 5.15).

Figure 5.16 shows the marine traffic in the seas adjacent to Labuan, Sabah, and Sarawak. The line of north-bound traffic through transiting through the Celebes and Sulu Seas is manifested in this graphic.

Within Brunei Bay and off the coast of Sarawak marine traffic (Fig. 5.17) on the stated date would appear to be that of vessels and oil rigs (blue dots) engaged in the exploration and exploitation of hydrocarbons from the substratum of the seabed of the southern sector of the South China Sea within the geologically termed Luconia Province.

The symbology employed on the online maps are as follows: the arrowhead indicates that the vessel is underway (on the way or moving) in the direction of the arrowhead is pointing; the circle indicates that the vessel has stopped (not underway). This could indicate that the vessel is possibly at anchor or made fast to the shore (wharf) or in attendance as a supply or work boat. It is also possible that the signal indicates that the vessel may be on patrol duties.

The categories of the vessels and their colour codes are as follows: General Cargo (yellow); Tanker (orange); Military ship (red); Passenger transportation (light green); Fishing boat (blue); Speed boat (dark green); Yacht/Sailing vessel (purple); Undefined ship, a category that is undetermined (grey); and other objects (dark blue). Refer to Fig. 5.18.

Navigation in the Malacca Strait

Viewed without the clutter of the previous maps, Fig. 5.19 shows the southern sector of the Malacca Strait and the location of aids to navigation, the Traffic Separation Scheme, the sites of the radar and their 40-m range overlaid on the limits of the reporting zones. The Deep-Water and Close Inshore routes are also depicted.

As an example of the coordinated approach to cooperation in managing the Straits of Malacca and Singapore the littoral Straits States—Indonesia, Malaysia, and Singapore—have representatives (technical experts) who meet at regular intervals to discuss various issues relating to the safety of navigation within these confined waters. At the 29th Tripartite Technical Experts Group (TTEG) Meeting of the Straits States, in Jakarta, from 6 to 8th December 2004 discussions focused on several issues agreed in the agenda, namely Aids To Navigation; Marine Casualty Affecting Traffic Movement; Mandatory Ships Reporting System (STRAITREP); Contravention of Rule 10 of The Collision Regulations (COLREG); Voluntary Pilotage Service in The Straits of Malacca and Singapore; Marine Electronic Highway (MEH); Signal To Be Displayed By Vessel Crossing The TSS in the Straits of Malacca and Singapore; Transit Anchorage Area (TAA); and Preparation of the 1st TTEG Familiarization Meeting With User States.[13]

Since that meeting in Singapore, in July 2004, Indonesia has not built new aids to navigation in and around the Straits of Malacca and Singapore and adjacent seas.

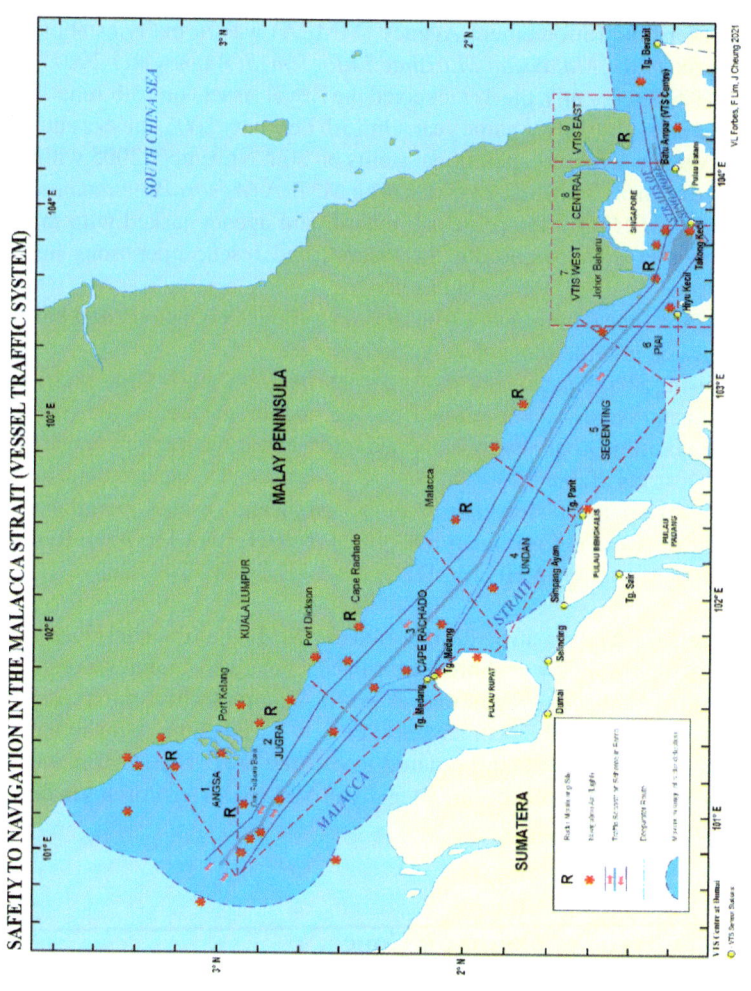

Fig. 5.19 Marine electronic highway and vessel traffic separation scheme. Marine casualties within the seas and straits of the Southeast Asian region can be found in literature and reports issued by international, national and regional agencies and authorities

Only repairing, maintaining and improvement measures were taken to the existing aids to navigation. During 2005, Indonesia owned 17 lighthouses, 14 buoys, 9 floating buoys and 5 RLB (Radar Reflector Buoy) to serve marine safety within the Straits.

5.5 Maritime Districts of the MMEA

The Malaysian Maritime Enforcement Agency (MMEA) was formally established with the enactment of the *Malaysian Maritime Enforcement Agency Act 2004 (Act 633)* in May 2004. Subsequently, the *Act* received the Royal Assent on 25th June 2004 and was *Gazetted* on 1st July of the same year. On 15th February 2005, the *Act* entered into force. The Agency achieved operational status on 30th November 2005 with the commencement of patrols by MMEA vessels. The MMEA (*Agensi Penguatkuasaan Maritim Malaysia; APMM*) is the principal government agency tasked with maintaining law and order and for coordinating search and rescue operations in the Malaysian Maritime Zone and on the high seas.[14]

MMEA is in effect the Coast Guard of Malaysia but does not belong and has no plans to be integrated into the Malaysian Armed Forces.

Figure 5.19, using a base map which shows the international maritime boundaries, depicts the administrative divisions of MMEA. The operational area of the Agency is the Malaysian Maritime Zone which is divided into 5 Maritime Regions consisting of 18 Maritime Districts (Table 5.3).

The MMEA is responsible for coordinating search and rescue operations. A network of Maritime Rescue Coordination Centres (MRCC) and Maritime Rescue Sub-centres (MRSC) monitors VHF, MF DSC, 2182 kHz, and VHF channel 16 for distress traffic.

On 31st March 2021, Malaysian media reported that MMEA introduced *Operation Jangkar Haram*, to spot illegally anchored ships in Malaysia's sovereign waters. The operation urges merchant ship owners to seek permission to anchor in the 'eastern waters' off the Johor coast in the Natuna Sea (See Fig. 5.4). Failure to do so would result in the Master of the ship being ordered to exit the area. The MMEA would not hesitate to act against ships anchoring illegally in Malaysia's territorial sea limits (Fig. 5.20).

5.6 Maritime Areas of Common Concern

A cooperative agreement between Indonesia, Malaysia, and the Philippines was established in June 2017. Its objective was to identify and define Maritime Areas of Common Concern and associated ship reporting systems in the vicinity of the southern Sulu Sea, the Sulu Archipelago and north-western Sulawesi Sea. Within this area multiple transit Corridors, established to mitigate the threat to shipping, have

Table 5.3   Operational areas of MMEA

Region 1	Operational area	Regional headquarters
Northern Pen. Mar. Region	P. Langkawi—S. Bernam	Langkawi, Kedah
District	*Operational area*	*Headquarters*
Maritime district 1	P. Langkawi-Kuala Muda	Bukit Malut, Langkawi
Maritime district 2	Kuala Muda-Parit Buntar	Batu Uban, Pinang
Maritime district 3	Parit Buntar-S. Bernam	Lumut, Perak
Region 2	*Operational Area*	*Regional Headquarters*
Southern Pen. Mar. region	S. Bernam-Endau	Johore Bahru, Johore
District	*Operation Area*	*Headquarters*
Maritime district 4	S. Bernam-Sepang	Port Klang, Selangor
Maritime district 5	Sepang-Kuala Kesang	Port dickson, N. Sembilan
Maritime district 6	K. Kesang-Johore C'way	Johore Bharu, Johore
Maritime district 7	Johore C'way-Endau	T. Sedili, Johore
Region 3	*Operational Area*	*Regional Headquarters*
Eastern Pen. Mar. region	Endau-Tumpat	Kuantan, Pahang
District	*Operation Area*	*Regional Headquarters*
Maritime district 8	Endau-Tanjong Geliga	Kuantan, Pahang
Maritime district 9	T. Geliga-Besut	K. Terengganu, Tereng
Maritime district 10	Besut-Tumpat	Tok Bali, Kelantan
Region 4	*Operational Area*	*Regional Headquarters*
SarawakMar. region	T. Datu-T. Baram	Kuching, Sarawak
District	*Operational Area*	*Headquarters*
Maritime district 11	T. Datu-Igan	Kuching, Sarawak
Maritime district 12	Igan—T. Payong	Bintulu, Sarawak
Maritime district 13	T. Payong—T. Baram	Miri, Sarawak
Region 5	*Operational Area*	*Regional Headquarters*
SabahMaritime region	T. Baram-Pulau Sebatik	KotaKinabalu, Sabah
District	*Operational area*	*Headquarters*
Maritime district 14	T. Baram—Kuala Penyu	Labuan
Maritime district 15	K. Penyu-Kg. Mendawang	KotaKinabalu, Sabah
Maritime district 16	Kg. Mendawang-Beluran	Kudat, Sabah
Maritime district 17	Beluran—Kunak	Sandakan, Sabah
Maritime district 18	Kunak—Pulau Sebatik	Tawau, Sabah

Note On the present version of the map Maritime Districts 11–18 inclusive are not shown. MMEA's websites do not have a graphic that depicts this information (As of 31.8.2023)

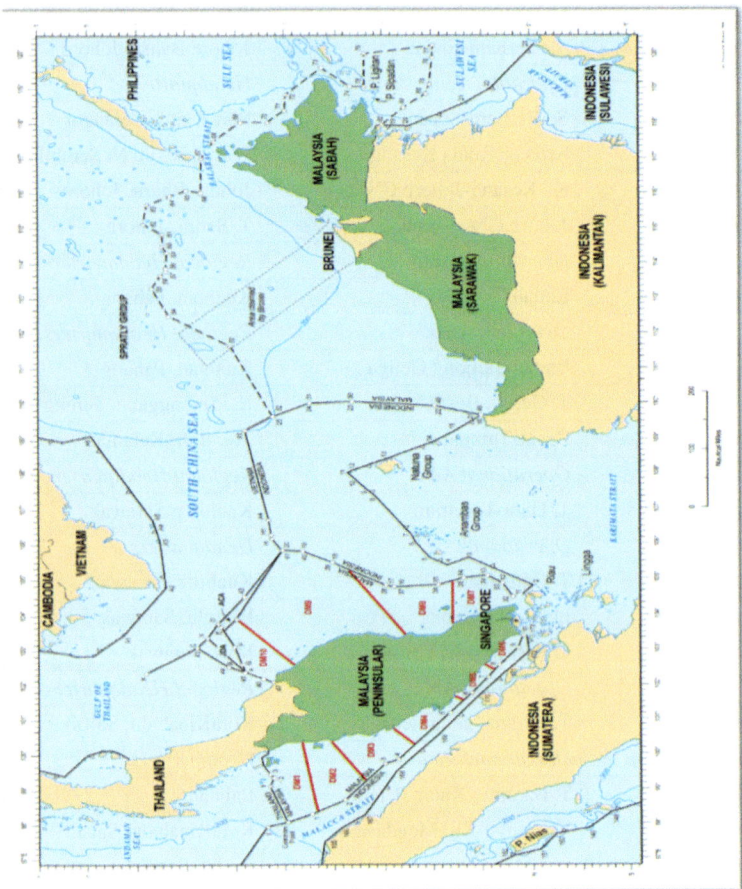

Fig. 5.20 Maritime Districts of the Malaysian Maritime Enforcement Agency

been established in these areas, including Basilan Strait, Moro Gulf, Alice Strait, and Sibutu Passage.

Recommended Transit Corridors (RTC) have been established between the east coast of Sabah, the Sulu Archipelago, and the west coast of Mindanao. The RTCs are not Traffic Separation Schemes and are not marked by any aids to navigation. For further information, mariners are advised to consult the relevant documents available from the Marine Department of Malaysia, Sabah Region as advised in *Notice to Mariners* No. 14/2017 and the webpages of the Malaysia Marine Department.[15]

Areas of Concern Ship Reporting System

The security situation since 2017 in the vicinity of Sabah (Malaysia), the Pangutaran Group (Philippines), and the Sulu Archipelago (Philippines) necessitated the establishment of a Ship Reporting System in conjunction with the Sulu Archipelago Transit Corridor. The area covered by the reporting system is the boundary of the Maritime Area of Common Concerns, which is encompassed within the following geographical coordinates:

a.	7° 11′ 00″ N,	118° 32′ 00″ E
b.	5° 48′ 00″ N,	120° 30′ 00″ E
c.	4° 48′ 00″ N,	120° 30′ 00″ E
d.	3° 11′ 33″ N,	119° 23′ 52″ E
e.	3° 56′ 00″ N,	118° 22′ 30″ E
f.	5° 21′ 00″ N,	119° 21′ 30″ E
g.	6° 21′ 00″ N,	117° 57′ 00″ E

Ships' Masters are required to report to the relevant authorities at least 24 hours prior to entering the Marine Area, detailing the vessel's complete routeing information.

National Heritage Zones Areas designated as such are protected from unauthorized interference. Local authorities should be contacted for further information. A National Heritage Zone of the NW coast of Brunei surrounds the offshore islands of Pulau Keraman (5° 13′N., 115° 08′E), Pulau Rusukan Kecil, and Pulau Rusukan Besar, as well as the E and S coasts of Pulau Labuan.

5.7 Submarine Operating Areas

Three designated submarine exercise areas have been established: off the Northwest coast of Sabah and Sarawak (Fig. 5.21) and off the East coast of Malaysia (Fig. 5.22). To ensure the safety of submarine operations, Masters of ships and other craft must notify the Malaysian National Security Council and the Submarine Control Centre at least seven days prior to conducting any of the following activities in the designated Malaysia Submarine Exercise Areas.[16]

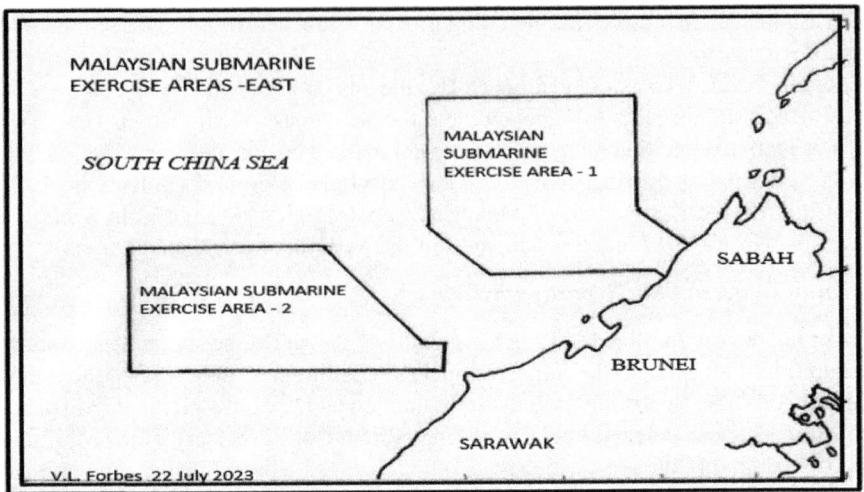

Fig. 5.21 Submarine exercise area off Sabah and Sarawak, Malaysia

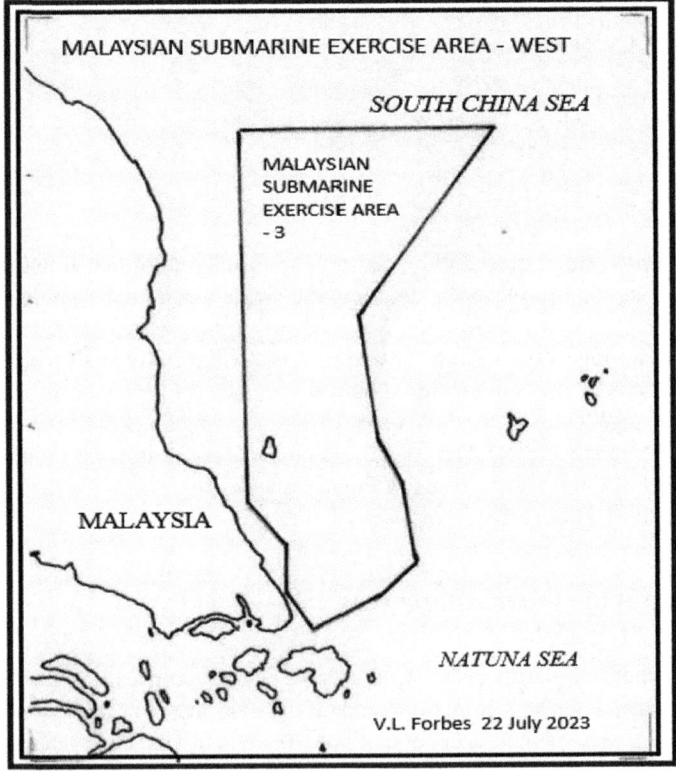

Fig. 5.22 Submarine exercise area off Peninsula Malaysia

The Government of Malaysia has stated that it will not be responsible for damage to, or loss of vessels, equipment, and life caused by the failure to notify the appropriate Malaysian authorities prior to conducting the indicated operations.

5.8 Landfall Lighthouses of Malaysia

A landfall lighthouse is so named as it is the first light to be seen by an observer (mariner) approaching the coast from the open sea. It is so situated and has luminous range and geographical range so great that it can be identified at a vast distance. Numerous lighthouses were erected during present-day Malaysia's rule by the Portuguese, the Dutch, and the British administrators (who oversaw the greatest number of new lighthouses built), and, later, the government of Malaysia, to provide safe navigation in and out of ports or through dangerous seas.[17] One of the earliest features was the Muka Head lighthouse structure on Penang (Pinang) Island. The elevation of the focal plane of the light is 242 m. The lighthouse was established in 1883. The lighthouse at Kuala Selangor, Selangor is featured in Fig. 5.23.

Figure 5.24 portrays Pulau Angsa and the lighthouse was established in 1887. The geographical location is Lat. 03° 11.2′ N, Lon.101°11.2′E. The structure, a white concrete building is 11 m, and the light characteristic is Flashing (3) White and Red 10 s with a range of 22 nautical miles.

Figure 5.25 depicts the original lantern and lens of P. Angsa Lighthouse as displayed at the World Maritime Day in Putra Jaya. This author, pictured next to the lantern, participated in this event.

A list of the landfall lighthouses of Malaysia, as depicted in Fig. 5.26, is in the Annex IX of this volume.

Malaysia's Marine Department functions to develop and implement systems pertaining to safety of navigation, safety of ships and all type of marine craft, management of seafarers, port facility security and to conduct examinations for maritime training in its effort to provide quality and satisfactory services to its clients which comply to international standards and national legislation and rules. The Department is charged with the responsibility for ensuring the safety of Malaysian-flagged vessels and foreign flagged vessels in the ports of Malaysia,

Aids to Navigation (AtN) in the Malacca Strait are regulated by the Malacca Strait Council, a consortium of the maritime authorities of Indonesia, Malaysia, and Singapore.

To ensure the effectiveness of aids to navigation the availability and reliability must meet the standards of International Association of Lighthouses Authorities (IALA). The aids to marine navigation must attain certain standards: lights in landfall lighthouses—availability of at least 99%; light beacons—availability of at least 98%; and light buoys—availability of at least 97%. IALA *Guideline No. 1116* of December 2016 offers a Selection of Rhythmic Characters and Synchronisation of Lights for Aids to Navigation.

Fig. 5.23 Image of lighthouse at Kuala Selangor, Selangor. This structure was built in 1907 and is 27 m tall. Elevation of light facility is 73 m. *Source* Author's collection

This document applies to marine aids to navigation signal lights on fixed and floating fixtures. It is intended for provision of integrated guidance on the following topics: selection of colours; flash duration; character length; use of simultaneous fixed and flashing signals; synchronisation and sequencing; and sharing of good practice. International and national conventions and regulatory frameworks provide the standards enforced by the administering authority. Guidance is offered by IALA relating to the technical aspects of selecting the rhythmic characters as defined in Recommendation E110(1). It includes temporal considerations, selection of colours, the use of the fixed and flashing character, user considerations, synchronisation, and sequencing.

Fig. 5.24 Pulau Angsa Lighthouse. Built in 1887; elevation 36 m

5.9 Summary

The discussion in this chapter focused on marine traffic and maritime safety within Malaysia's maritime jurisdiction. A brief mention was made of the effects of the COVID-19 Pandemic as it related to the maritime trade and aviation movement in the air space over Malaysia's maritime jurisdiction at a specific moment in time and the plight of Flight MH370 of March 2014.

Maritime safety is of prime importance especially within the Straits of Malacca and Singapore. The following chapter focuses on the establishment of marine parks, protected zones, and marine biotic and mineral resources.

Fig. 5.25 Original lantern of Angsa Light and the author. *Source* Author's collection

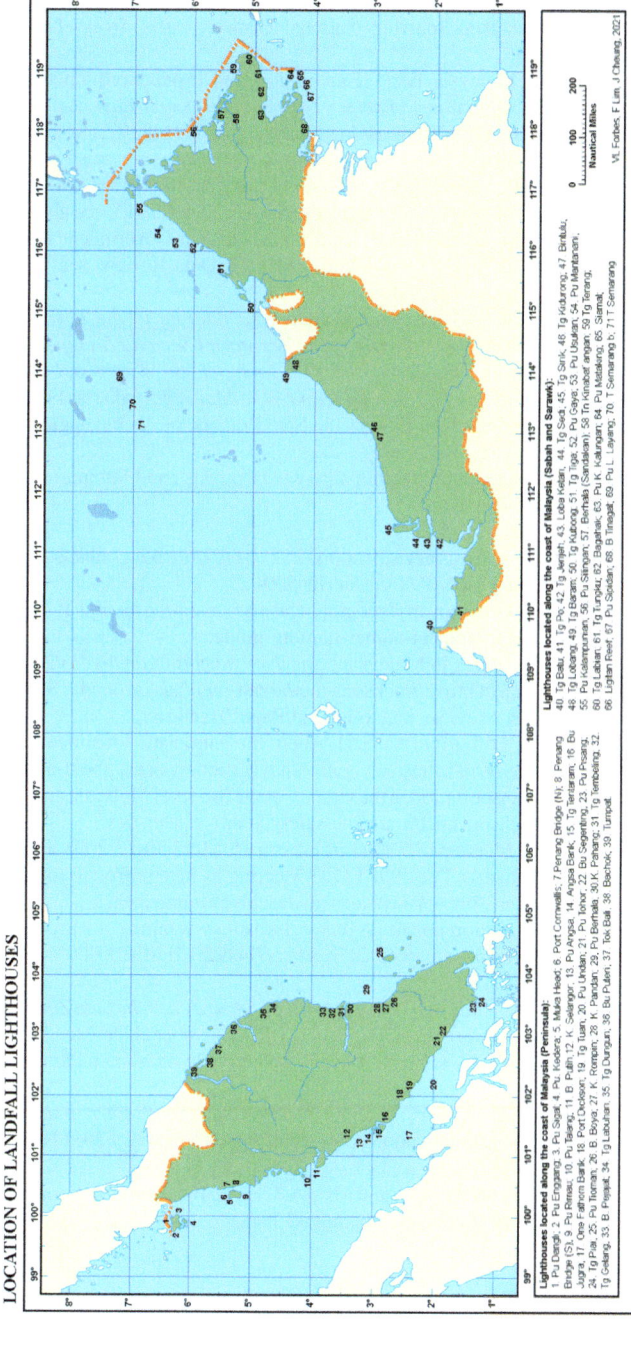

Fig. 5.26 Location of landfall lighthouses of Malaysia

Notes

1. Marine Department of Malaysia, Spurring National Growth, 2018; *StrasseLink* web site <strasselink.com/straits.php>.
2. PETRONAS Overview of Malaysia E&P (Exploration and Production) <petronas.com>; Publication 120, *Sailing Directions, Pacific Ocean and South-East Asia Malaysia*, p. 264 https://msi.nga.mil/api/publications/download?key=16694492%2FSFH00000%2FP ub120bk.pdf&type=view.
3. CANSO—Civil Air Navigation Services Organisation https://canso.org; see also Texas A&M University 'Mathematician theorizes what happened to MH370, 8/6/2015; see also Australian Transport Safety Bureau, 'About the Search' and the Malaysian Government's Ministry of Transport Report on MH370 Safety Investigation; MH370 Data Review, see https://www. atsb/gov/au/.../mh370-data-review-2022-final.
4. *Treaty between Malaysia and the Republic of Indonesia relating to the Legal Regime of Archipelagic State and the Rights of Malaysia in the Territorial Sea and Archipelagic Waters as well as in the Airspace above the Territorial Sea, Archipelagic Waters and the Territory of the Republic of Indonesia lying between East and West Malaysia*, Jakarta, Feb. 25, 1982, entered into force May 25, 1984, UN, Law of the Sea: Practice of Archipelagic States 144-155 (UN Sales No. E.92.V.3, 1992).
5. International cable Protection Committee https://www.iscpc.org': and Submarine cable map https://www.submarinecablemap.com.
6. GEBCO https://download.gebco.net.
7. See, for example, USNI News https://news.usni.org/2021/10/07/breaking-attack-submarine...; and https://www.nytimes.com/2005/05/18/us/adrift-500-....
8. PDNG https://www.mygeoportal.gov.my/geographical-names-database; and JUPEM https:// www.jupem.gov.my/page/geographical-names-naming-guide.
9. The Naming of Undersea Features in Malaysia https://www.geoinfo.my/hydro/2019....
10. Background of the MEH Project https://mehsoms.net/about-meh/background... See also: IMO https://www.imo.org/MediaCentre/SecretaryGeneral/Pages/Handover.
11. Australian Maritime Safety Authority (AMSA) 'About the Automatic Identification System' refer to: https://www.amsa.gov.au/safety-navigation/navigation-systems/about-automatic....
12. Observations of maritime transportation in the vicinity of the Straits of Malaysia and Singapore. See, for example, https://mehsoms.net/maritime-safety/straitrep-statistics....
13. See, for example, The 29th Meeting of TTEG in Jakarta 15:14:55 30-12-2004 Report of The 29th Meeting of TTEG in Jakarta The 29th TTEG Meeting in Jakarta from 6 to 8 December 2004 and for subsequent years. https://www.nas.gov/archivesonline/data....
14. The Malaysian Maritime Enforcement Agency https://www.mmea.gov.my/index.php..., https://msi.nga.mil/api/publications/download?key=16694492%2FSFH00000%2FPub1 20bk.pdf&type=view.
15. Hamzah B.A and Forbes, V.L., (Editors) (2019) *Maritime Security and the Sulu Zone: Readings on History, Peace Making and Terrorism*, Publisher: UPNM, National Defence University of Malaysia, Kuala Lumpur, pp. 537, see in particular, the article by Prashanth Parameswaran at pp. 108–110.
16. Publication 120, *Sailing Directions, Pacific Ocean and South-East Asia Malaysia*, p. 264 https://msi.nga.mil/api/publications/download?key=16694492%2FSFH00000%2FP ub120bk.pdf&type=view.
17. See Note 1; Marine Department Malaysia and MIMA Team (2017) *Spurring National Growth*, KL, 140pp; and https://www.detailedpedia.com/wiki-list....

Chapter 6
Marine Parks, Protected Zones, and Marine Resources

> *The most important thing that we can do to save our oceans is to dramatically expand our efforts to establish new marine protected areas and make sure that critical fish spawning sites and ecosystems remain undisturbed.*
> Serge Dedina

Abstract Protection of the marine environment and establishment and management of marine parks and protected zones off the coast and distant offshore islands are major issues of discussion in the light of sustainable development of resources on the one hand, whilst promoting concepts of benefiting from a Blue Economy and promoting Green Technology. The narrative in this chapter may be considered a controversial mix of topics; however, policy makers and administrators are confronted with unifying themes and issues. A basic principle in the development of policies is the recognition of multiple uses and users within reserved areas and public spaces and the conflictual issues between economic development and society's needs.

Keywords Marine park · Protected zone · Vulnerable marine ecosystem · Oil and gas tenement blocks · Marine cadastre · Tectonic structure

6.1 Introduction

In this chapter the discussion focuses on the establishment of marine parks, protected zones, and marine biotic and mineral resources of Malaysia. It may appear to be a controversial mix of topics; however, there are unifying themes and issues. Whilst the concept of sustainable development is being promoted, and the Blue Economy and Green Technology are publicized there is deep concern for a warming Earth and Climate Change. According to the World Meteorological Organisation's Report of 2023, the global mean temperature during 2022 was 1.15 °C (1.02–1.28 °C) above the 1850–1900 average. The years 2015–2022 were the eight warmest in the 173-year instrumental record. The year 2022 was the fifth or sixth warmest year on record, despite ongoing La Niña conditions. Indeed, on Monday 3 July 2023, the world's average temperature reached a new high of 17.01 °C (C) for the first time according to the US National Centre for Environmental Prediction. The previous record of 16.92 °C was deduced in August 2016.

The concentrations of the three main Greenhouse Gases (GHG) are carbon dioxide, methane, and nitrous oxide. The annual increase in methane concentration for the was the highest on record. Real-time data from specific locations show that levels of the three greenhouse gases continued to increase in 2022 and warmer temperatures during 2023.[1]

The above two factors, namely rise in average temperature and concentration of GHG, are of concerns to humans, animals and marine life and planet Earth as a whole. Protection of the ecosystems and environment—marine, terrestrial, and atmospheric—must be a priority of all governments and political ideology. An ecosystem is that of a community or group of living organisms that live and interact with each other is a specific environment. There are many types of ecosystems. They are terrestrial, forest, grassland, desert, tundra, aquatic, freshwater and marine. (Armstrong 1986).

Protection of the ecosystems and environment have been voiced by the United Nations and other international fora and by research scholars and scientists. Within the maritime sphere establishing areas of special significance and purpose has been promoted for many decades.

Marine Protected Area

A Marine Protected Area (MPA) is a region of the ocean designated for long-term conservation. There are no official criteria; however, there are guidelines proposed by the International Union for Conservation of Nature and Natural Resources (now the World Conservation Union) (hereinafter referred to as IUCN). In 2023, under 6.5% of the ocean is protected, however, less than 2% of this surface area is covered by exclusively 'no-take' policy, namely removal of resources from the MPAs is prohibited.[2]

Marine Protected Areas (MPAs) are established under national legislation to conserve natural diversity of native species in their habitats and provide some measure of insurance against environmental degradation or management uncertainty. They play a key role in protecting and conserving the global ocean ecosystems. MPA effectiveness depends on adequate water quality to a considerable extent. Most pollution comes into MPAs from land-based and ship sources; however, the maritime and fishing industries also play a role. Therefore, MPAs located across shipping lanes or near cities, river mouths, industrial, or port facilities were found to experience substantial water pollution.

Malaysia possesses over a half a million hectares of mangrove area and tropical rainforest covering about 65% of the total land area of which nearly 11% is pristine.

6.2   Marine Parks off Peninsula Malaysia

The first marine protected area for Peninsula Malaysia was the Fisheries Prohibited Area established in the waters eight kilometre (km) from the island of Redang in 1983. The designation of this area was made under the *Fisheries (Prohibited Area) Regulations*, 1983 under the *Fisheries Act* (1963). Subsequently the waters three-km width off twenty-one islands in the States of Kedah, Terengganu, Pahang, and Johor were added to the list. Like Redang these areas were initially declared as Fisheries Prohibited Areas before being gazetted as Marine Parks in 1994. Since then, the waters of 38 islands have been declared as marine parks in Peninsula Malaysia under the *Establishment of the Marine Parks Order* of 1994 (as amended) of the *Fisheries Act*, 1985. Two more islands were added to the list in 1998[3] (See Table 6.1).

The establishment of the marine parks was made with the goal of creating multiple-use areas for the protection, conservation, and management of the marine environment. The specific objectives of these parks as identified in the "National Marine Parks Malaysia—policy and concepts" documents. These MPAs fall under different jurisdictions: off Peninsular Malaysia, the 42 MPAs are managed by the Marine Park Department of the Ministry of Natural Resources and Environment. These are mostly islands declared as marine parks and comprise both terrestrial and marine components.

Some of the reefs in the Marine Parks of Peninsula Malaysia were extensively surveyed in 2000 by researchers involved in collaborative project between the then Marine Parks Section of the Fisheries Department and Coral Cay Conservation, a conservation group based in the United Kingdom.

Three marine parks are gazetted under Sarawak's *National Park and Reserve Ordinance*. These MPAs are not necessarily established for the conservation of biodiversity.

Five MPAs within Sabah's jurisdiction (see Sect. 6.3) are administered by the *Sabah Parks Enactment*. The first marine park was established in 1974. The Sugud Islands Marine Conservation Area (SIMCA) was the first private managed marine

Table 6.1 Legislation and statistics relating to MPAs

Establishment of Marine Parks Malaysia Order 1994		
Fisheries Act 1985 (No. 317 of 1985)		
Establishment of Marine Parks Malaysia (Amendment) Order 1998		
Establishment of Marine Parks Malaysia (Amendment) Order, 2008		
Number of MPAs: 51; Total surface area: 15,661 km^2; Maximum size: 10,200 km^2		
Minimum size: 11.8 km^2; Mean: 270.4 km^2; Malaysia's EEZ area: 468,167 km^2		
MPAs within EEZ: 3.5% of area		
Coral reefs	Total area: 1585 km^2	Percentage within MPA: 32.5
Sea grass	Total area: 125 km^2	Percentage within MPA: 32
Mangrove	Total area: 5591 km^2	Percentage within MPA: 1.2

conservation area in Malaysia. Administered by Reef Guardian, it was established in 2001. The Tun Mustapha MPA, an area of about one million hectare, is home to green sea turtles and dugongs.

Pulau Sipadan, an oceanic island formed over millennia, off the south-east coast of Sabah in the Celebes (Sulawesi) Sea is registered as MPA.[4]

The concept of Ecosystem Approach to Fisheries Management (EAFM) is a major shift in the way that marine biotic resources and their sustainability are perceived and practiced in Malaysia.

Figure 6.1 portrays the protected zones around the Groups of Johor Islands off the east coast of Peninsula Malaysia in the vicinity of Lat. 2° 14′ and 2° 45′ N, Lon 104° 09′ E.

The 13 islands of the Pulau Tinggi Marine Park Group (Table 6.2) contain 155 species of coral and 219 species of fish. The islands are scattered, and the marine park covers a surface area of 765 km². Surveys carried out in 2000, indicated that the corals in the Johor Islands were impacted by the 1998 coral bleaching event. Human impact, fortunately, is low given the small population on the islands and its distance from the mainland.

The six islands in the Pulau Perhentian Marine Park cover an area of almost 570 km² (Fig. 6.2). Of the six islands, Pulau Perhentian Besar and Pulau Perhentian Kecil are the most popular tourist destinations and have experienced the most human impact. They are located off Kuala Besut, east coast of Peninsula Malaysia (Table 6.3).

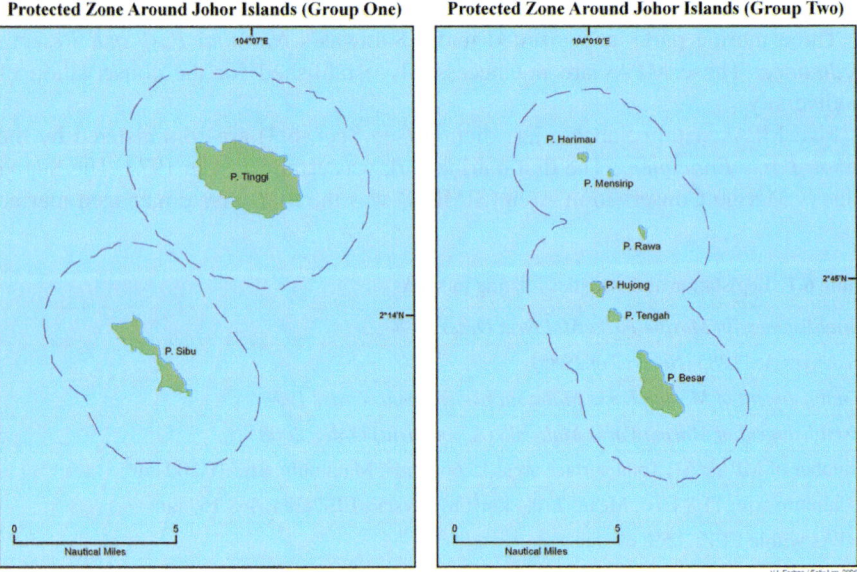

Fig. 6.1 Protected zones around the Johor Group of islands

Table 6.2   Islands within the Johor Group Marine Park

Name	Size (km²)	Longitude E	Latitude N
P. Goal	45.7	103° 58′ 12″	2° 32′ 12″
P. Harimau	49	103° 56′ 36″	2° 33′ 30″
P. Mensirip	46.6	103° 57′ 06″	2° 33′ 04″
P. Hujung	52.35	103° 57′ 12″	2° 29′ 30″
P. Tengah	51.49	103° 57′ 45″	2° 28′ 37″
P. Besar	84.14	103° 58′ 42″	2° 26′ 30″
P. Rawa	50.8	103° 58′ 39″	2° 31′ 12″
P. Tinggi	101.8	104° 07′ 12″	2° 17′ 54″
P. Mentinggi	43.99	104° 08′ 18″	2° 18′ 52″
P. Sibu	42.6	104° 04′ 21″	2° 12′ 51″
P. Sibu Hujung	11.83	104° 04′ 00″	2° 14′ 00″
P. Aur	97.45	104° 31′ 12″	2° 26′ 36″
P. Pemanggil	87.9	104° 19′ 36″	2° 34′ 26″
13 Islands	765.65		

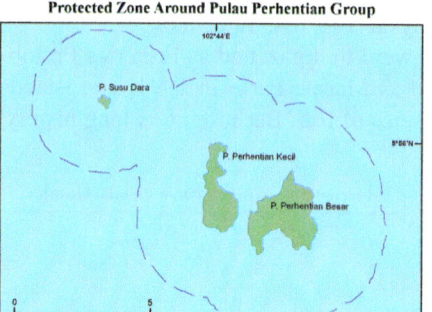 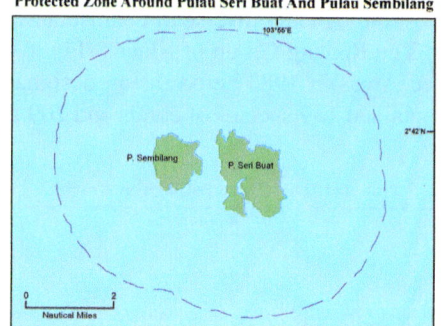

Fig. 6.2   Protected zones around Pulau Perhentian group

Table 6.3   Islands within the Perhentian Group Marine Park

Name	Size (km²)	Longitude E	Latitude N
P. Perhentian Kecil	81.07	102° 43′ 30″	5° 54′ 36″
P. Perhentian Besar	91.21	102° 45′ 30″	5° 54′ 00″
P. Susu Dara	14.28	102° 40′ 30″	5° 57′ 39″
P. Kapas	21.33	103° 16′ 00″	5° 13′ 15″
P. Tenggol	24.0	103° 40′ 00″	4° 48′ 24″
P. Nyireh	14.4	103° 40′ 00″	4° 50′ 41″
6 Islands	156.29		

Table 6.4 Pulau Seri Buat
and Pulau Sembilang

Name	Size (km²)	Longitude E	Latitude N
P. Seri Buat	77.2	103° 54′ 42″	2° 41′ 41″
P. Sembilang	60.6	103° 53′ 30″	2° 41′ 16″
2 Islands	137.8		

Pulau Seri Buat and Pulau Sembilang

These islands are located southwest of Pulau Tioman midway between Tioman and the east coast of the peninsular. The reefs of Pulau Seri Buat comprise of 68 species of corals and hosts 111 species of fish. The reefs around both islands have been subjected to prominent levels of sedimentation (Table 6.4).

The Pulau Payar Marine Park off Langkawi is the only marine park on the west coast of Peninsula Malaysia. (Fig. 6.3, left-hand map) Pulau Payar is one of the most heavily visited marine parks in Malaysia and experiences a significant level of human impact because of its popularity (Table 6.5).

The Tioman Group of islands is the largest marine park off Peninsula Malaysia and covers more than 676 km² of sea area (Figs. 6.3 and 6.5). A total of 183 species of corals and 233 species of fish were recorded in the group making it one of the most biologically diverse in Peninsula Malaysia (Fig. 6.4; Table 6.6).

The Redang Group of islands (Fig. 6.6) were first gazetted as Fisheries Prohibited Areas in 1983 before being designated as Marine Parks in 1994. The islands contained 149 species of corals and 209 species of fish. Because of its long history,

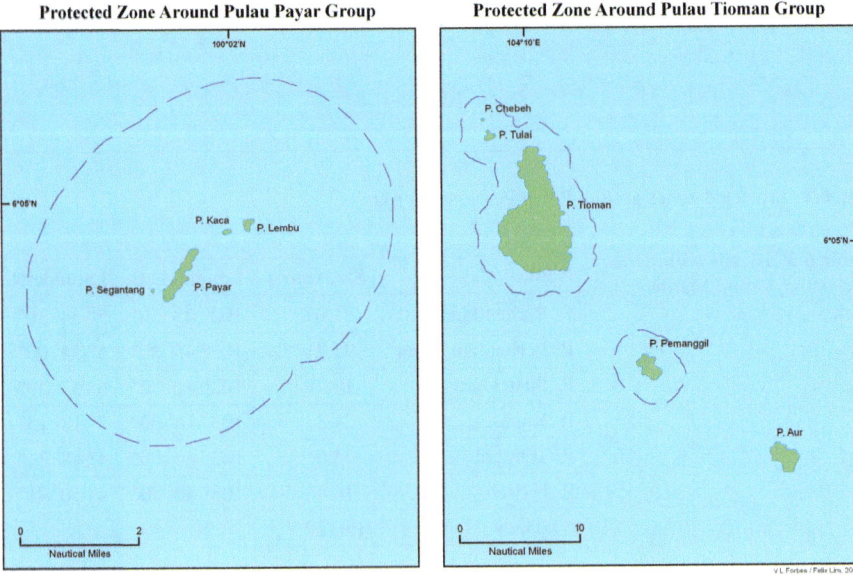

Fig. 6.3 Protected Zones around the Pulau Payar and Pulau Tioman Groups

Table 6.5   Islands of the Pulau Payar Group

Name	Size (km²)	Longitude E	Latitude N
P. Payar	54.91	100° 02′ 27″	6° 03′ 48″
P. Kaca	42.9	100° 03′ 06″	6° 03′ 22″
P. Lembu	46.13	100° 03′ 29″	6° 04′ 30″
P. Segantang	44.19	99° 55′ 37″	6° 02′ 36″
4 Islands	188.13		

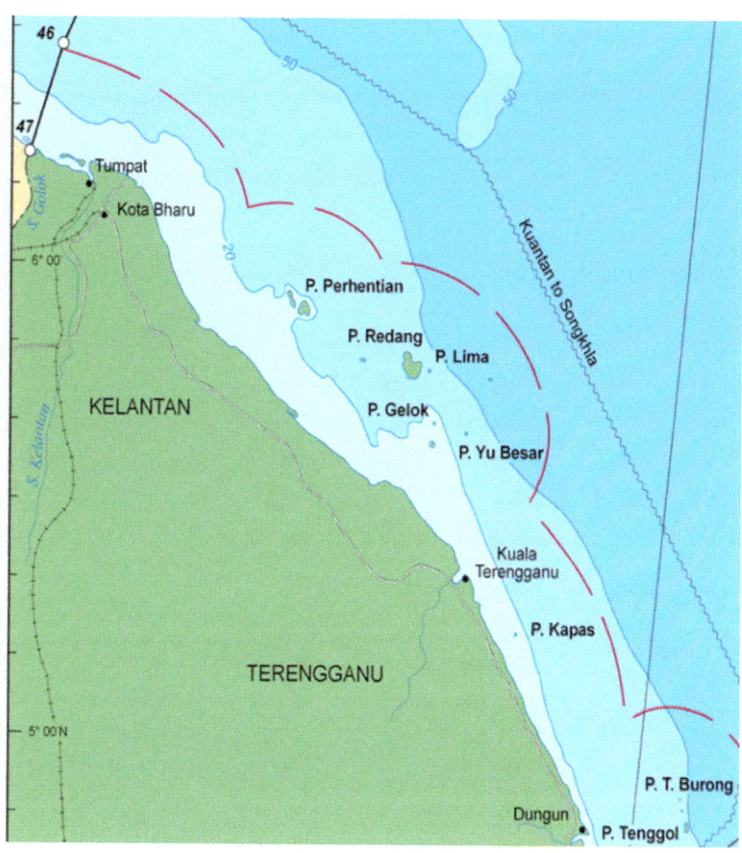

Fig. 6.4   Locational map for Fig. 6.2

the Pulau Redang marine parks have also seen some conflict between conservation and development policies and users including fishers (Table 6.7).

The Labuan Group of Marine Parks consists of three islands off the main island of Labuan. (Table 6.8) Because these islands are part of the Federal Territory of Labuan, the marine park is managed by the Marine Parks Section of Labuan Territory rather than by Sabah Parks.

Fig. 6.5 Locational maps for islands mentioned in Fig. 6.3

Table 6.6 Islands of the Pulau Tioman Group

Name	Size (km^2)	Longitude	Latitude
P. Chebeh	44.92	104° 06′ 30″	2° 56′ 00″
P. Seri Buat	77.2	103° 54′ 42″	2° 41′ 41″
P. Sembilang	60.6	103° 53′ 30″	2° 41′ 16″
P. Tioman	251.15	104° 10′ 21″	2° 46′ 48″
P. Tulai	63.06	104° 06′ 02″	2° 54′ 10″
P. Labas	44.78	103° 52′ 22″	2° 42′ 14″
P. Tokong Bahara	45.13	103° 54′ 28″	2° 41′ 01″
P. Gut	45.2	104° 10′ 00″	2° 40′ 00″
P. Sepoi	44.57	104° 06′ 22″	2° 55′ 47″
9 Islands	676.61		

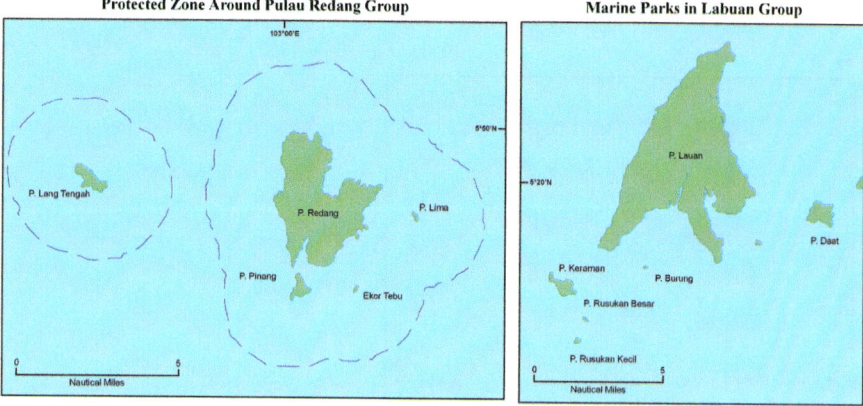

Fig. 6.6 Protected area around Pulau Redang Group; and Labuan Island

Table 6.7 Islands of the Pulau Redang Group

Name	Size (km^2)	Longitude	Latitude
P. Lang Tengah	61.5	103° 54′ 00″	5° 40′ 47″
P. Redang	127.5	103° 00′ 48″	5° 46′ 48″
P. Lima	43.9	103° 03′ 42″	5° 46′ 24″
P. Ekor Tebu	40.6	103° 01′ 52″	5° 44′ 52″
P. Pinang	48.9	103° 00′ 06″	5° 44′ 27″
5 Islands	322.4		

Table 6.8 Islands of the Labuan Group

Name	Size (km2)	Longitude	Latitude
P. Kuraman	66.95	115° 07′ 45″	5° 13′ 24″
P. Rusukan Besar	44.7	115° 08′ 12″	5° 11′ 18″
P. Rusukan Kecil	46.5	115° 08′ 36″	5° 12′ 10″
3 Islands	158.15		

6.3 Marine Parks off the Coast of Sabah

Turtle Islands National Park (Taman Negara Pulau Penyu is located about 40 km north of Sandakan in Sabah, East Malaysia (Fig. 6.7). The Group consists of three islands—Selingaan, Bakkungaan Kecil and Gulisaan including the surrounding coral reefs and maritime space. The Park is famous for its green turtles and hawksbill turtles which lay their eggs on the beaches of the islands. The Park covers an area of 17.4 km^2. The name Turtle Islands, however, refers to ten islands in the group, three of which are part of the National Park (Malaysia) and seven which belong to the Municipality of Turtle Islands, Tawi-Tawi, Philippines.

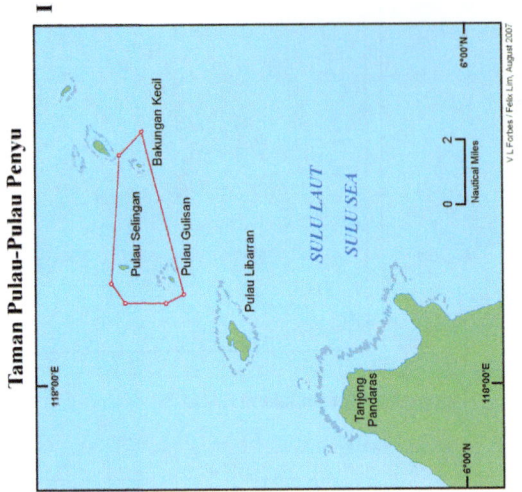

Taman Pulau–Pulau Penyu

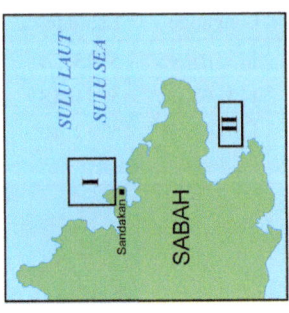

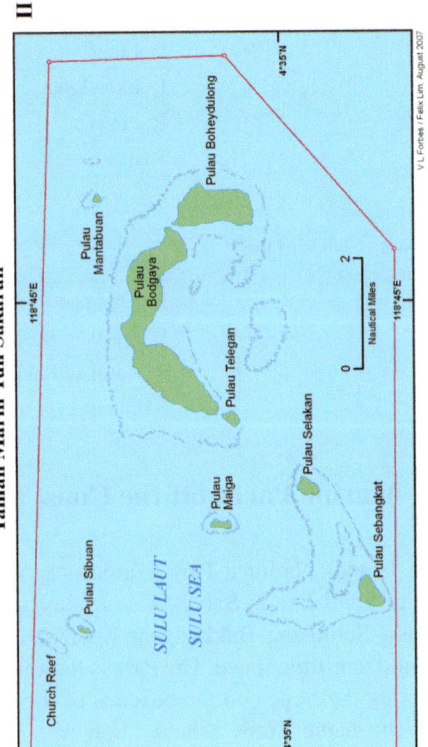

Taman Marin Tun Sakaran

Fig. 6.7 Marine Parks off the East coast of Sabah

On 1st August 1966, the first turtle hatchery in Malaysia was established on Selingaan, funded entirely by the Sabah State Government.[5] Turtle hatcheries on the remaining two islands followed shortly thereafter. In 1972, Selingaan, Bakkungaan Kecil and Gulisaan were designated as a Game and Bird Sanctuary. In 1977, this status was upgraded to that of a Marine Park.

On 13th June 2016, Malaysia formally established its largest marine park—the Tun Mustapha Park—off the State of Sabah in northern Borneo. At nearly one million hectares, the Tun Mustapha Park includes more than 50 islands and islets spread across Kudat, Pitas and Kota Marudu Districts. The Tun Mustapha Marine Park has more than 250 species of corals. It also harbours dugongs, endangered green turtles, and more than 300 species of fish (Fig. 6.8). The park will not ban fishing; however, it will allow local communities and commercial fisheries to operate in designated regions in a bid to ensure sustainable use of resources and undertake the sustainable management of the significant marine resources in the area that support jobs, livelihoods, and food security. The Park's official status should function as a model and an inspiration for marine conservation in the Coral Triangle and worldwide (Figs. 6.8, 6.9 and 6.10).

The Tunku Abdul Raham Park, off Kota Kinabalu, established in 1974, was the first Marine Protected Area in Sabah (Fig. 6.11). The Turtle Islands Park was established in 1975 and the Pulau Tiga Park was designated as a Marine Protected Area in Sabah now includes the Tun Sakaran Marine Park, off Samporna which was established in 2004.[6]

Fig. 6.8 Image of fish at Tunku Abdul Rahman M.P.

Fig. 6.9 Tunku Abdul Rahman Marine Park

Fig. 6.10 Mangrove ecosystem in T.A.R. Marine Park. *Sources* Figs. 6.8, 6.9 and 6.10 courtesy of Mr. Vee Chek

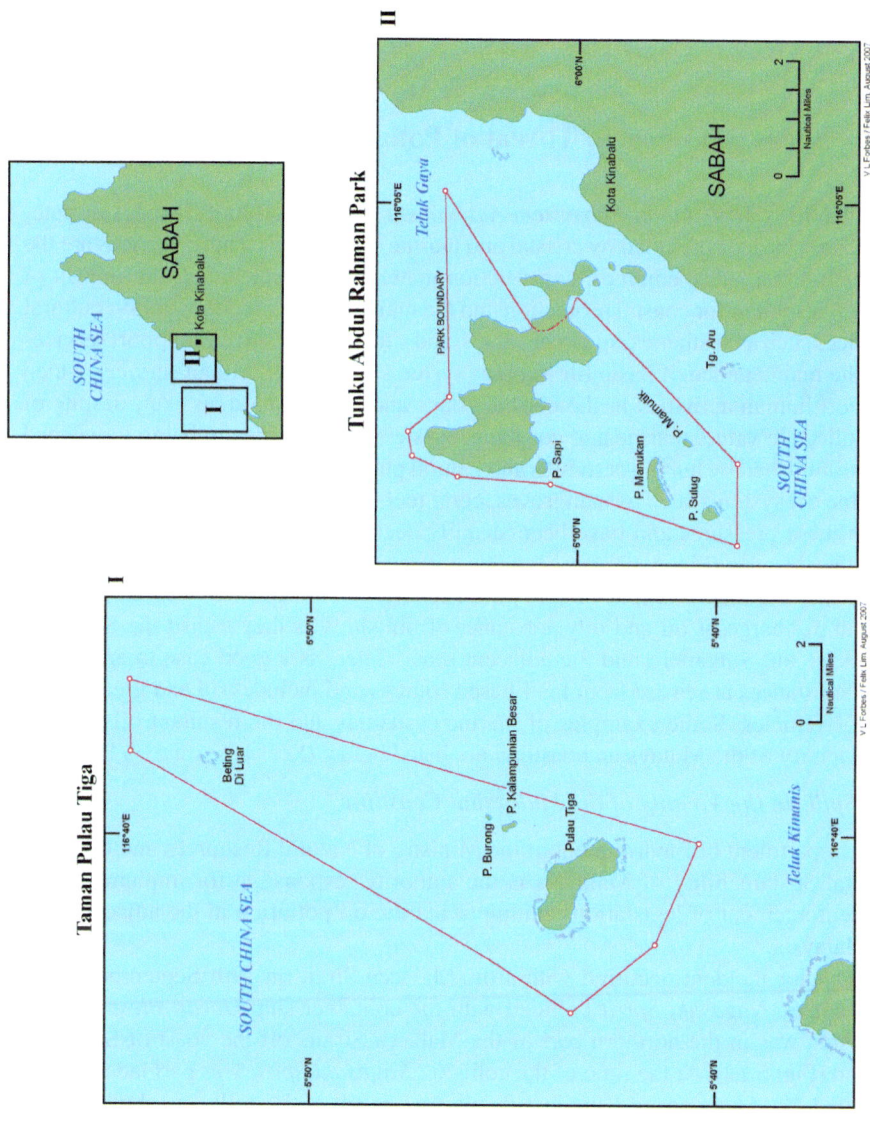

Fig. 6.11 Marine protected areas off the west coast of Sabah

The marine parks in Sabah were established with the aim of conserving the marine biodiversity of its coral reef ecosystems. In addition, park-specific objectives were also identified. The Pulau Tiga Park was established to protect its unique island ecosystem which includes mud volcanoes, coral reefs, and nesting habitat for sea snake. In comparison, Turtle Islands National Park was designed as a protected area for nesting sites for the green turtles and hawksbill turtles.

6.4   Areas of Concern: Threat of Pollution to Coastal Zone

The UNEP's *Millennium Ecosystem Assessment Reports* of 2005 listed four categories of services provided by coastal and marine ecosystems. These services are the provision of food, medicines, construction materials; and regulating the impact of the environment for coastal protection and the maintenance of water quality; cultural services such as aesthetics, spiritual values, and sites for tourism; and support services for the maintenance of basic life support systems. These are critical services which support human activities in the coastal zones and hinterland areas. Yet, despite of its collective value, marine and coastal ecosystems worldwide are being threatened and diminished by human activities and natural phenomenon. Malaysia's coastal and marine ecosystems such as mangroves, coral reefs and seagrass beds are not exempt from these pressures and have been steadily declining in quality and size over the last few decades (Fig. 6.12).

A major concern is that of oil spills because of collision between ships, accidental discharge of oil and oily substances from ships as they transit the Straits of Malacca and Singapore and the adjacent seas.[7] There is a good case to argue that oily substances at sea are from land-based sources and includes oil and grease from motor vehicles. Some examples of marine casualties and the resultant oil spills in the vicinity of the Malaysian coastline are listed below (See also Annex VI).

Oil Spills in the Vicinity of the Malaysian Coastline

The Department of Environment of the Ministry of Natural Resources and Environmental Sustainability of Malaysia is the authority responsible for implementation and enforcement of legislation relating to marine oil pollution in the adjacent seas of Malaysia.

A major incident occurred within the Malacca Strait on 19th September 1992. The tanker, *Nagasaki Spirit* collided with the container ship *Ocean Blessing*. The collision was in the northern part of the Malacca Straits off the coast of Sumatera Island (Fig. 6.12). At the time of the collision *Nagasaki Spirit* was part laden with a cargo of 40,154 tonnes of Khafil crude oil. As a result of the collision, about 12,000 tonnes of *Nagasaki Spirit's* cargo was released into the sea and caught fire.

The track of the oil slick was observed and recorded and portrayed in Fig. 6.13. The oil slick drifted as far north as the Langkawi Archipelago and eastward to the vicinity of Penang Island.

Its furthest southerly point was about Lat. 4° 30′ N.

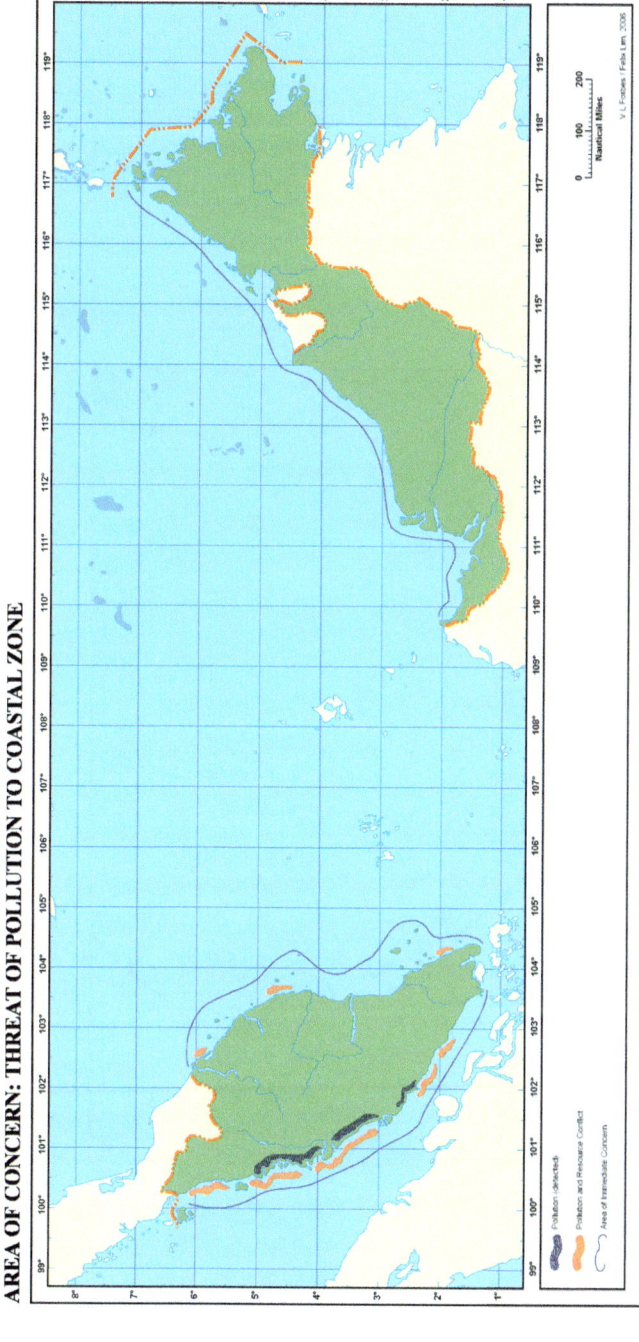

Fig. 6.12   Areas of concern: threat of pollution to coastal zone

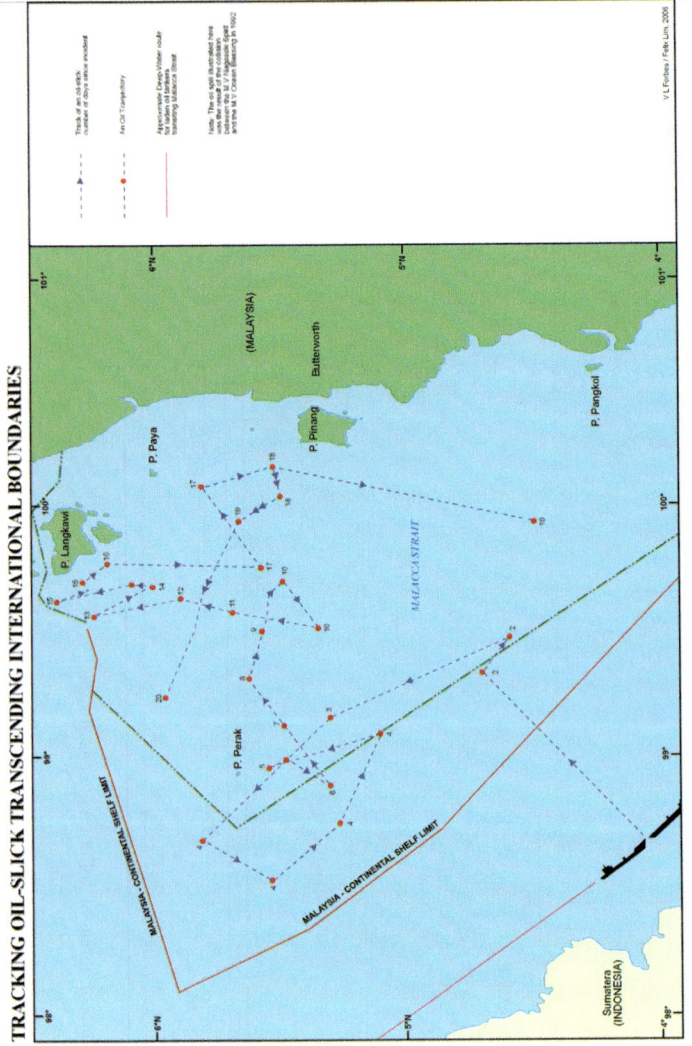

Fig. 6.13 Tracking an oil spill in the Malacca Strait

A review of oil spill incidents within Malaysia's jurisdiction between 2014 and 2016 was undertaken by Ishak and others (2019).[7]

Wastewater Discharge and Discard of Plastic Particles

Wastewater is defined as liquid wastes collected in a sewer system and conveyed to a treatment plant for processing. There are diverse types of wastewaters and here are a few of them: sanitary or domestic wastewater refers to liquid material collected from residences, business buildings, and institutions; industrial waste refers to wastes from manufacturing plants; municipal wastewater is a general term applied to liquid treated in a municipal treatment plant. The age of wastewater depends on the colour and odour.

A 2015 study noted that there were over 10,000 sea village homes in Sabah without proper sanitation. Collectively, about 50,000 residents had discharged an estimated 4.2 L of untreated wastewater directly into the adjacent seas.[8]

The concentration and yield of the total suspended solids from storm water runoff and sewage are similar. Streets and parking lots collect large amount of solid water, that runoff into streams and lakes, and Biochemical Oxygen Demand (BOD) in urban areas in much higher than in areas pristine, but less than raw sewage (Fig. 6.14).

Figure 6.13 depicted the conditions observed in 1994. The tabulated values below are results observed since in the following decade. Reports of pollution can be made to the Department of the Environment by telephone.

New laws to protect river and marine pollution were subsequently introduced and a National Water Resources Council was created in 1998. The Council was tasked with effective water management, including the implementation of interstate transfer of water. On 28th April 2006, the Government under its tough stance on Open Dump Sites and Rudimentary Landfills decided to close down with immediate effect 16 open dump sites located near water intake sites across the country.[9]

On 25th October 2021, the Government of Malaysia promulgated the Water Services Industry (Prohibited Effluent) Regulations 2021 (the Regulations). These Regulations stipulated that no person shall discharge or permit discharging any prohibited effluent as specified in the Schedule of the Regulations to a public sewage system or public sewage treatment plant without the approval of the National Water Services Commission (SPAN). The Regulations came into force on 1st December 2021.[10] Waste management has been given priority in many national and international agendas since this sector has tremendous impacts on climate change and loss of biodiversity. Improper waste disposal has caused water, air, terrestrial, and marine pollution. Marine debris or marine litter has been identified as a major factor that has deteriorated marine health and impact on biodiversity negatively. MARPOL 73/78 was introduced to combat marine pollution from ship-source and land-based facilities such as ports and oil refineries. Malaysia's vast coastlines abutting the regional seas of Southeast Asia is home to a rich marine biodiversity which could be severely affected by marine debris, particularly from plastic particles. Malaysia as a nation is a major manufacturer of plastic material.

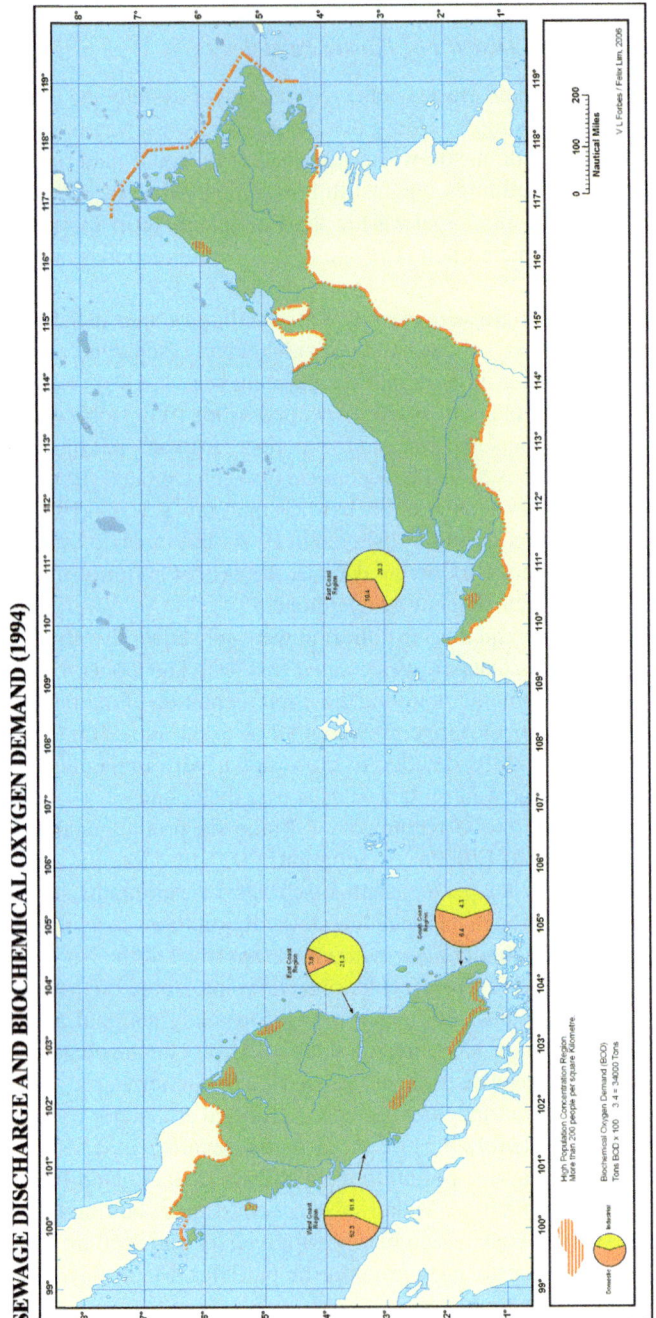

Fig. 6.14   Sewage discharge and biochemical oxygen demand (1994)

6.5   Vulnerable Marine Ecosystems: Seagrass Beds

Seagrass beds are the least officially known of the three ecosystems. Their importance, however, should not be understated as like coral reefs they also function as nursing and feeding grounds for commercially valuable fish. They are unique flowering plants which can live submerged in the shallow marine ecosystems. Within Malaysia, seagrass beds are found in association with shallow inter-tidal pools, semi-enclosed lagoons, coral reef flats and subtidal zones.

At least ten sea grass species are to be found along the coast of Peninsular Malaysia and Sabah and Sarawak. Seagrass grow on substrates ranging from sand and muddy/sand to coral rubble.

Seagrass is important as a source of food for the critically endangered dugongs and seahorses. Unfortunately, because of the lack of understanding of its significance and lack of awareness of its existence many seagrass beds are now threatened by development activities such as coastal land reclamation and construction of ports. Important seagrass areas in Sungai Pulai, Tanjung Adang and Merambong Shoal in Johore are threatened by port development activities while sand mining threatens the seagrass areas at the Sungai Paka estuary in Terengganu. Researchers have observed that at least 100 fish species and 20 prawn species are known from sea grass beds on the west coast of Peninsular Malaysia.[11] (Fig. 6.15).

Unlike mangroves and corals, seagrasses are not directly protected by law. No specific law exists to protect seagrass and seagrass beds and those which are protected are accorded such protection by virtue of being inside the boundaries of marine parks. Seagrass protection is a significant gap in coastal and marine ecosystems conservation in Malaysia. Because of its significance and the threats, it faces, seagrasses could no longer be left unmanaged. Table 6.9 lists major seagrass areas in Malaysia, their management status, and the threats that they face.

Malaysia's mangroves are among the most diverse in the world. A two-year MIMA survey of sixty major mangrove sites found the presence of sixty-nine species belonging to twenty mangrove families including rare species with limited distribution. Besides the mangroves, there are also habitats for many endangered species including migratory birds and the Proboscis monkey signifying the diversity in Malaysia. More importantly, the mangroves support Malaysia's burgeoning fisheries industry by serving as nursing and feeding grounds for juvenile fish and prawns.[12]

Mangrove areas in Malaysia have declined steadily from 800,000 hectare in the 1950s to approximately 695,000 hectares in 1973 and 575,000 hectares in 2003. This represented a twenty-eight percent decline in coverage over a period of fifty years. The decline resulted primarily from the conversion of mangroves for other land-use such as human settlements, agriculture, aquaculture, industries, and infrastructure. Most of the mangroves lost were from state land forests which are not protected by law. The decline of mangroves from forest reserves was gradual and totalled only 21,411 hectares from 1980 to 2003, representing a loss of 4.24% of total national mangrove areas. On the other hand, this is also alarming as it means that even mangroves designated as protected areas could be converted for other uses.

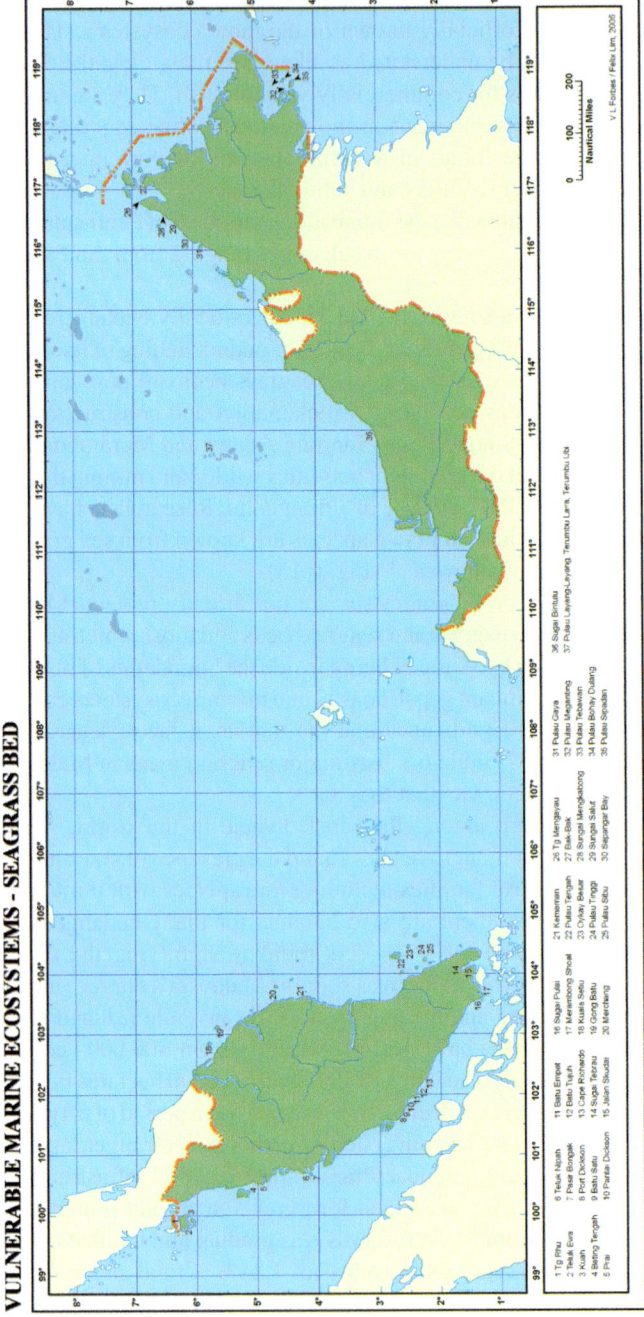

Fig. 6.15 Vulnerable marine ecosystems: seagrass beds at nominated locations

Table 6.9 Threats through human activities to selected seagrass beds in Malaysia

State	Location of seagrass beds	Conservation status	Threats
Kedah	Pulau Langkawi (Tanjung Rhu and Teluk Ewa)	None	• Land reclamation for tourism facilities • Pollution from cement industry • Impacts from boating and recreational activities
Negri Sembilan	Port Dickson	None	• Reclamation for tourism facilities • Sand/coral mining • Pollution from solid wastes and sewage • Uncontrolled tourism and recreational activities
Johor	Sungai Pulai estuary, Tanjung Adang and Merambong shoal	Mangrove Forest Reserves and RAMSAR site	• Land reclamation for port development and expansion (Tanjung Pelepas Port), and industrial parks • Massive ship navigation/movement • Ship-sourced pollution • Potential pollution from petrochemical industries • Heat water and wastes from Tanjung Bin power plant (coal) • Clearing of mangroves • Impacts from harvesting of fisheries resources
	Sungai Johor estuary and adjacent areas (Straits of Johor, Pulau Tekong and Pulau Ubin, Singapore)	Mangrove Forest Reserves	• Land reclamation (Pulau Tekong, Pulau Ubin and Changi area) • Sand mining • Industrial wastes from Pasir Gudang, Tebrau and Woodlands (Singapore) Industrial Parks • Massive ship navigation/movement • Ship-based pollution • Domestic wastes and sewage

(continued)

Table 6.9 (continued)

State	Location of seagrass beds	Conservation status	Threats
	Pulau Sibu, Pulau Tinggi, Pulau Besar, Pulau Rawa and adjacent islands	Johor Marine Parks and Mersing Islands National Park	• Sedimentation from the impacts of illegal trawling at marine park • Impacts from boating and recreational activities • Untreated wastes
Terengganu	Sungai Paka Estuary and Paka Shoal	Mangrove Reserve	• Sand mining • Impacts from harvesting of fisheries resources
Sarawak	Kuala Lawas	Mangrove Reserve	• Impacts from harvesting of fisheries resources
Sabah	Tunku Abdul Rahman Park	National Park	• Land reclamation at Kota Kinabalu and adjacent areas • Destructive fishing cyanide and fish bombing • Direct discharge of wastes from illegal settlement from Pulau Gaya and mainland of Kota Kinabalu • Impacts from boating and recreational activities • Ship-based pollution
	Karambunai, Sepangar Bay, Sungai Salut and Sungai Mekabong	None	• Land clearing for Kota Kinabalu Industrial Park, naval base, and settlements • Destructive fishing (cyanide and fish bombing) • Impacts from boating and recreational activities • Pollution from petrochemical industries • Ship-based pollution (Sepangar and Kota Kinabalu Ports)

(continued)

Table 6.9 (continued)

State	Location of seagrass beds	Conservation status	Threats
	Sulaman Lake		• Impacts from boating and recreational activities • Aquaculture development • Illegal cutting of mangroves
	Pulau Banggi and Pulau Balambangan	Gazetted as Tun Mustapha Marine Park and Mangrove reserves	• Sand and coral mining • Destructive fishing (cyanide and fish bombing) • Illegal trawling activities • Illegal clearing of mangroves • Impacts from harvesting of fisheries resources
	Darvel Bay	Mangrove reserves (Lahad Datu, Kunak, and Semporna) and some islands proposed as Tun Sakaran Marine Park	• Sand and coral mining • Destructive fishing (cyanide and fish bombing) • Illegal trawling activities • Illegal clearing of mangroves • Impacts from harvesting of fisheries resources

Eighty-five percent of Malaysia's 575,000 hectare of mangroves is protected under several protected area categories which include forest reserves, wildlife sanctuaries, state parks, national parks and RAMSAR Convention sites. This constitutes the bulk of Malaysia's mangroves which will be protected in 'perpetuity' as envisaged in the National Forestry Policy of 1985. (Figs. 6.16 and 6.17).

Malaysia's coral reefs are mostly shallow fringing reefs in the offshore islands and are an extension of the 'Coral Triangle'. The rest are small patch reef, coral atoll, and barrier reef. The United Nations Environment Programme's (UNEP) *World Atlas of Coral Reef* estimated the size of Malaysia's coral reef area at 3600 km^2 which is 1.27% of world total coverage.

Table 6.10 illustrates the type of threats and their scale values that degrade and destruct coral reefs in Malaysia. Over 85% of the corals reefs in Malaysia were threatened; however, management of the reef systems have improved in the past decade.[13]

Fig. 6.16   An area of mangroves, Malaysia. *Source* Courtesy of Ms. Sue Suri Hati

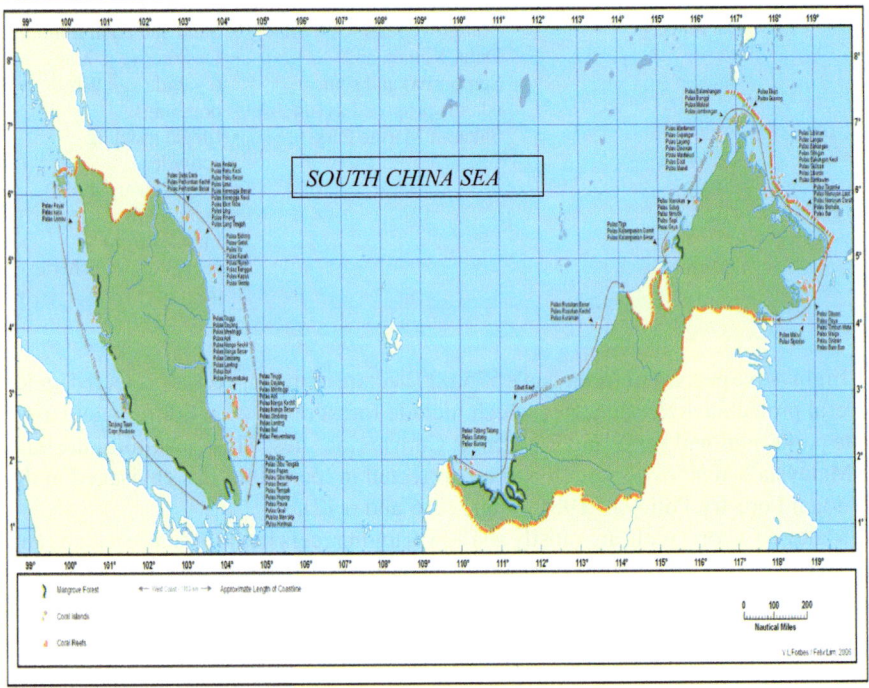

Fig. 6.17   Vulnerable marine ecosystems: mangrove forest and coral islands

Table 6.10 Threats to corals and coral reefs in Malaysia

Threats	West Coast of PeninsularMalaysia	East Coast of PeninsularMalaysia	East Malaysia
Fishing intensity	4	3	5
Fishing damage	3	3	5
Fish blasting	2	2	4
Gleaning	2	1	3
Boat scouring	2	3	4
Population pressure	4	3	4
Sedimentation	5	3	3
Domestic and agriculture pollution	3	2	4
Industrial pollution	3	1	1
Oil spill	2	1	2
Disease and predation	2	4	3
Dredging	2	1	2
Coral mining	1	1	3
Tourist activities	1	2	2
Coral bleaching	1	1	1

The scale values
1 = None to rare
2 = Very low concentration
3 = Some damage, some stress
4 = Medium to high damage
5 = Very high, high stress, very damaging
Source UN UNEP/GEF South China sea global environment programme project facility reversing environmental degradation trends in the South China Sea and Gulf of Thailand NATIONAL REPORTS on Coral reefs in the Coastal Waters of the South China Sea

Two areas for coral conservation in Malaysia are the islands of Redang and Tioman. Recent reports have suggested that the corals of Tioman Island have come under pressure from a marina development project and may be further threatened by a proposal to construct a new airport on the island. Despite these threats, Malaysia's corals are among the better managed in the region and have been parts of the country's marine protected area system since 1984. (See Fig. 6.18).

6.6 Commercial Fishing Ports and Fishery Infrastructure

The Malaysian fisheries sub-sector has for decades played a significant role in contributing to the nation's blue economy. According to the official portal of Malaysia's Department of Fisheries the sub-sector produced nearly 1.79 million

Fig. 6.18 Marine ecosystem: coral and fish. *Source* Courtesy of Ms. Sue Suri Hati

metric tonnes of edible fish during 2023. Additionally, 241 million pieces of orna-
mental fish and 26 million bunches of aquatic plants worth RM 16.5 billion. The
total production value of the fishery industry increased by 3.6% since 2022.[14]

Malaysia is a member of several international organisations relating to fisheries
management.

Although prices have soared (10–30%) over the past two decades the *per capita*
consumption in Malaysia has been around 50 kg annually, which ranks second in the
Southeast Asian region. Using latest scientific findings and technologies developed
from research activities the national fishing industry could become more competitive.

Shortage of labour, price of diesel fuel, adverse weather conditions and depletion
of fish stock and the adverse effect of illegal and un-regulated fishing activities in
the regional seas affect the price of fish for consumption in Malaysia.

According to the official source, namely the Fisheries Department, in 2023, there
were 41,282 boats operating within Zone A (0–5 nautical miles from the coast); 4.161
vessels working in Zone B (5–12 M); 2997 boats fishing within Zone C (12–30 M);
716 engaged in trawling in Zone C2 (30 M to limit of EEZ); and 17 vessels operating
on the High Seas (Fig. 6.19).

There are four species of turtle found in Malaysia. The species, listed on the
IUCN Red List as endangered, include the green turtle (*Chelonia mydas*), leatherback
(*Dermochelys coriacea*), hawksbill turtle (*Eretmochelys imbricata*) and olive ridley
turtle (*Lepidochelys olivacea*)[15] (Fig. 6.20).

On Peninsular Malaysia, most of the recorded landings are at Kedah, Penang,
Perak Melaka, Johor, Terengganu, Pahang, and Kelanatan. Along the coast of
Sarawak most of the turtle landing recordings are at the Talang-Satang National
Park. Landings also occur on the beaches of Tanjung Dato National Park, Telok
Melano, Samunsam Wildlife Sanctuary, Sematan, and Similajau National Park. The
nesting on the coast of Sabah and adjacent islands that include the archipelago of the
Turtle Islands Park and on Sipidan Island.

The green turtle is extensively distributed in Malaysia. The hawksbill turtle nesting
sites are to be found at six sites on the coast of Peninsular Malaysia as well as long
parts of the coasts of Sabah and Sarawak. The leatherback is found on the beaches
of Rantu Abang and Paka in Terengganu and occasionally at Chendor and Johor.

COMMERCIAL FISHING PORTS AND FISHERY INFRASTRUCTURE

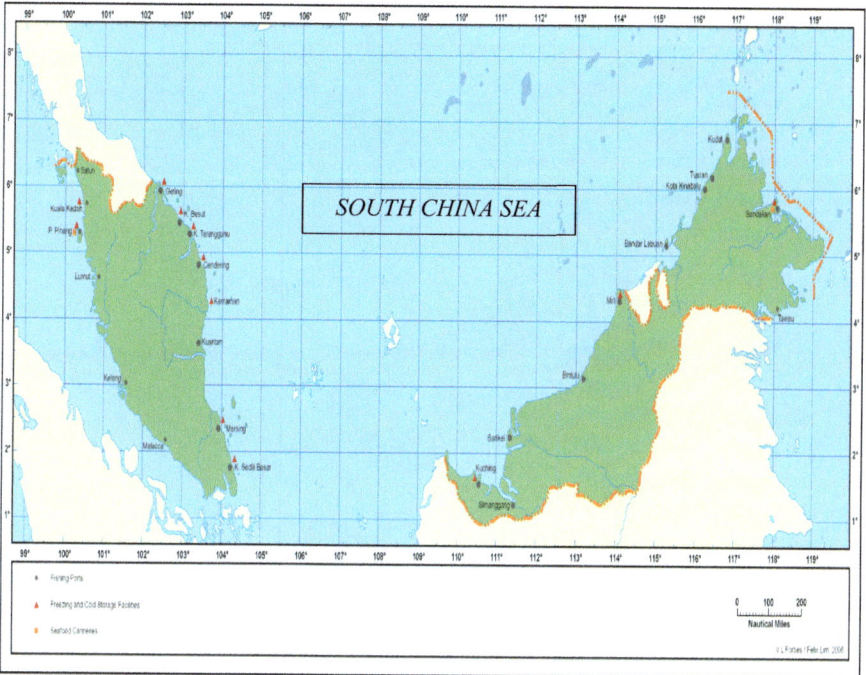

Fig. 6.19 Commercial fishing ports and fishery infrastructure

Nesting sites of the olive ridley turtle is not precisely known as records are generally not accurate nor consistent.

The substantial decline in turtle population was due to over-exploitation of turtle eggs over several decades. The continuing loss of nesting habitat is also a contributing factor why marine turtles throughout the world are deemed an endangered species. The loss or reduction of a single nesting beach, due to resort development or establishment of recreational facilities has contributed to the decline of the turtle population.

The main threats to the turtle population are due to the by-catch in the trawl fisheries operations and the uninterrupted long-term harvesting of the eggs and adults and the destruction of their habitat. Other threats include the accidental and deliberate discarding of fishing nets overboard which trap turtles who may eventually drown if the turtle cannot untangle and swim away. Other natural and human-induced problems such as erosion, sedimentation, and other land-based and vessel-source pollution also contribute to potential loss of turtle nesting sites.

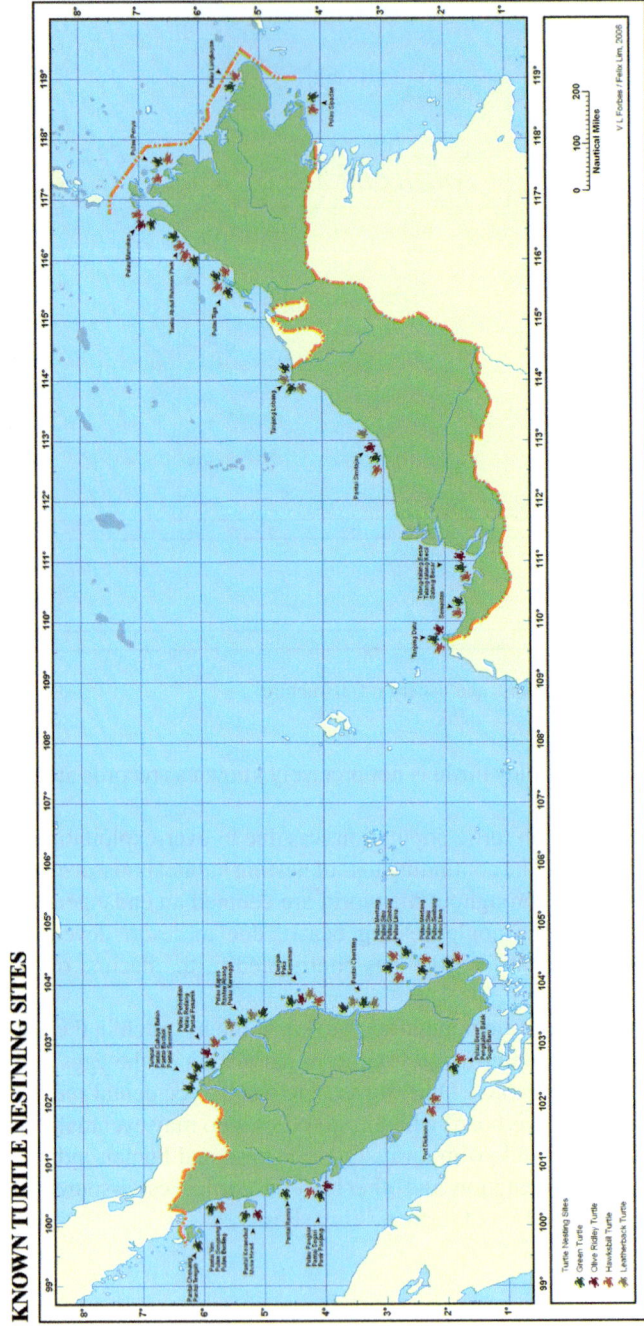

Fig. 6.20 Known turtle nesting sites

6.7   Oil and Gas Tenement Blocks in Offshore Malaysia

Figures 6.21 and 6.22 illustrate the extent of activity over many decades of exploration and exploitation of hydrocarbon operation in the offshore regions of Malaysia.

In its Media Release in November 2023, PETRONAS, the national hydrocarbon exploration and exploitation agency of Malaysia, recorded 19 exploration discoveries and two exploration-appraisal successes. It represented over one billion barrels of equivalent (bboe) of new hydrocarbon resources for the nation during the year. According to the report 25 new wells were drilled during the year.[16]

The majority of the discoveries were made in the Sarawak Basin in particular within the Balingian and West Luconia geological provinces. Three discoveries were made during 2023 in the Northwest Sabah Basin in shallow-water as well as deep-water and ultradeep-water areas in the South China Sea.

The offshore exploration blocks—areas permitted by the appropriate authority of the Government of Malaysia—within which companies undertake exploration and exploitation of hydrocarbon reserves are known to exist in the substratum of the seabed within Malaysia's maritime jurisdictional space are depicted in Figs. 6.21, 6.22 and 6.23.

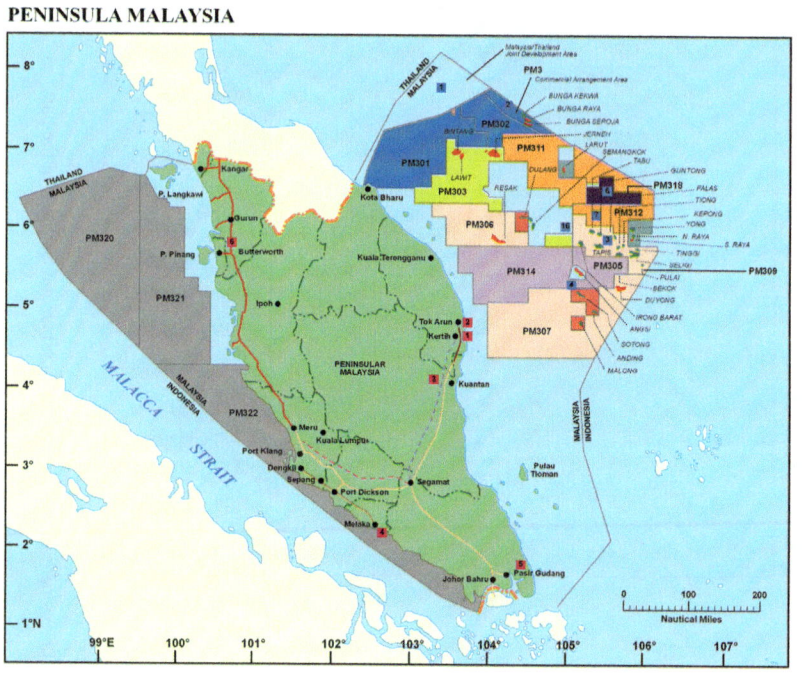

Fig. 6.21   Offshore exploration tenement blocks granted in the past (c.2009)

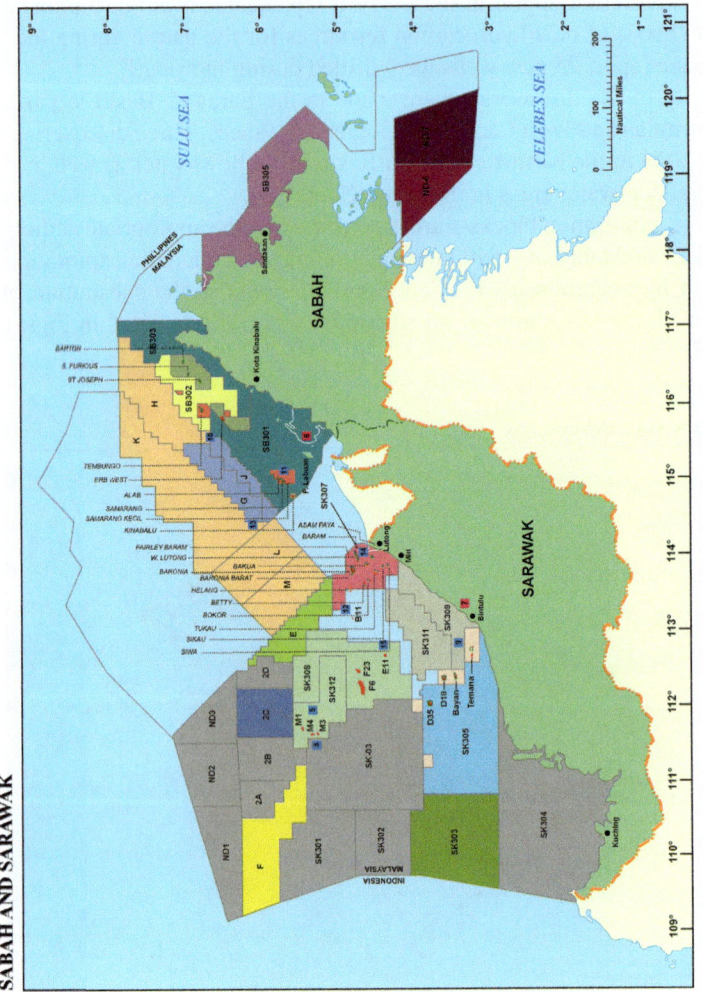

Fig. 6.22 Offshore tenement blocks granted in Sabah and Sarawak in the past

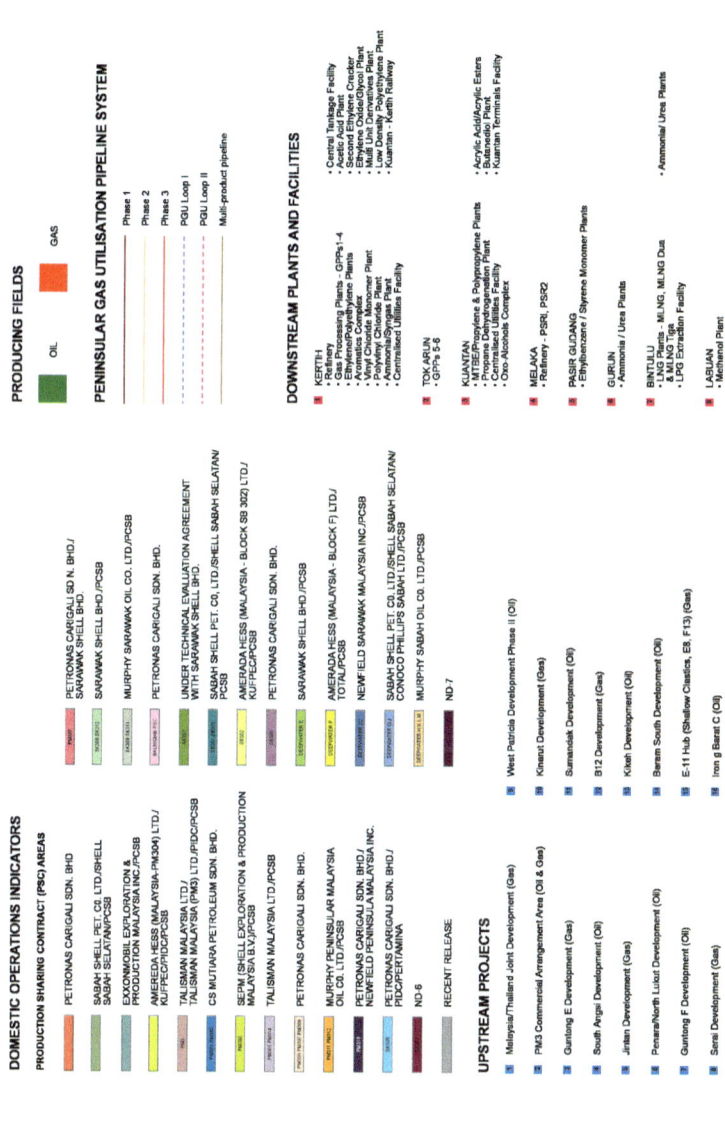

Fig. 6.23 Tenement holders and operators

World demand for oil is expected to rise between 2023 and 2028 to reach 107 million barrels per day due to the needs of the petrochemical and aviation sectors, according to the authoritative sources, IEA (International Energy Agency) and OPEC World Oil Outlook (2022).

Figure 6.23 is a map legend to accompany Figs. 6.21 and 6.22. It lists the operators and/or permit holders of the various exploration blocks shown on the preceding maps; and shows the upstream projects and downstream plants and facilities. Holders of tenements may have changed in subsequent years. The names of the companies listed may have changed over the years as the industry is economically dynamic, complex, and secretive to say the least. Only selective information is offered, especially in the electronic media.

The Malaysia-Thailand Joint Development Area (JDA) covers an area of about 7250 km^2 (Fig. 6.24).[17]

The Malaysia/Thailand JA currently consists of three blocks, namely A-18, B-17, and B-17–01. The companies entered contracts with MTJA.[18]

A Transmission Pipeline was completed in December 2004 and the Malaysia-Thailand Gas Separation Plant (GSP-1) was completed in December 2005. Details of the operations of the JDA are available in their annual reports. (Fig. 6.25).[19]

The names and numbers of the hydrocarbon operations within the M/T JDA (Fig. 6.25) are as follows: Muda—1; Muda—2; Jenga—3; Jenga East—4; Jenga—5; Jenga South sits below Jenga; Bumi East—6; Bumi—7; Sunya—8; Cakerwala—9; Bulai—10, Wira—11, Sumudra S—12; Sumudra N—13; Senja—14; Mali—15; Tapi—16. (Fig. 6.26):

On 1st December 2021, PTTEP announced yet another discovery of contaminant-free gas at the Nangka-1 exploration well in Block SK417, in offshore Sarawak. The well was drilled to a total depth of 3758 m in September 2021. The company operates the shallow-water block, 90 km from the coast of Sarawak. (Fig. 6.27). Figure 6.28 illustrates the configuration of a cluster of platforms linked for hydrocarbon exploitation.[20]

6.8   Tectonic Structure of Seabed: Offshore Sabah and Sarawak, South China Sea

Figure 6.29 is a generalised version of the sedimentary basin lying offshore off the northern coastline of Borneo. It also shows the reef systems that are collectively termed the Spratly Archipelago (southern sector), various fault lines, sediment thickness (as brown isolines with their respective thickness in kilometres) and bathymetry (as blue isobaths, in metres). A cross-section profile of the substratum on either side of the Palawan Trench is portrayed in an inset. The edge of the natural continental shelf is the 200-m isobath. The limiting line of the juridical continental.

The North-West Borneo Trough is floored by an abyssal plain at 2.9–3.4 km. The basin is crossed by several northeast-southwest submarine ridges, reflected by a

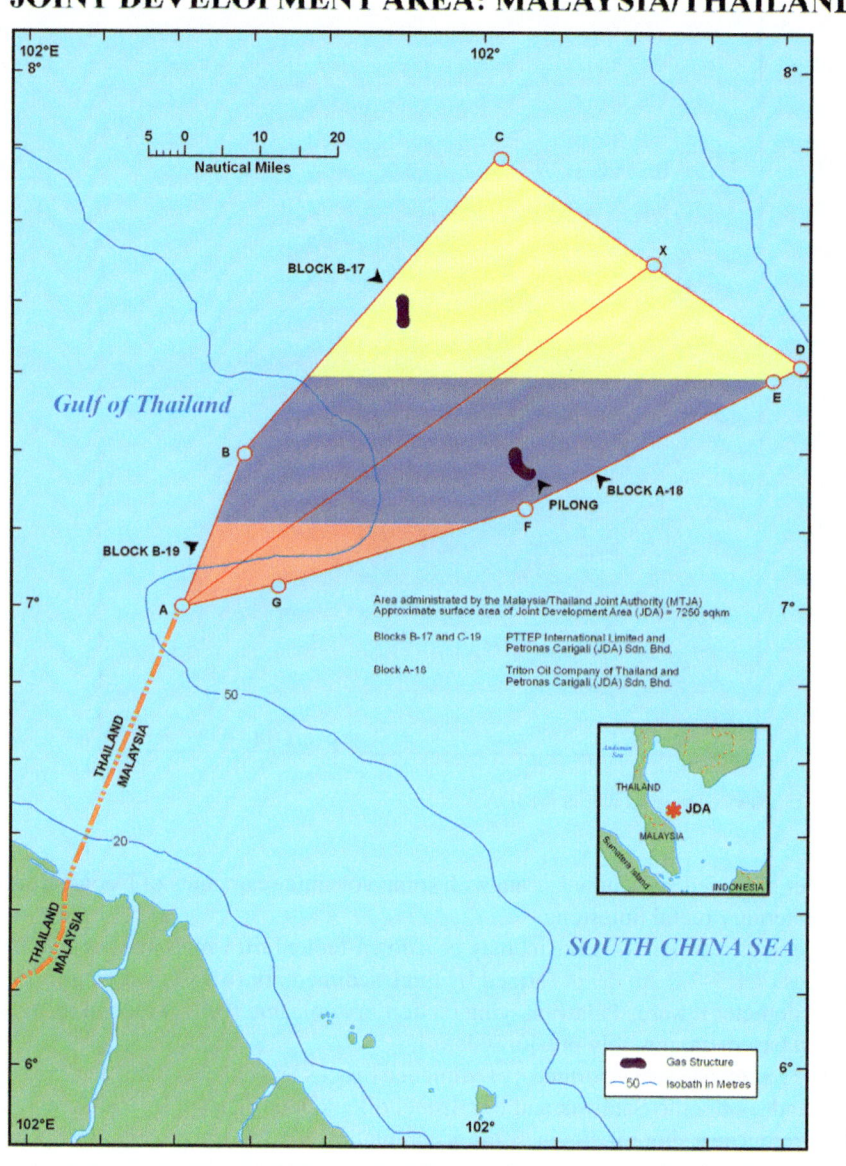

Fig. 6.24 The Malaysia/Thailand JDA (Joint Development Area) 2009. *Source* Author, 2009

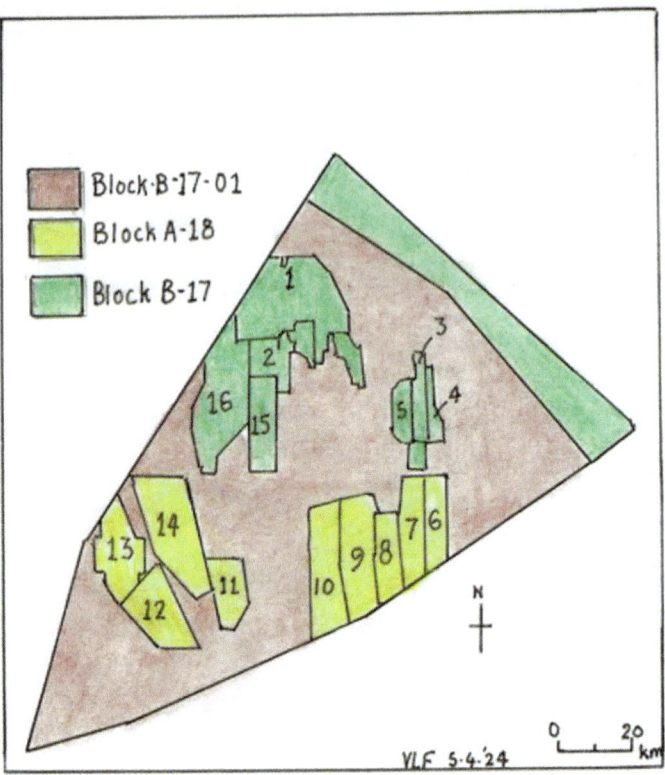

Fig. 6.25 Revised permits within the JDA

number of folded and faulted strata with some volcanic seamounts of Neogene and/ or Quaternary reefal limestone.

The Southeast Palawan Basin, may be a down-faulted rift which was occupied by terraces of 1.5–2.0 km deep, formed by thick sedimentary rocks thickening to over four kilometer towards Palawan, lying on Tertiary volcanic rocks, which in turn rest upon a basement, possibly pre-Jurassic.

The Sarawak Basin is formed by sedimentary rocks layered over 8 km that extends west and south onto Sarawak and the Brunei oilfield area. The biological activity is high around near shore areas where the accumulation of oil and gas potentially occurs. Huge quantities of organic matter are buried and protected from oxidation in many offshore sedimentary basins. With increasing burial over millions of years, chemical reaction gradually transforms some of the original organic matter into hydrocarbon reserves—oil and natural gas.

Jurisdictional Limits to Fisheries

To regulate the fishing activities the offshore waters are divided into fishing zones, A Conservation Zone has been established which is within one nautical mile (M) from

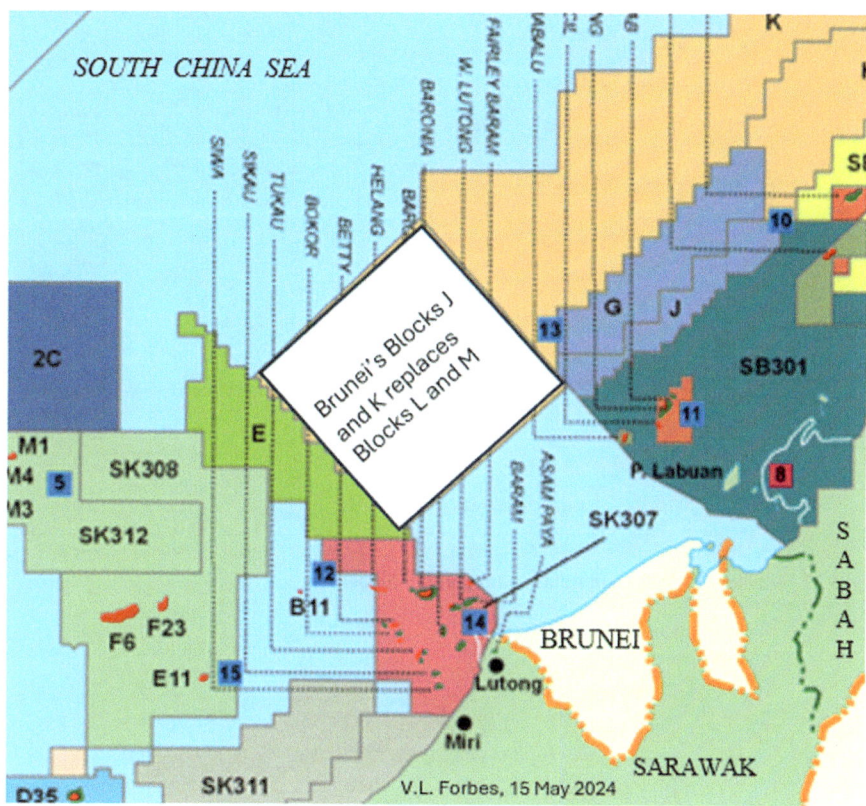

Fig. 6.26 Brunei's blocks 'J' and 'K'

the Malaysian coastline. Fishing within 2 M of the marine parks is prohibited with an exception made for the tourism sector, for example, diving groups.

Zone A is reserved for small vessels operating traditional fishing gears within one to five M from the shore. Commercial fishing operation (trawling and purse seining) are not permitted to operate within this zone.

Commercial fishing activities (trawling and purse seining) with boats below 40 GRT are allowed to operate within Zone B which has a defined marine space of 5–12 M from the coast.

Trawlers with some purse-seiners below 70 GRT are permitted to operate within Zone C1 which has a width of 12–30 M) from the Malaysian coast. Vessels engaged in commercial fishing operations over 70 GRT are allocated Zone C2 which extends from 30 to 200 M from the coast. (See Fig. 6.30; Tables 6.11 and 6.12).

For each Zone, the optimum number of fishing vessels has been determined based on estimation of maximum sustainable yield. There has also been a freeze on the issuance of new fishing licences for the inshore waters. Limited number of licences for offshore vessels is still being issued annually.[21]

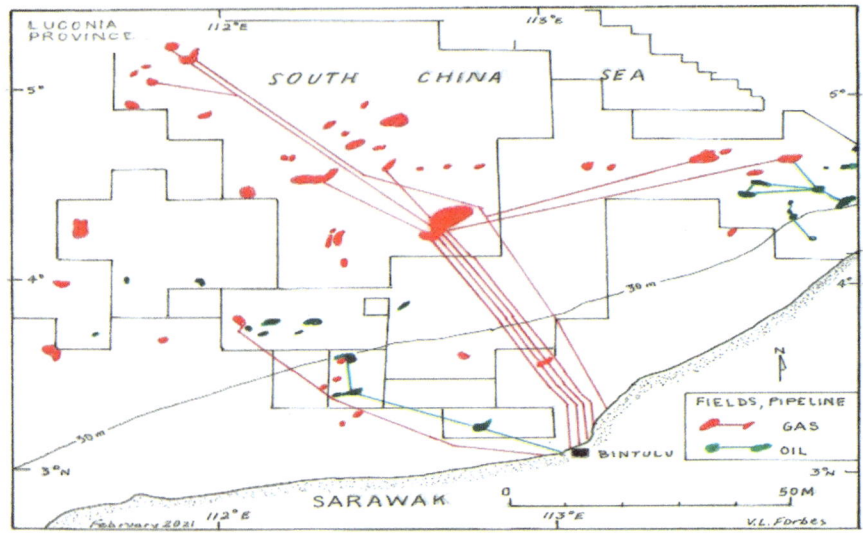

Fig. 6.27 Offshore hydrocarbon fields and pipelines, Sarawak

Fig. 6.28 A cluster of platforms operating in Malaysian waters. *Source* Mr. Mohd Muzakir

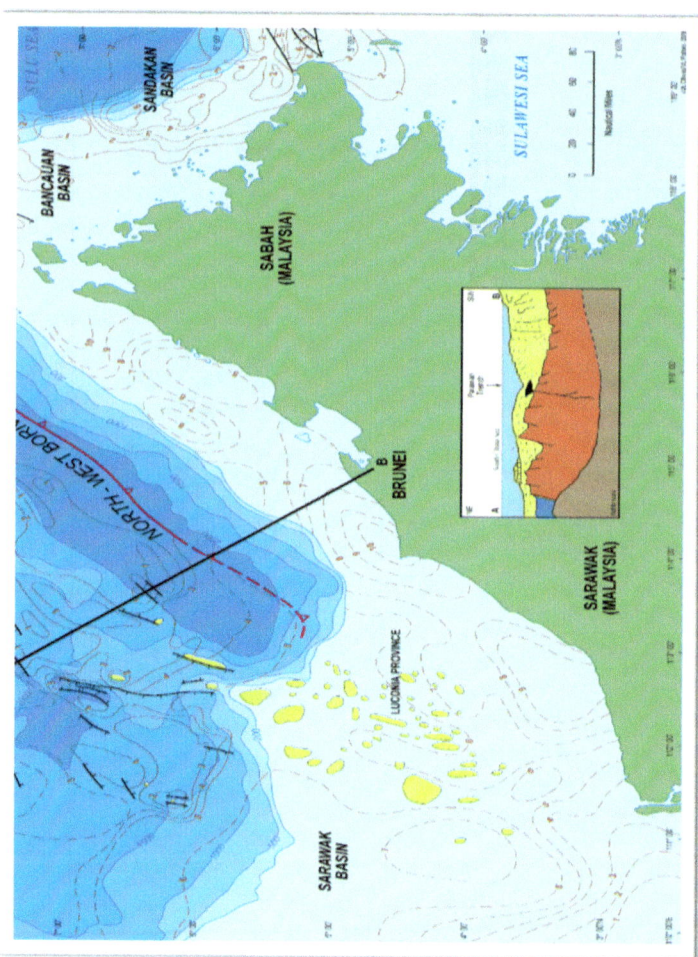

Fig. 6.29 Tectonic structure of South China Sea seabed

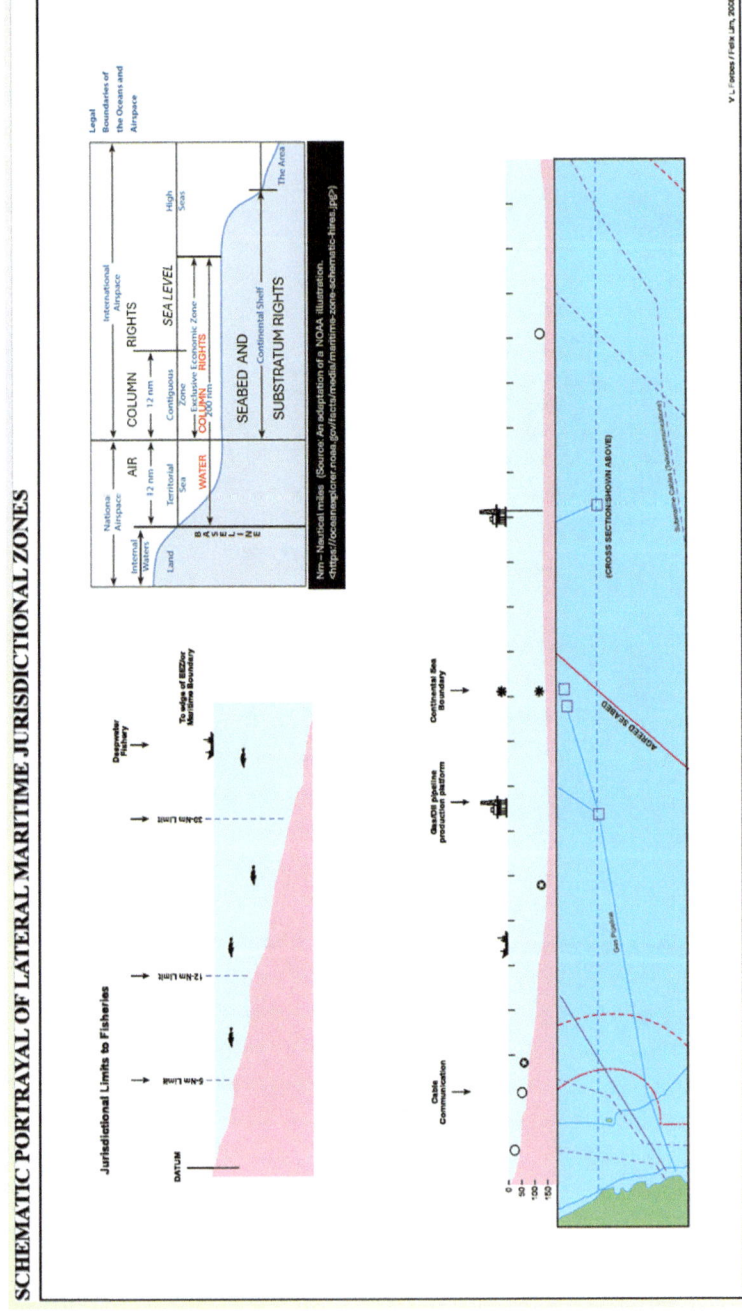

Fig. 6.30 Schematic portrayals of lateral and vertical maritime jurisdictional zones

Table 6.11 Number of fishers by state and zone, 2023

State	Zone A	Zone B	Zone C	Zone C2	Zone C3	Total
Perlis	504	14	480	415	–	1413
Kedah	6345	164	1007	84	–	7609
P. Pinang	4528	215	225	56	288	5312
Perak	6028	2656	5864	3669	–	18.217
Selangor	3787	1405	1428	0	–	6620
N. Sembilan	480	–	–	–	–	480
Melaka	997	–	–	–	–	997
Johor	6429	895	1626	1278	–	10,228
Kelantan	2728	118	210	371	–	3403
Terengganu	5590	688	1444	589	–	8311
Pahang	1950	515	3154	2196	–	7815
Total Pen	*39,375*	*6648*	*15,438*	*8658*	*288*	*70,405*
Sarawak	7307	236	2.092	604	–	10,239
Labuan	400	–	–	–	–	400
Sabah	22,740	7925	587	48	–	31,300
Total E. Mal	*30,447*	*8.161*	*2679*	*652*	*0*	*41,939*
Total	69,822	14,807	18,117	9310	288	112,344

Table 6.12 Number of fishing vessels by state and zone, 2023

State	Zone A	Zone B	Zone C	Zone C2	Zone 3	Total
Perlis	340	7	100	39	–	486
Kedah	3307	158	333	17	–	3815
P. Pinang	2181	49	31	1	17	2279
Perak	3056	979	974	290	–	5299
Selangor	2124	496	237	–	–	2857
Sembilan	359	–	–	–	–	359
Melaka	828	–	–	–	–	828
Johor	3649	296	189	51	–	4185
Kelantan	1.137	22	72	86	–	1317
Terengganu	2329	102	191	56	–	2678
Pahang	846	122	426	137	–	1531
Peninsula	*20,156*	*2231*	*2553*	*677*	*17*	*25,634*
Sarawak	4355	54	362	35	–	4806
Labuan	280	–	–	–	–	280
Sabah	16,491	1876	82	4	–	18,453
E Malaysia	*21,126*	*1930*	*444*	*39*	*–*	*23,539*
Total	41,282	4161	2997	716	17	49,173

In the second illustration of Fig. 6.30 the utilisation of ocean space is shown as one viewed on a plan as against the reality across the sea floor and water column.

6.9 Summary

The narrative in this chapter might appear to be a controversial mix of topics; however, in the context, there is a unifying theme throughout. On the one hand the discussion focuses on Malaysia's marine parks and protected zones and marine ecosystems. It spotlighted on the issue of the threat to the coastal zone and marine life and seagrass beds, mangrove forests and coral reefs caused by marine pollution and other human-induced interactions.

The attention is then directed to the offshore exploration within Malaysia's maritime area and commentaries on the joint exploration ventures between Malaysia and Thailand, Brunei, and Malaysia.

The following chapter focuses on the major and minor commercial ports of Malaysia and some of the maritime security issues experienced in the seas, for example, actual and potential acts of armed robbery and piracy and threats of terrorism, in the vicinity of the Malaysian coast.

Notes

1. UN World Meteorological Organisation's Report of 2023. See also Greenhouse Gases at the Climate Portal https://climate.mit.edu/explainers/greenhouse-gases.
2. International Union for Conservation of Nature and Natural Resources (IUCN).
3. Malaysia's Marine Park and Conservation < https://marinepark.dof.gov.my/en/ >.
4. < ctatlas.coraltriangleinitiative.org/Country/Index/MYS > ; and Marine Park, Department of Fisheries https://marinepark.dof.gov.my/en/; https://www.scribd.com/document/203628845/Pub-120-Pacific-Ocean-and-Southeast-Asia-Planning-Guide-10th-Ed-2013.
5. Refer: Turtle conservation in Malaysia < https://www.turtleconservationsociety.org.my >.
6. Malaysia Department of Environment, Ministry of Natural Resources and Environmental Sustainability official portal < https://www.doe.gov.my/en/2021/10/26/oil-spill-contingencies >.
7. I. C. Ishak, A, M, Arof and M.R. Zoolfakar (2019) The Review of the Oil Spill Incidents in Malaysia between 2014 and 2016. https://www.reserachgate.net/publications/338036690; The Nagasaki Spirit Explained https://everything.explained.today/The_Nagasaki_Spirit....
8. Refer: UNEP 'Enhancing Wastewater, Nutrient Management and Sanitation' available at https://www.unep.org./topics/ocean-seas-and-coasts/ecosystem....
9. National Water Resources Policy, 2012 https://www.doe.gov.my/wp-content/2021/09/.
10. See Note 8, above.
11. Seagrass, Malaysia https://www.seagrasswatch.org/malaysia/
12. UNEP/GEF South China Sea Global Environment Programme Project Facility Reversing Environmental degradation Trends in the South China Sea and Gulf of Thailand National Reports on Coral Reefs in the Coastal Waters of the South China Sea.
13. T.S. Wan 'Monitoring, Control, and Surveillance of Fisheries in Malaysia', pp. 34–44; see also FAO *The State of World Fisheries and Aquaculture Blue Transformation in Action*, 2024. https://www.fao.org/publications/home'fao-flagship/publications/th.

UNEP's World Atlas of Coral Reef UNEP-WCMC Resources.

14. Department of Fisheries Malaysia https://www.dof.gov.my
    Fisheries Industry Scenario—Department of Fisheries Malaysia Official Portal (dof.gov.my).
    World Bank Document—World Bank Group *Malaysia Economic Monitor* Raising the Tide, Lifting All Boats, The World Bank, Washington, October 2023; see also Annual Fisheries Statistics 2023, Volume 1, available as a pdf document 9 at Perangkaan Perikanan I—Portal Rasmi Jabatan Perikanan Malaysia (dof.gov.my).
15. Gavin Jolis and Lau Min Min (2015) Sea Turtle Conservation in Malaysia: Issues, Challenges and Recommendations https://www.researchgate.net/publication/322752509....
    Turtle Conservation < https://marinepark.dof/gov.my/en/lo....
16. Petronas Media Release of 30 November 2023 'Malaysia Records Over I Billion Barrels of Oil Equivalent of Exploration Discoveries in 2023', available: Petronas' webpage 14/07/24.
17. Auburn, F.M., Ong, David and Forbes. V.L. 'Dispute Resolution and the Timor Gap treaty', *Occasional Paper* No. 35, 1994, IOCPS, UWA, Perth, 74 pp.
18. See Auburn, Ong and Forbes, 'Dispute Resolution and the Timor Gap treaty', *Occasional Paper* No. 35, 1994, IOCPS, UWA, Perth, 74 pp. Ong offers a comparative analysis of the JDA and Timor Gap Treaty, pp.28–40.
19. Refer: Malaysia Thailand Joint Authority https://www.mtja.org.
20. https://www.offshore-mag.com/regional-reports/asia/article/14198881/malaysia-adapts-terms-for-latest-offshore-bid-round.
    Refer to Annual Report of the M/T JDA < dmf.go.th/bid19/annual/08.html > .
21. Forbes and Basiron (2009); Fisheries Country Profile: Malaysia (2018) < https://www.sea fdec.org/fisheries-country-profile-malaysia-2018 >.

Chapter 7
Major Commercial Ports of Malaysia, Sea Lanes and Security

It is not the going out of port, but the coming in, that determines the success of a voyage.
Henry Ward Beecher

Abstract Malaysia's commercial ports play an important and vital role in spurring the national growth and contributing to the nation's Blue Economy. The ports of Malaysia are strategically located to take full benefits from the concepts of the Belt and Road Initiative adopted by the People's Republic of China in 2013. The narrative in this chapter is supplemented with maps that depict the limits of Malaysia's major commercial ports; the prescribed sea lanes to select ports of Sabah; areas of acts of armed-robbery and piracy over a period of many years. A brief discussion is offered on Indonesia's Archipelagic Straight Territorial Sea Baselines and Archipelagic Sea Lanes as these concepts partially impact of Malaysia's maritime jurisdictional limits.

Keywords Port limit · Prescribed sea lane · Archipelagic straight baseline · Archipelagic sea lane · Acts of armed-robbery and piracy

7.1 Introduction

In this Chapter the narrative focuses on the major commercial ports of Malaysia and security of the sealines of communications within Malaysia's maritime jurisdiction vital components for an enhanced blue economy.

Since time immemorial, the waters around Malaysia supported not just the movement of the people but also functioned as facilitator for trade to the surrounding communities, and regions, as well as internationally. This resulted in a fascinating mix of peoples, cultures, and commercial activities that match any major cosmopolitan centres found around the world past and present.

In terms of trade movements, the waters and the ports of the country have contributed significantly to the growth of the Malaysian Blue Economy as conduits for the movement of exports to earn much-needed foreign exchange and for imports used in consumption and inputs for the industries of this country. Trade has grown significantly since the country's independence in 1957 and especially since the 1990s when the foundations of modern industrial Malaysian nation had its beginnings.

Malaysia's international trade experienced tremendous growth over the last four decades rising from RM9.45 billion in 1970 to RM158.76 billion in 1990 to RM13,739 billion by 2019. Except for a handful of years, the trade balance has always been in Malaysia's favour and stood at RM91.58 billion in 2015. Malaysia is firmly positioned as a major world transhipment hub, with a regular port of call for diverse types of vessels. Very Large Crude Carrier (VLCC), tanker vessels, LNG (Liquified Natural Gas) carriers, cargo vessels, container-loaded vessels, bulk carriers, passenger vessels, and fishing vessels are among the ships, boats, and submersible craft that navigate the Malacca Strait according to the analysis of the vessel characteristics conveyed to the ship reporting system.[1]

However, between March 2020 and mid-2022, the maritime sector of Malaysia, as with other economic sectors and countries worldwide were affected by the COVID-19 Pandemic and its aftereffects.[2]

Logistics is the mainstay of the supply chain, and is key to stimulating trade, facilitating business efficiency and economic growth to strengthen the nation's competitiveness. Malaysia's strategic location, good regional linkages, steady annual economic growth, and relatively stable government organisation and electronic digital (paperless) transaction systems in place and strong transport infrastructure, it is well placed to face the challenges and the competition offered by regional rivals. A slight glitch, in the form of COVID-19 Pandemic during 2020/21, had disrupted trade and affected the economy of the country; however, Malaysia, was not alone and more importantly was resilient and has returned positively.

Malaysia's Third Industrial Master Plan (2006–2020) had projected that marine cargo managed by its ports will have increased from 252.6 million tonnes in 2005 to over 751 million tonnes in 2020. However, marine cargo that is carried onboard liner ships must be able to move efficiently across the hinterland to and from ports since most of the businesses and consumers that are buying and selling goods are located outside the port area. The continued efficiencies for global trade gained by use of liner shipping are dependent upon an inland transportation network that permits for the timely and efficient overland transfer and transport of cargo.[3]

The ports of Malaysia are ideally located to capture a fair share of the maritime trade that transit the seas of Southeast Asia. This fact is assisted by the indicator that Malaysia readily adopted the initiative taken by the People's Republic of China (PRC) concept of *One Belt, One Road* (OBOR) or the *Belt and Road Initiative* (BRI). The BRI or OBOR is a global infrastructure development strategy announced by the President of the PRC in October 2013 whilst on a visit to Indonesia. Its aim is to promote economic development and inter-regional connectivity between countries who sign up to the initiative. Finance for infrastructure development can be sourced through a dedicated banking system that is backed by the PRC and other financial institutions. Malaysia was an early supporter of the initiative especially due to its ideal location and port facilities at the maritime crossroads.[4]

However, competition from Singapore's Marine and Port Authority, Indonesia's Tanjung Priok and Thailand's Laem Chabang challenge for share of the trade that

Malaysia has the potential to capture. The container-handling ports of Port Klang complex and Port of Tanjung Pelepas (PTP) are within the top 20 container-handling facilities in the international rankings.

Plans are in place to expand the Port of Tanjung Pelepas' (PTP) total volume container capacity of 10.5 million twenty-foot equivalent units (TEUs) to 15 million TEUs. The cost of expansion is estimated at around RM5 billion (Fig. 7.1). In 2015, the Government of Malaysia announced an allocation of RM300 million to improve connectivity along the final stretches of road network to access the three ports in the vicinity to boost exports.[5]

The freight railway network in Malaysia remained generally under-utilised (in 2022), with the share of railway freight volume among the total domestic freight volume only amounting to about 0.5%. Transportation capacity will not be easily augmented as most rail sections being single-tracked and non-electrified. Safe and properly constructed dual-track rail lines could relieve the pressure now placed on the overstretched road networks. Rail-based freight transportation on Peninsula Malaysia covers more than 2600 km. This figure will increase when the east–west link, Kuantan to Port Klang, which is currently under construction, and an upgrade in the north/south transportation system is completed and in operation.

Malaysia's logistics sector contributed around 3.6% to the country's GDP as of 2016 and anticipated to increase growth to 4.3% by 2020 in that sector, however, collaboration with China is considered essential. However, planning for growth also requires consideration to determine ways to curb bottlenecks at customs and quarantine checkpoints, and ensuring cargo is cleared efficiently and expeditiously. The six Malaysian ports that were earmarked for further developments are Penang, Port Klang, Kuantan, Malacca, Tanjung Pelepas on Peninsular Malaysia, and Bintulu in Sarawak.

The Maritime Silk Road (MSR) revised concept of the Government of China in October 2013 which is part of the 'One Belt, One Road' (OBOR) initiative was to increase investment and foster collaboration across the historic Silk Road, places Malaysia's ports in an ideal position to capitalise on the huge trade volumes, investments, and socio-economic and infrastructure developments that OBOR is poised to bring to the global and regional economies. Indeed, many projects are earmarked to take advantage of the tremendous opportunities that are already on the drawing board or at least, the groundwork prepared.

The China-funded East Coast Rail Link (ECRL) on Peninsular Malaysia, launched in 2017, was expected to be completed by 2023 at an estimated cost of RM55 billion. The envisaged dual-track railway, commencing at Kuantan, would cater for passenger and freight and intended to connect ports on the east coast of Peninsular Malaya with those of the west coast, particularly, Port Klang, and was expected to alter the present trade routes which ply between the Malacca Straits and South China Sea via the Straits of Singapore. By participating in the PRC-initiative OBOR (One Belt, One Road), Malaysia's economy, trade and logistics services will be further enhanced. During 2016, container throughput of Port Klang recorded an increase of 10.8% to 13.17 million TEUs. The port is Malaysia's largest and the 12th busiest in the world ranking.

Fig. 7.1 Satellite-derived imagery of the West Johor Strait and of PTP (top, left of centre) and Tuas (Singapore). This is an area of rapid development and busy with shipping movements at the two ports. *Source* An extract of graphic courtesy of Google Earth Map

Further south, along the west coast of the peninsular, north of Malacca, a RM12.6 billion (US$2.8 billion) Kuala Linggi International Port (KLIP) is under construction on reclaimed land of an area of 250 ha, with funds from the OBOR concept. The new port will offer a planned 1.5 million cubic metres oil storage facility; repair and dry-docking enterprises; and re-fuelling services for large tankers.

7.2 Ports of Malaysia

Figure 7.2 depicts the approximate location of four major commercial ports and numerous minor ports sited along the coastline of Malaysia. A list of cargo and passenger jetties of Malaysia is tabulated in Table 7.1. The limits of the 41 ports administered by the Malaysia Marine Department are described on the website of the Department.[6] Further details of select major ports are provided in maps and narratives, below.

Section 7.3 discusses the limits of the Malaysian ports within the Malacca Strait; Sect. 7.4 focuses on the limits of the of East and West sectors of the Johor Strait and ports along the northern littoral of Singapore Strait; and Sect. 7.5 examines the ports of Malaysia located along the southern littoral of the South China Sea and western littoral of the Sulu Sea.

7.3 Malaysian Ports Within Malacca Strait

Ports of Telok Ewa and Langkawi (Kuah Port)

The prescribed limits of the Port of Langkawi are defined by geographical coordinates. The details are listed in the Marine Department's website (Table 7.2).

Langkawi's main town and port, Kuah, attracts ferries from the mainland and Pulau Pinang. Duty free shops have been established and many new hotels and chalets have been constructed and operational on Langkawi Island (Fig. 7.3).

The twin ports at Penang and Butterworth lie further south from Langkawi Port complex. The area encompassed by the Port of Penang complex is delineated on maps and a satellite-derived image as depicted in Figs. 7.4 and 7.5.

The Three Ports of Klang: Port Klang

All facilities and services developed by Port Klang Authority have been privatised although Port Klang Authority retains regulatory authority and oversees the operations of the companies.

The defined limits of Port Klang in accordance with the powers conferred by subsection 6(3) of the *Merchant Shipping Ordinance* 1952 [Ord. 70/1952], the Minister altered the port limits of Klang (Fig. 7.6) as defined in the Legal Notification No. 73 of 1953 by substituting for item 79.[7]

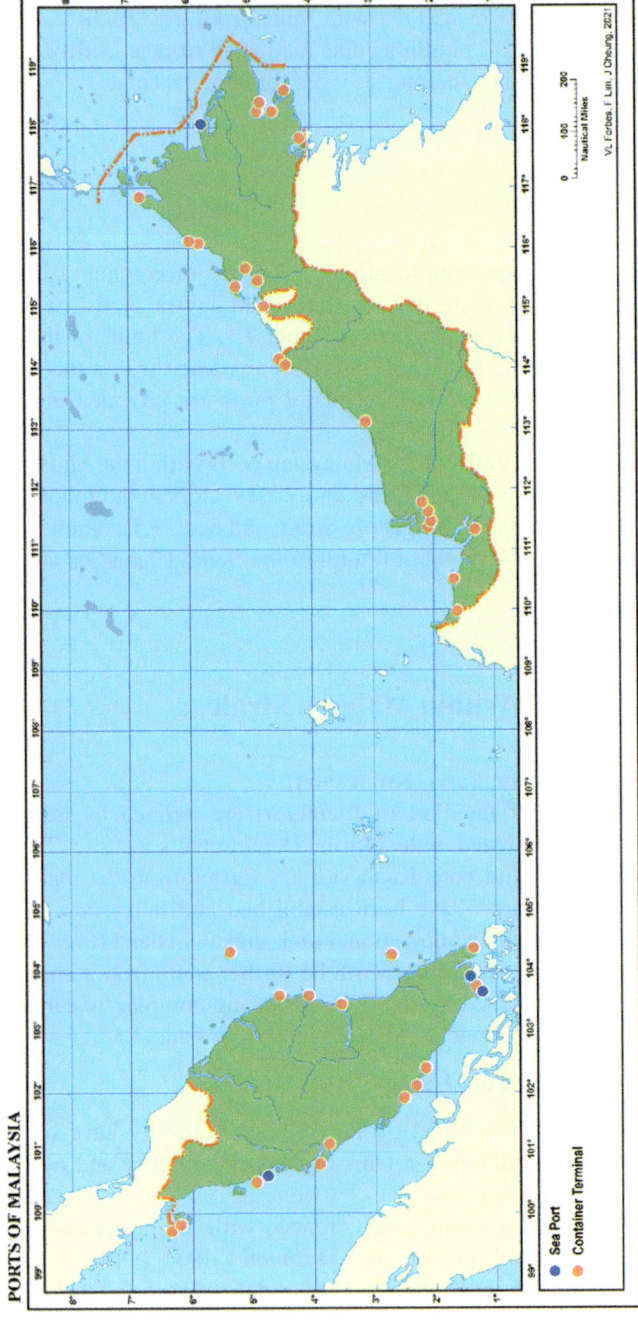

Fig. 7.2 Major ports of Malaysia

Table 7.1 Cargo and passenger jetties of Malaysia

Region	Cargo Jetty	Passenger Jetty
Northern	Jeti Lama Jabatan Laut, Kuala Perlis	Terminal Jeti Penumpang Kuala Perlis, Perlis
	Jeti Jabatan Laut Tg. Lem bong, Langkawi	Terminal Jeti Penumpang Kuah, Langkawi
	Terminal Jeti Kargo Kuala Kedah, Kedah	Terminal Jeti Penumpang Kuala Kedah, Kedah
	Jeti Kargo Ramp Kg. Acheh, Perak	Terminal Jeti Penumpang Pangkor, Perak
		Jeti Penumpang Sg. Pinang Kecil, Perak
		Jeti Penumpang Sg. Pinang Besar. Perak
		Jeti Penumpang Damar Laut Perak
		Terminal Jeti Penumpang Lumut Perak
		Terminal Jeti Penumpang Bagan Datoh, Perak
Central	Jeti Jabatan Laut Pulau Ketam, Selangor	Terminal Jeti Penumpang Port Dickson, Negeri Sembilan
	Jeti Penumpang Sg. Lima, Pulau Ketam, Selangor	
	Jeti Penumpang Pulau Ketam, Selangor	
	Jeti Institut Pusat Latihan, Pulau Indah, Selangor	
Southern	Jeti Jabatan Laut Johor Bahru	Terminal Jeti Penumpang Muar, Johor
	Jeti Jabatan Laut Tg. Pelepas, Johor	Jeti Penumpang Mersing, Johor
		Jeti Penumpang Tanjung Pengelih, Johor
Eastern	Jeti Jabatan Laut Kuantan, Pahang	Terminal Jeti Penumpang Kg. Tekek, Pulau Tioman, Pahang
	Terminal Jeti Kargo Rompin, Pahang	Jeti Penumpang Mukut, Pulau Tioman, Pahang
	Jeti Kargo Kuala Terengganu, Terengganu	Jeti Penumpang Kg. Juara, Pulau Tioman, Pahang
	Jeti Kargo Kuala Besar, Kelantan	Terminal Jeti Penumpang Tg. Gemok, Pahang
	Jeti Jabatan Laut Tok Bali, Kelantan	Terminal Jeti Penumpang Kuala Terengganu, Terengganu
		Jeti Penumpang Kuala Besut
Labuan	Jeti Feri Kenderaan. Labuan	Dermaga Merdeka
Sarawak	Jeti Jabatan Laut Muara Tebas, Kuching, Sarawak	
	Jeti Jabatan LautTanah Puteh, Sarawak	
	Jeti Jabatan Laut Sarikei, Sarawak	
	Jeti Jabatan Laut Sibu, Sarawak	

Table 7.2 Port of Telok Ewa

Type of berth	Length (m)	Depth (m)	Max. vessel size
General cargo	480	5	5000 dwt
Dry bulk cargo	165	5	5000 dwt

The limits of South Port and the remaining facilities in North Port (after the privatisation of container terminal to Klang Container Terminal (KCT) in 1986) and the dockyard were privatised to Kelang Port Management Sdn. Bhd. in 1992. The company, as the second private port operator in Port Klang, manages over 21 berths including a five-berth container terminal. With the recent equity restructuring exercise carried-out by the shareholder, KPM became wholly owned subsidiary of Northport (Malaysia) Bhd., (formerly known as KCT) (Figs. 7.7, 7.8 and 7.9). Tables 7.3, 7.4 and 7.5 inclusive offer some simplified statistics of the facilities at Port Klang.

During 2023, Port Klang was rated the 11th busiest port in the world. Domestic and international traffic flow for container handling in the export, import and transshipment categories at Northport and Westport in the Port Klang complex recorded a total number of TEUs of 14,061,022.

On 8th December 2023, the Government of Malaysia, Port Klang Authority and Westport Malaysia Sdn. Bhd. signed the Third Supplementary Agreement of Privatisation. Westport has a 58-year concession. It is expected to begin the second expansion phase of the planned development sometime during the third quarter of 2024. The new concession will include the existing facilities in Westport and the new facilities to be developed during the concession period which will involve an investment of RM 39.6 billion (or about US$83 million). The expansion of the Westport Terminal—CT10 to CT17—is expected to increase the capacity from 14 million TEUs to 27 million TEUs based on projections until 2028. The planned development will be undertaken on sustainability principles, in accordance with green and smart concepts, including the use of automation and digitalisation that will not only reduce greenhouse gas emissions but also future proof the port.

Port of Kuala Linggi International Port (KLIP)

Sungai Linggi International Floating Trans-shipment and Trading Hub, also known as the Lift-Hub, handles liquid bulk transfer operations. Vessels moor alongside each other at anchor to conduct ship-to-ship (STS) operations (Fig. 7.10). In addition to the Lift-Hub a barter trade port operates at the mouth of the Linggi River, small coastal vessels and barges can be accommodated. The Designated Transfer Area (DTA) is located approx. 3.5 M off the west coast of the Malay Peninsula, 16 M NW of Malacca and 18 M SE of Port Dickson.

The designated port limits are bounded by lines connecting the following geographical coordinates: 1. 02° 23.3′ N, 101° 58.2′ E; 2. 02° 17.4′ E; 3. 02° 16.5′ N, 102° 04.3′ E; 4. 02° 14.7′ E; 5. 02° 19.4′ N, 101° 55.2′ E.

This private port facility's, currently under construction, primary aim is to foster economic growth, enhance regional connectivity and be a global green industrial hub for energy source, port, and maritime services. The facility will provide tank

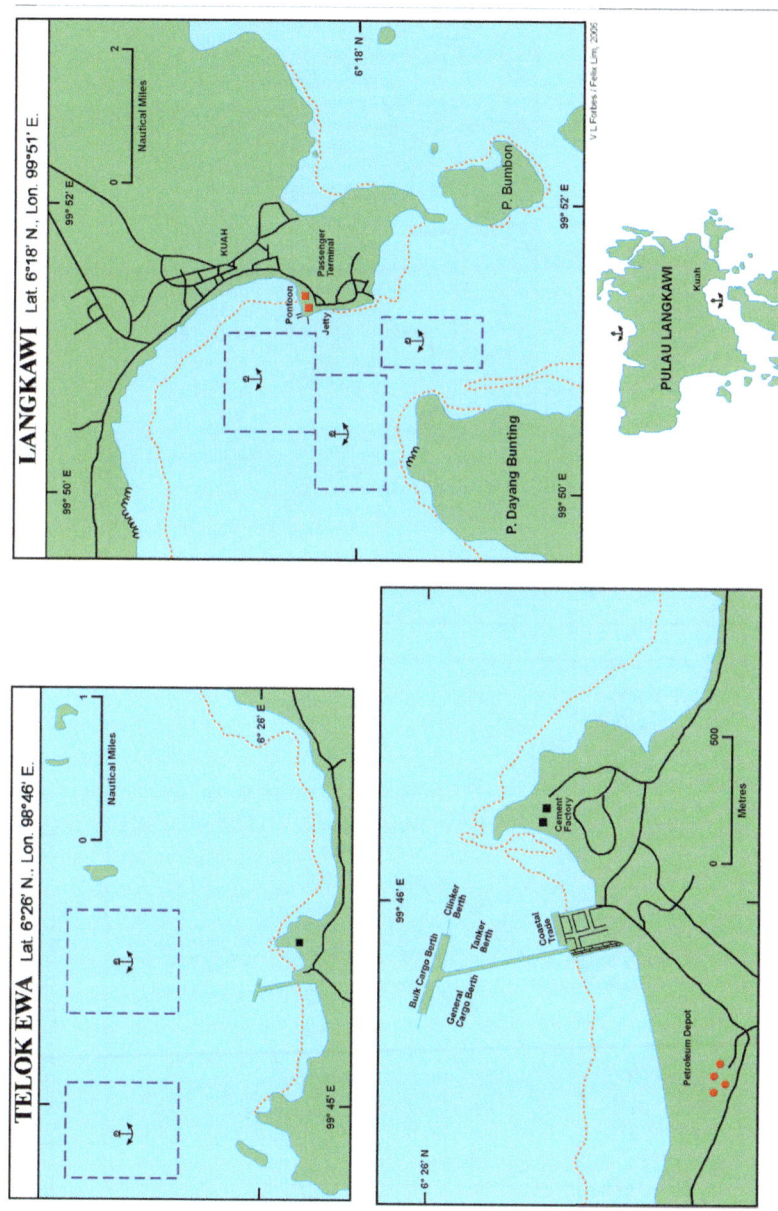

Fig. 7.3 Port Langkawi and Telok Ewa

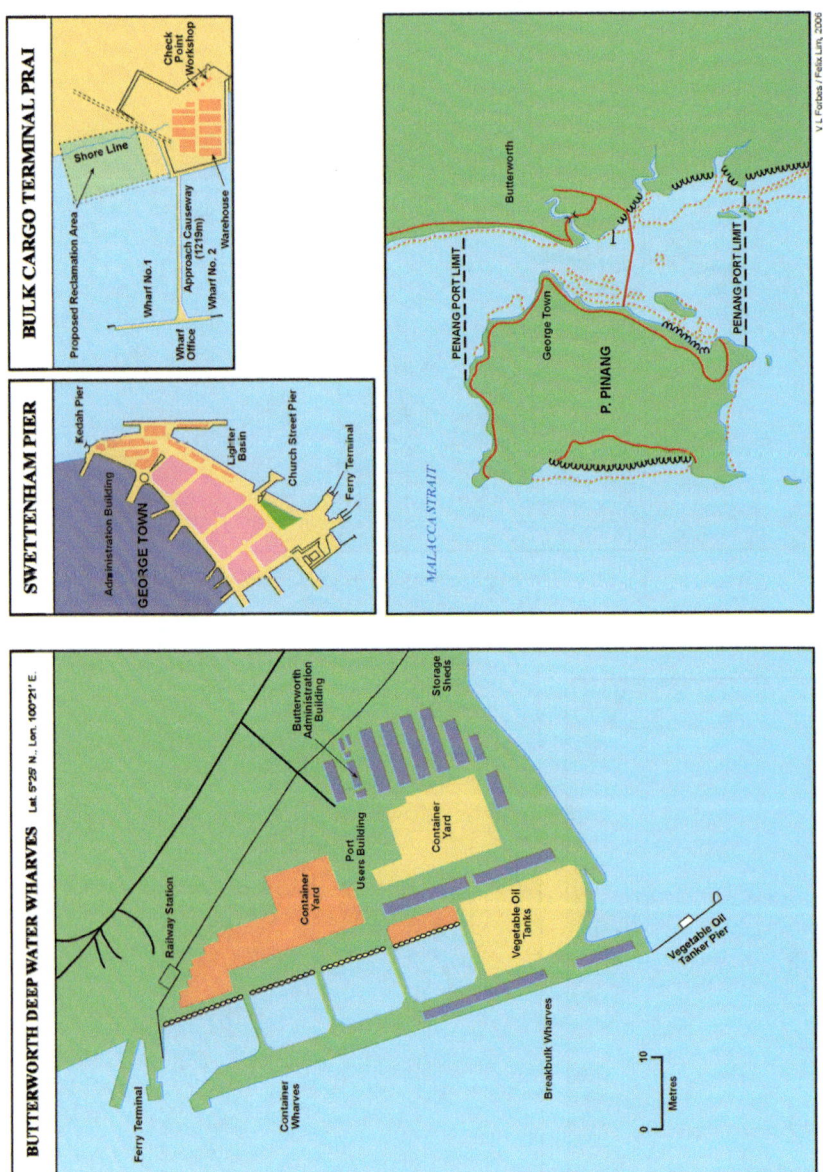

Fig. 7.4 Butterworth deep water port and Penang port

Fig. 7.5 Satellite-derived imagery of George Town and Butterworth, Malaysia. Note the reclamation work off T. Tokong. *Source* An extract of Google Earth Map

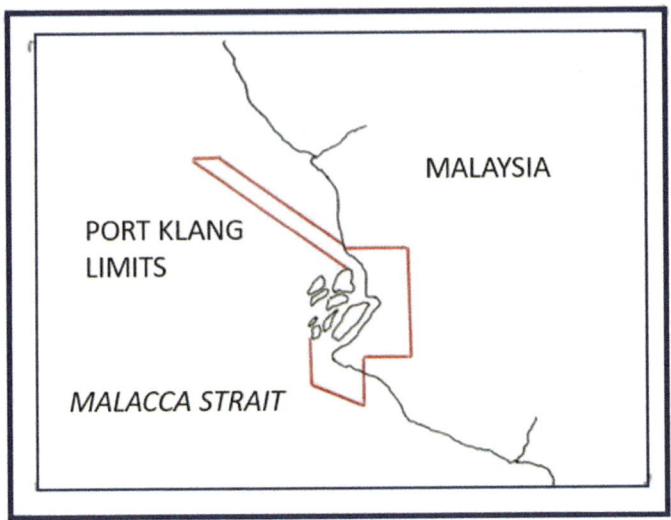

Fig. 7.6 Limits of Port Klang, Malaysia

storage for liquid bulk cargoes, liquified petroleum gas (LPG) and liquified natural gas (LNG) as well as dedicated areas for ship maintenance, repair, and overhaul.

Tanjung Bruas and Anchorage, Malacca Port

Malacca Port (Figs. 7.11 and 7.12) was administered by Klang Port Authority until 2 November 1992. Following a request from the Malacca State Government, the Port facilities and services were privatised to Syarikat Perkhidmatan Pelabuhan Gabungan Sdn Bhd. For ocean-going and coastal vessels an anchorage of 3–6 m depth is available between 1 and 1.5 M off Malacca Harbour. There is no specific prohibited anchorage. Discharge of cargo into lighters (barges) is common at the anchorage (Tables 7.6 and 7.7).

7.4　Malaysian Ports Within the Straits of Johor and Singapore

Port of Tanjong Pelapas (PTP), Johor, is sited on approximately 783 ha of prime land and currently has the capacity to handle up to 6 million TEUs per annum (Fig. 7.13). Phase 2 of the port involved dredging and reclamation of land for an additional eight berths, while widening and deepening the shipping access channel to receive the Super Post Panamax container vessels of the future. The first two berths of Phase Two are now in operations, remaining berths will be constructed in line with demand.

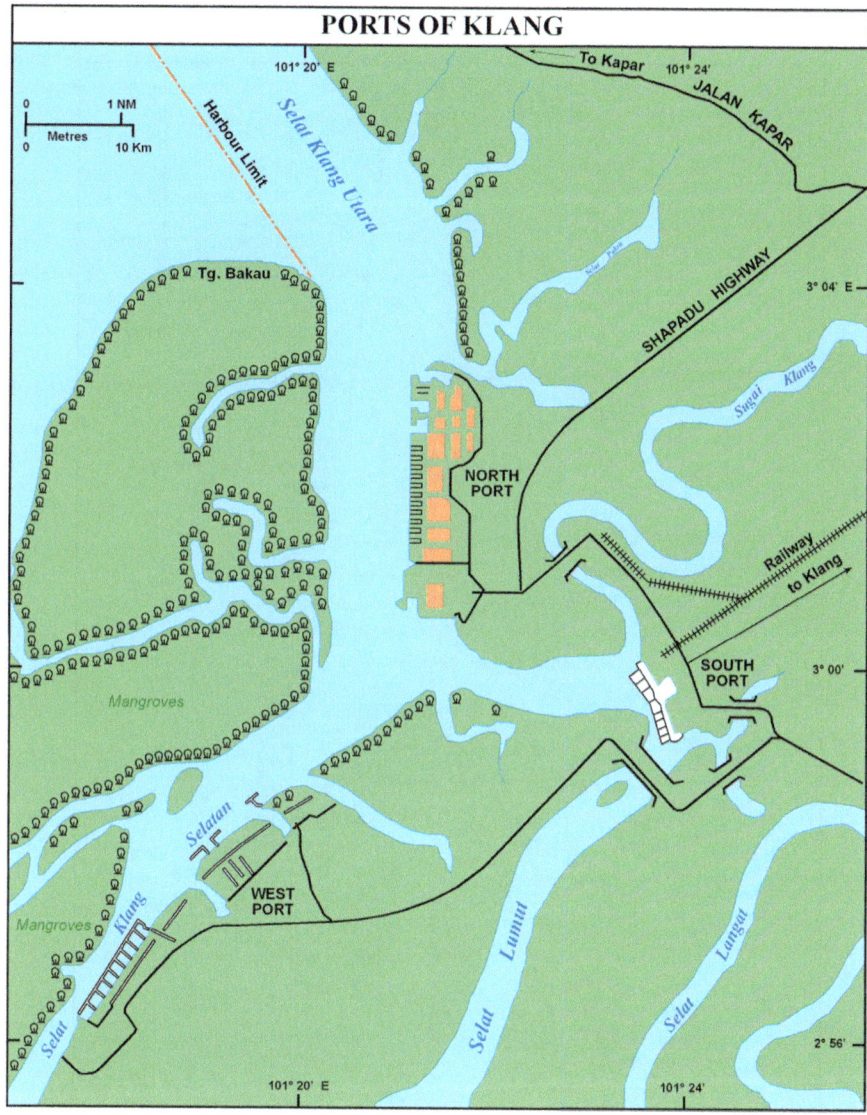

Fig. 7.7 The Three Ports of Klang

It currently offers eight berths of linear quay, directly behind which, is the port's container yard (one of the largest in the region). The terminal is equipped with Super Post Panamax quayside cranes that can cater to the next generation of Super Post Panamax vessels.

PTP's key advantage is that it is a mere 45 min from the confluence of the world's busiest shipping lanes. Easily accessible from the Malacca Strait, PTP is situated

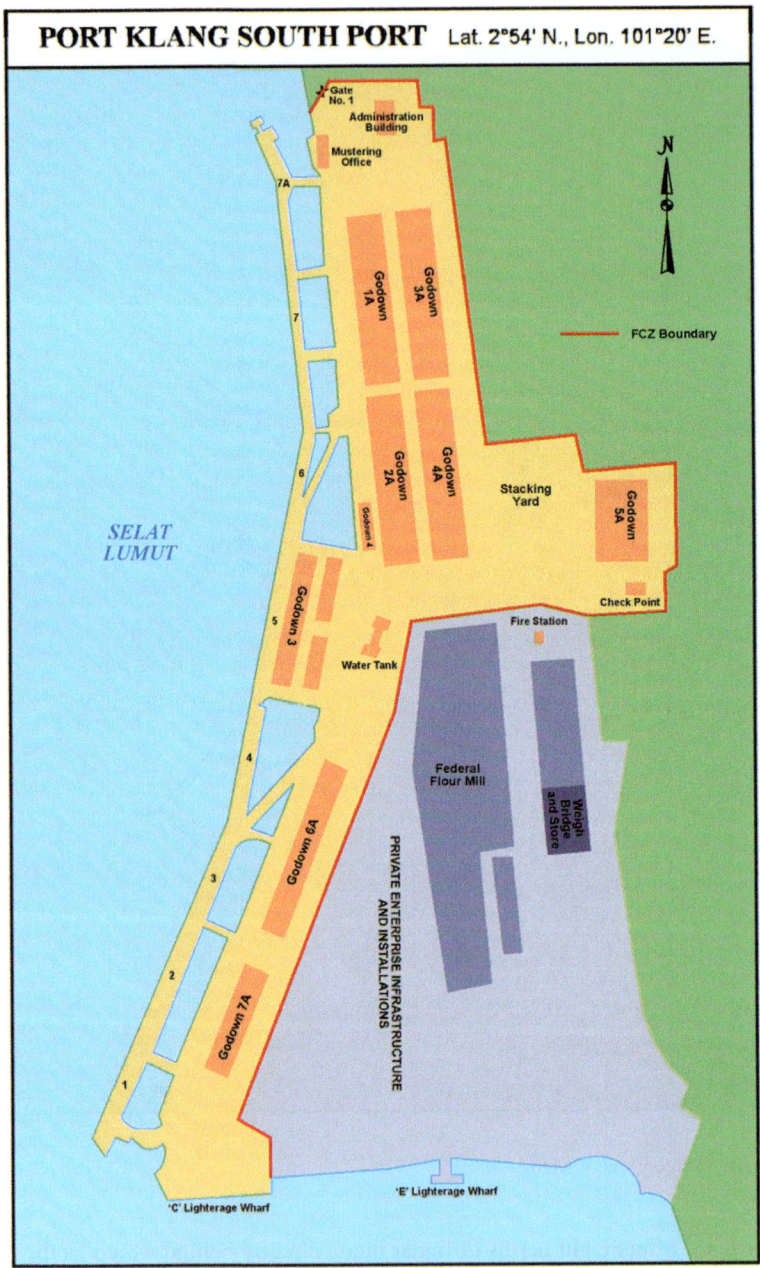

V L Forbes / Felix Lim, 2006

Fig. 7.8   South Port, Port Klang

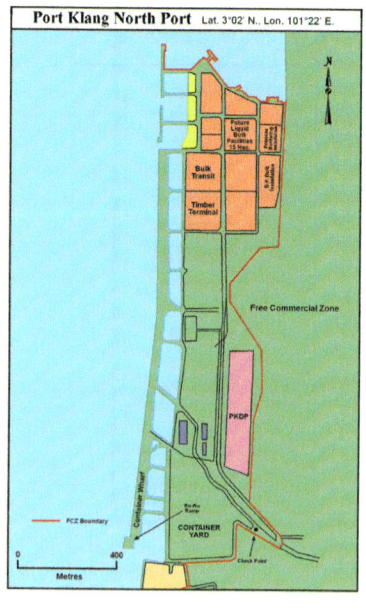
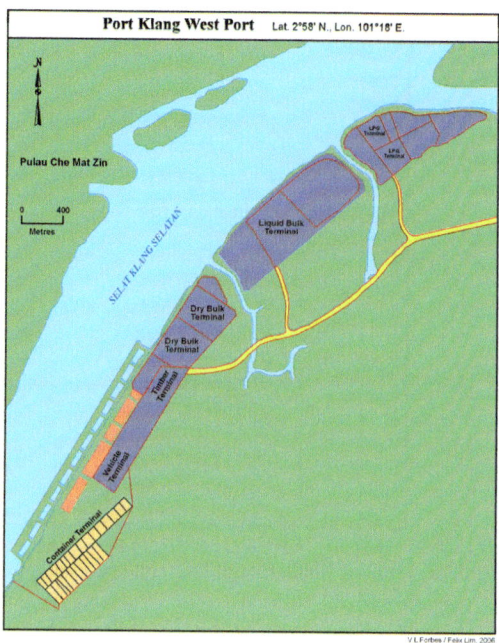

Fig. 7.9   North Port and West Port, Port Klang

Table 7.3   Break bulk terminal, Port Klang

Number of berths	4
Berth length (m)	800
Handles steel and scrap metal	

Table 7.4   Dry bulk terminal, Port Klang

Number of berths	3
Cement jetty	1
Berth length	800
Grab unloaders	2
Continuous unloader	1
Conveyor belt	10,000 m length

Handles sugar, grains, slag, fertiliser, and cement

Table 7.5   Liquid bulk terminal, Port Klang

Number of berths	3
Bunkering jetty	1
Berth length (m)	818
Handles 36 types of cargo	

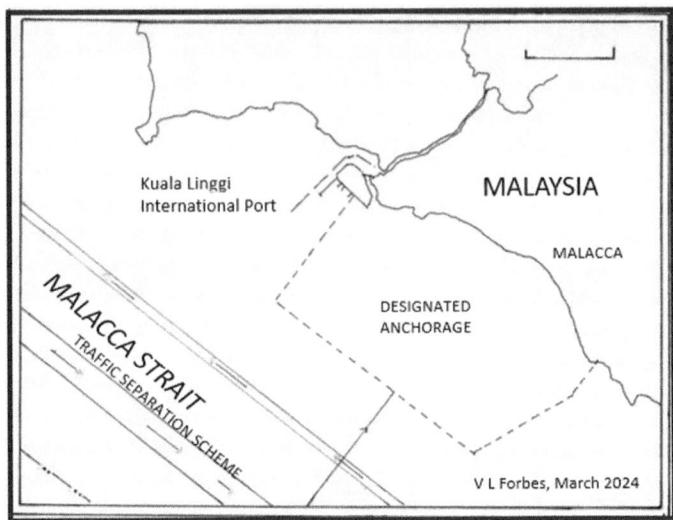

Fig. 7.10 Limit of Kuala Linggi International Port

on the eastern side of the mouth of the Pulai River in South-West Johor. PTP occupies 783 ha. Development of the port has been in stages. On the map: the colour coding employed in the utilisation of land is purple for 'Heavy Industrial Area'; orange for light to medium industries; pink for distribution park 'phase B'; yellow for distribution park 'phase A'; and light green for the cocoa transhipment centre.

The substantial-sized Forest City Malaysia, a multi-use $150-billion project, located in West Johor Strait, just to the east of PTP, was being developed by one of China's largest real estate developers, Country Garden Holdings, Ltd.[8] Possible environmental, social, and economic impacts of the project have generated substantial controversy, within Malaysia, and internationally, specifically with neighbouring Singapore. The Forest City Study will provide the basis for a role-play simulation that can be used to help real estate students learn how to better account for sustainable development concerns and interact with angry stakeholders at the local, regional, national, and international levels. The management of the project are also interested in learning about the ways that Chinese development companies see their corporate social responsibility (especially when they operate outside of China). The findings and analysis of the team has produced a new massive open online course (MOOC), an innovative multimedia and interactive case study, and a series of journal articles pertaining the evaluation and valuing of ecosystem services.

Covering a total of 1386 ha, the four artificial-island Forest City project was reclaimed in four phases: Island 1 (396 ha), Island 2 (767 ha), Island 3 (164 ha) and Island 4 (59 ha acres). Hall Contracting was engaged by the contracted builder JBB to undertake the dredging component of Phase 1. This involved the dredging of approximately 3 million m³ of sand which was brought to the dredge site by sand-carrier barges. A combination of Hall Contracting staff and locally sourced labour

Fig. 7.11 Reclamation work in progress off Malacca. *Source* Google Earth Map

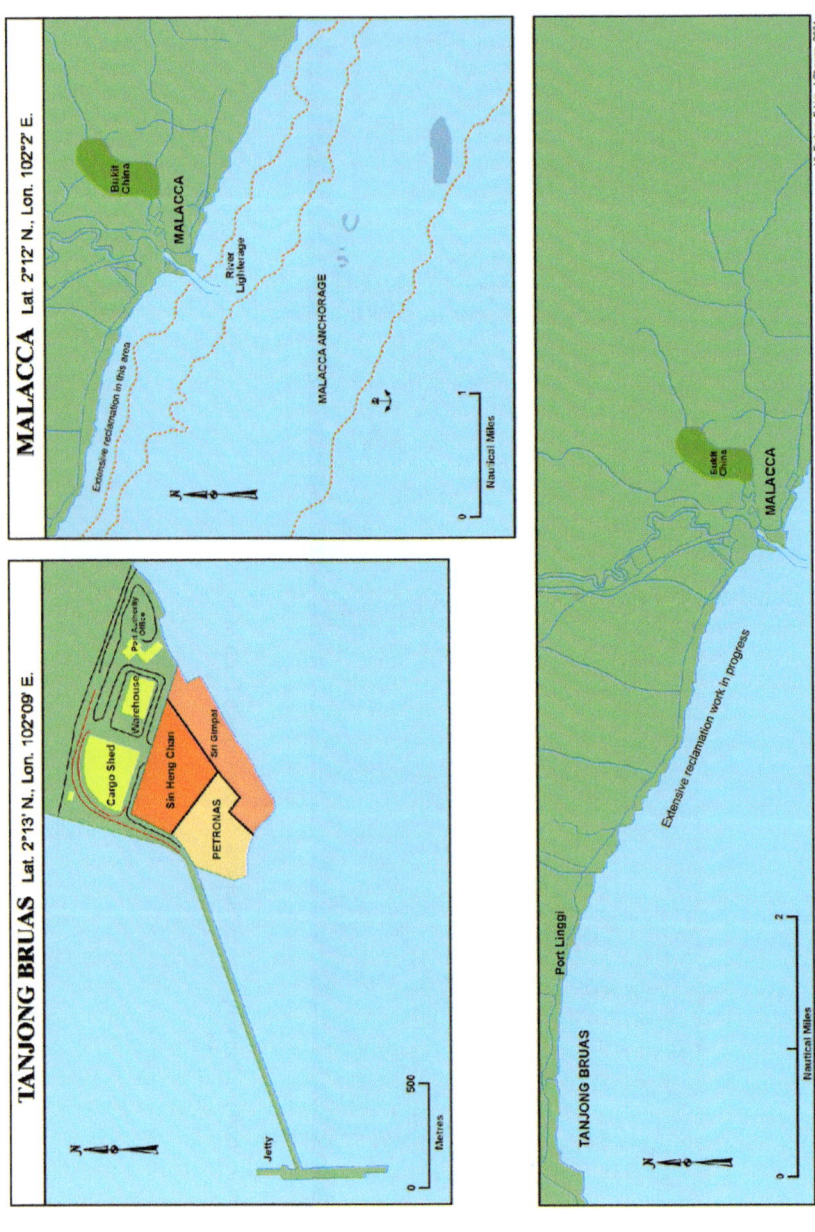

Fig. 7.12 The Ports of Tanjong Bruas and Anchorage, Malacca Port

Table 7.6 Berthing facility

Location	Berth	Length (m)	Depth (m)	Vessel size (dwt)
Tanjung Bruas	T-shaped	170	9	10,000

Table 7.7 Storage and warehousing

Type of storage	Capacity
Warehousing	3000 sqm
Liquid tank farm	6500 m/tonnes
Private godown (storage shed)	24,722 sqm

and sub-contractors was used to undertake the task according to the developer's statement. The areas of reclamation within West Johor Strait are evident in Fig. 7.14. This body of water had, in 2023, at least four major reclamation developments at various stages of construction and development.

Port of Johor Bahru

The revised port limits of Johor Bahru in accordance with Legal Notification P.U. (B) 393 of 1999, was published on 21 October 1999 (Fig. 7.15).

Johor Port Sdn. Bhd., a wholly government-owned government enterprise, took over, in January 1993, all port facilities and services from Johor Port Authority which was established in 1973. The port was fully privatised in August 1995 by Seaport Terminal (Johore) Sdn. Bhd. which became the holding company of Johor Port Bhd. (Refer to Figs. 7.16 and 7.17).

7.5 Malaysian Ports Along the Southern Littoral of S.C.S.

The ports along the east coast of Peninsula Malaysia include Desaru, Endau (Mersing), Pulau Tioman, Tok Bali, Kerteh, Dungun, Terengganu, Kuala Besut, Setiu, and Tumpat.

Kuantan Port (Fig. 7.18—map on the left) is one of the busiest ports in part due to the Belt and Road Initiative. A railway link to the west coast ports is under construction.

The Kuching-Pending Centre (KPC) services the petrochemical industries, spawned by the gas and petroleum and petrochemical hub of Malaysia. KPC has excellent berthing facilities totalling 2981 m in length to cater to the various cargo compositions managed by the port.

Kemaman Port and Kota Kinabalu Port

In exercise of the powers conferred by Section 6 (3) of the *Merchant Shipping Ordinance* 1952, the Minister amended item 85 in the First Schedule of Legal Notice No. 73 of 1953 by removing the item in the second column on the Kuala Kemaman port

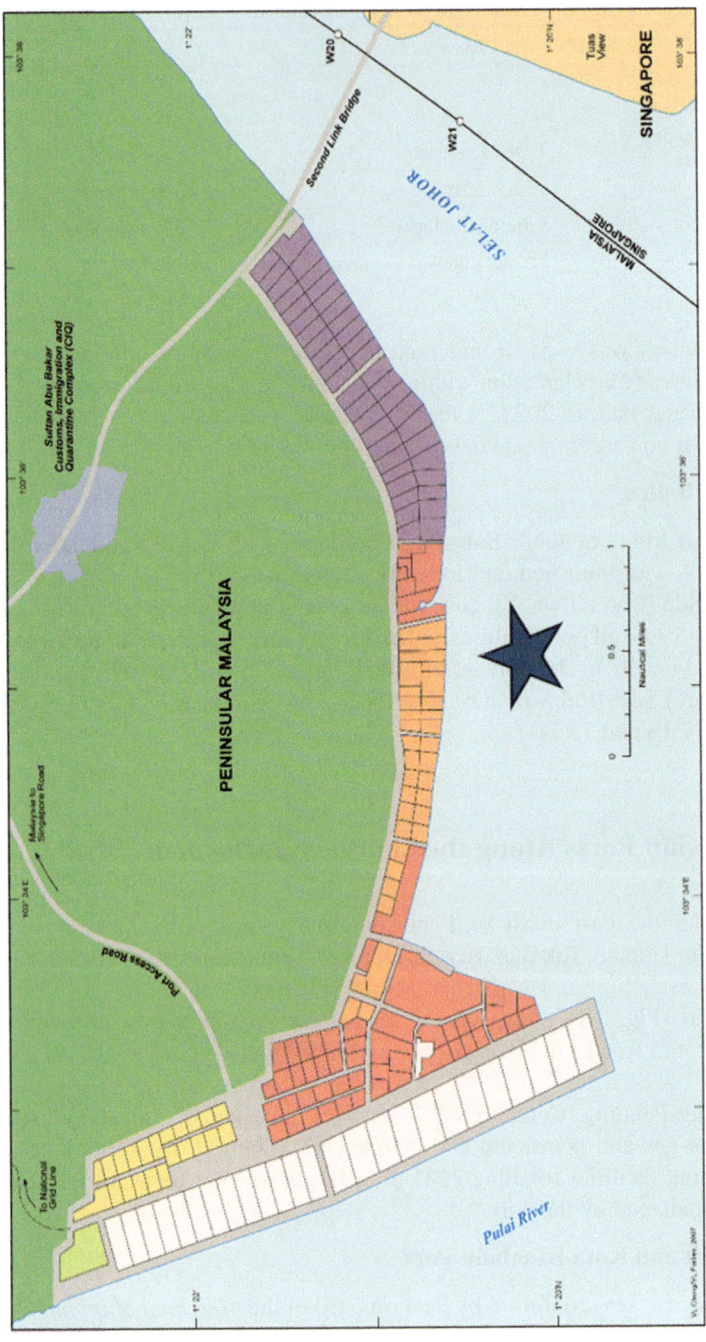

Fig. 7.13 Port of Tanjong Pelapas (PTP), Johor. *Note* Development of *Forest City*'s location (blue star) depicted in Fig. 7.22 in an area just above the scale bar and the existing coastline—centroid Lat.1° 20′ N, Lon.100° 34.5′ E

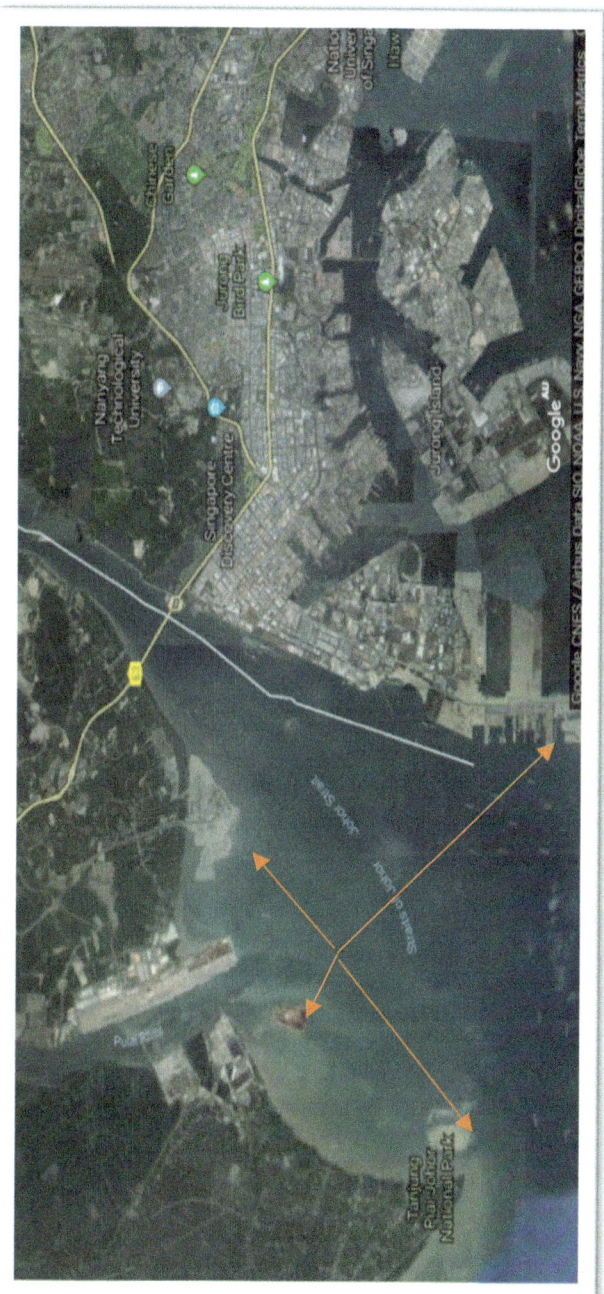

Fig. 7.14 Satellite-derived image of reclamation works (orange arrows) in West Johor Strait. Three features belong to Malaysia and the fourth, Tuas, relates to Singapore

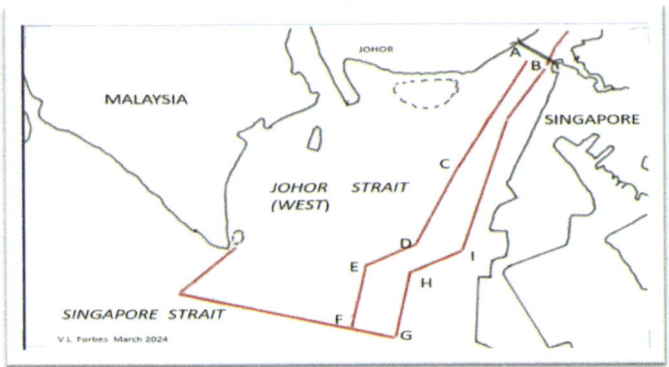

Fig. 7.15 Delineation of Johor Port limits (western section)

area; and amended the port limits by including in the second column, the following: The area enclosed by the following notional lines: from a point on the coast at Lat. 4° 16.4′ N., Lon. 103° 28.8′ E due East to Lat. 4° 16.4′ N., Lon. 103° 34.0′ E., thence South to Lat. 4° 08.0′ N., Lon. 103° 34.0′ E., thence due West to the coast and including the coast line and extending for a distance of two nautical miles from the mouth of the River Kemaman and returning to position (A); including all piers, jetties, landing places, wharves, quays, docks and other similar works whether within or without the HWM and any position on the shore or bank within about 45.72 m of the HWM subject to any rights of private property therein (Fig. 7.19).

The centroids for Kota Kinabalu Port are Lat. 5° 59′ N., Lon. 116° 0′ E., and Sepangar Bay: Lat. 6° 5′ 0.2 N Lon. 116° 7′ 0.3 E. The limits of these port facilities include Gaya Bay and waters to the South-West that are bounded by a line joining Tanjong Aru and the Northern point of Pulau Gaya thence a line to Gaya Head including rivers entering the same for 200 m, inland from the mouth of the river encompass a vast area. Sepangar Bay Container Port Terminal is experiencing a surge in development due to the expansion at the Port of Kota Kinabalu (Table 7.8).

Bintulu Port and Bintulu Fishing Port

The Bintulu Port Authority was established on 15 August 1981 under *Bintulu Port Authority Act* 1981. The port started operations on 1 January 1983. Bintulu Port Sdn. Bhd., a fully government owned company, took over all port facilities and services from Bintulu Port Authority in January 1993. Bintulu Port operated by Bintulu Port Sdn. Bhd. and Bintulu Port Authority remains as the regulatory body. Bintulu Port has a total land area of 320 ha. Its water limits extend in an arc seaward as far as 18 km (about 9 M) (Fig. 7.20).

Bintulu port is the only export gateway for Malaysia's export earner liquefied natural gas produced from the Central Luconia field, off the Sarawak coast. In addition, it manages a growing volume of a variety of general cargo, liquid and dry bulk, and containers.

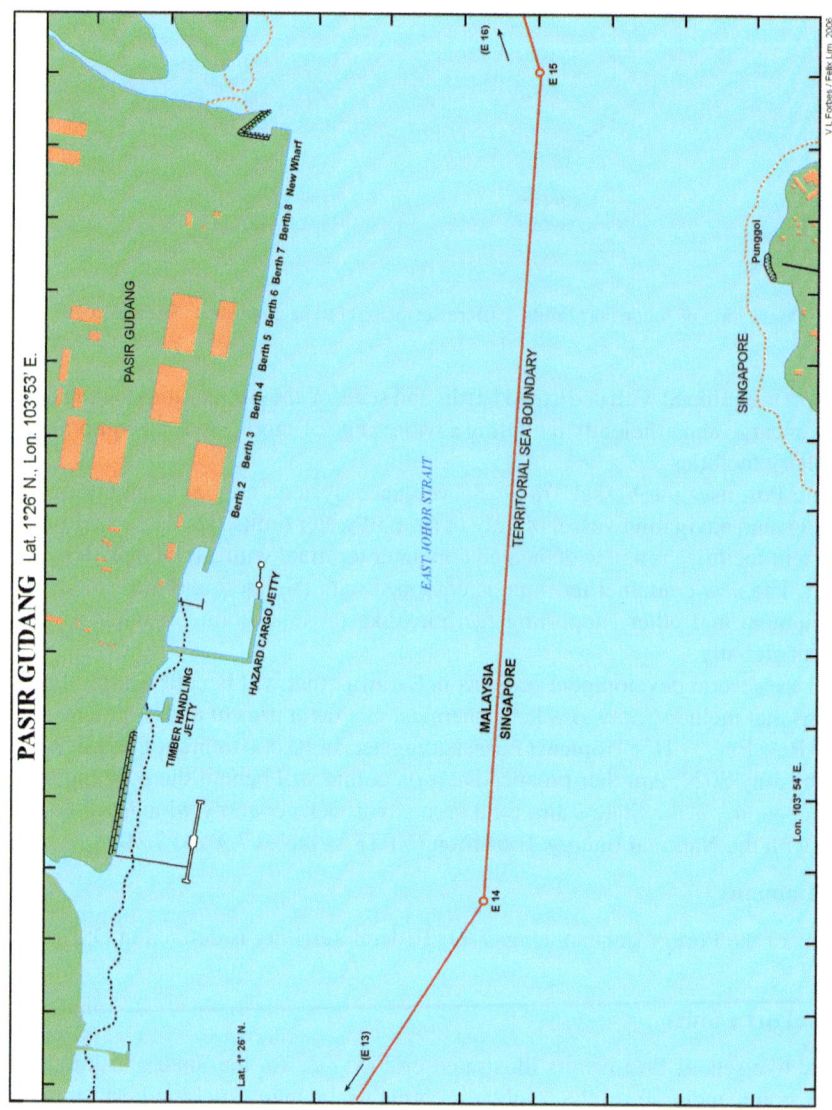

Fig. 7.16 Port of Pasir Gudang, Johor

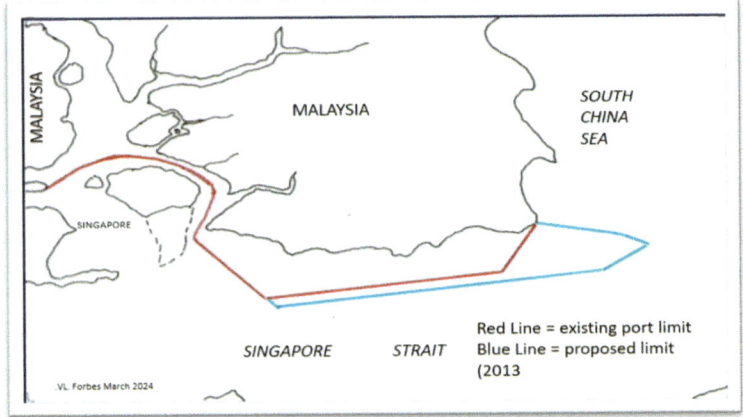

Fig. 7.17 Delineation of Johor Port limits (eastern sector). Yet to be confirmed: 30 June 2024

The port is equipped with dedicated berths and state-of-the-art facilities to manage a variety of cargo and efficiently including a wide range of cargo handling equipment and ancillary facilities.

Bintulu Port uses the Vessel Traffic Surveillance System, crucial in monitoring, controlling, and navigating vessels safely in the port water limits. The latest addition to its current facilities is a new dedicated container terminal with a total quay length of 450 m. The new container terminal is equipped with the latest and most modern port equipment and other supporting facilities like those available at other world class ports globally.

There are several development projects in Sarawak that will benefit Bintulu Port operations that include Sarawak's Petrochemical methanol project and shipments of Biomass Raw Energy Hot Tropical Grass, adding nearly 10,000 tonnes of liquid bulk volume by July 2024. Another prospective project that will benefit the port and the national economy is the State's aim for a green hydrogen economy which will be in keeping with the National Energy Transition (NETR) (Tables 7.9 and 7.10).

Port of Labuan

The limits of the Port of Labuan, (Fig. 7.21) Federal Territory is defined in Ord. 70/1952.[9]

Sarawak Port Limits

There are five sets of Port Limits illustrated on Fig. 7.22 for the State of Sarawak. They are, commencing from the southwest corner of the map: Kuching Port Limits, Tg. Manis Port Limits, Rajang Port Limits which encompasses all the riverine ports in the region, the Samalaju Port Limits, Miri Port Limits which is comprised in three parcels of land (locations).

An announcement on 24th May 2023, by the Government of Sarawak, noted that there will be a centralised Sarawak Ports Authority to oversee the operation of all

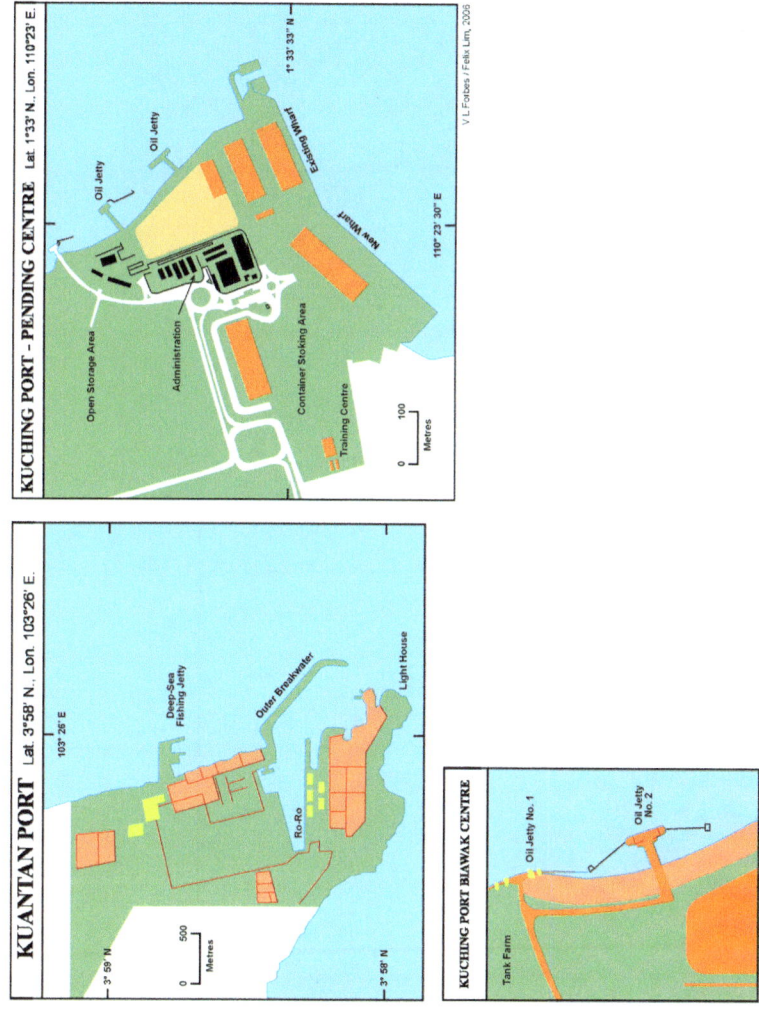

Fig. 7.18 Ports of Kuantan and Kuching-Pending, and Biawak Centres

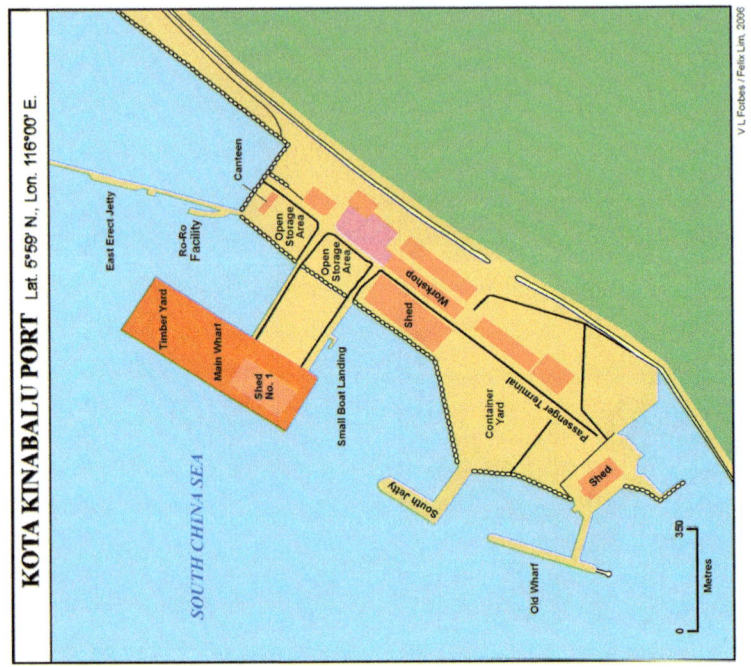

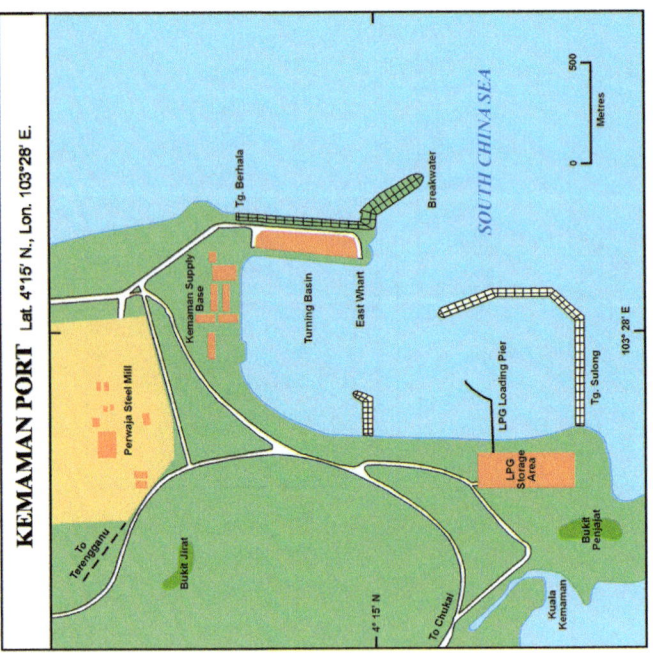

Fig. 7.19 Ports of Kemaman and Kota Kinabalu

Table 7.8 Berthing facilities at Port of Kota Kinabalu

Number of Berth	Length (m)	Depth (m)	Vessel size (dwt)
Main wharf 7	122–350	10	16,000
Berths 1 to 3	350	9.5	16,000
4	120	9.5	16,000
5 and 6	200	6.6	16,000
7	120	8.0	16,000
North jetty (2)	90–180	7.0	10,000
Berths 8 and 9	181	7.0	10,000
South jetty (3)	98–120	–	–
Berth 10	120	8.5	6000
Berth 11	120	6	6000
Berth 12	100	7	6000

ports. The aim was to regulate, streamline and coordinate the management operation, planning and development of all ports to ensure that each of the ports play its niche role in promoting business and trade. Bintulu Port will be one of the clusters of the state's ports. Its future development and growth will be harmonised and streamlined with the other ports in Sarawak. Bintulu Port is de-federalised and now regulated under Sarawak's authority—a subject matter that was pursued under the *Malaysia Agreement 1963* (MA63) Agenda.

Sabah Ports

The Sabah Ports Authority (SPA) is a state statutory body established in 1968 by the *SPA Enactment* 1967 which was repealed and replaced by the *SPA Enactment* 1981. The Authority is under the jurisdiction of the Ministry of Infrastructure Development, Sabah. Sabah Ports Authority acts as the regulatory authority responsible for matters related to port activities and ensures the terminal operator who operates the server ports in Sabah with the privatization agreement fully and set standards plus benchmarks for port operations comparable with ports in the region (Fig. 7.23).

There are eight designated commercial ports within the state of Sabah. Along the west coast abutting the South China Sea they are Kota Kinabalu (formerly Port Jesselton) (Lat. 05° 59′ N, Lon. 116° 00′ E.), Sapanagar Bay Oil Terminal (Lat. 06° 05.2′ N., Lon. 116° 07.3′ E.) and Sapanagar Bay Container Port (Lat. 06° 05.2′ N., Lon. 116° 07.3′ E.). On the north coast within Marudu Bay (*Telukan Marudu*) is Kudat Port (Lat. 06° 53′ N., Lon. 116° 51′ E.). Along the east coast of Sabah, abutting the Sulu Sea there are Sandakan Port (Lat. 05° 48′ N, Lon. 118° 04.8′ E) Lahad Datu Port (Lat. 05° 20′ N., Lon. 118° 02′ E.) and Kunak Port (Lat. 04° 41′ N., Lon. 118° 51′ E.) The Port of Tawau (Lat. 04° 14′ N., Lon 117° 52′ E) is located on the shores of Cowie Bay and facing the Sulawesi Sea just north of Pulau Sebatik.

The limits of the Port of Kota Kinabalu are stated as comprising Gaya Bay and waters to the southwest bounded by a line joining Tanjong Aru and the northern point

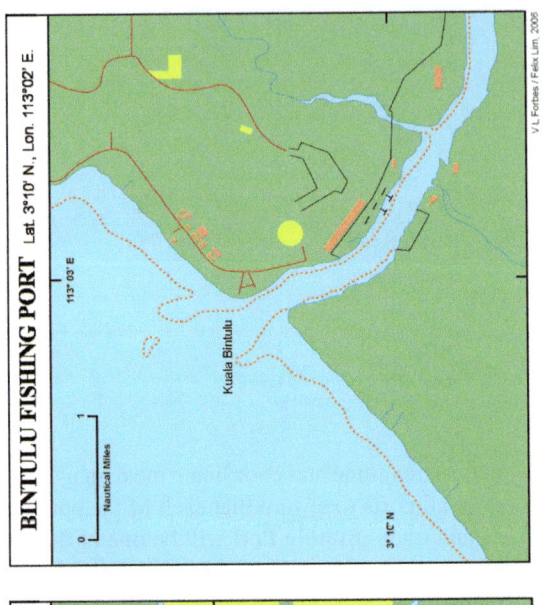

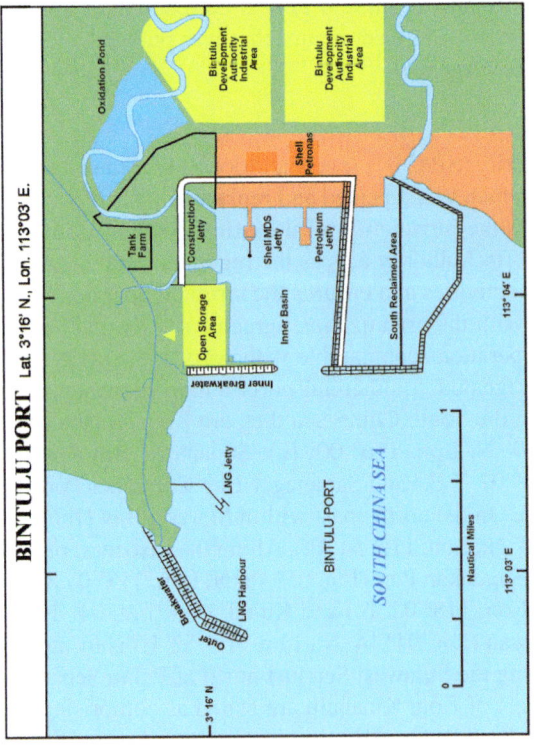

Fig. 7.20 Bintulu Port and Bintulu Fishing Port

Table 7.9 Port Facilities at Labuan

Type of berth	Length (m)	Depth (m)	Vessel size (dwt)
New Liberty wharf	244	10	16,000
Labuan passenger	20	–	–
Terminal (pontoon)	8	–	–
Victoria wharf	83.6	4.6	

Table 7.10 Private Jetties at Labuan

Type of berth	Length (m)	Depth (m)	Vessel size	Purpose
Shell jetty	213	9.4	6000	Petroleum
Iron ore jetty	220	18	150,000	Iron ore
Methanol jetty	650	13	35,000	Methanol
Asian supply base	120	8	6000	Offshore
Sabah flour mill	–	–	–	Wheat and maize

of Pulau Gaya thence a line to Gaya Head including rivers entering same for 200 m inland from the mouth of the river.

The limits of the Port of Kudat are deemed as a line drawn from Tigasmil Point to Sandilands Rock and thence to Tanjong Kapor including all passages and rivers entering the same for six hundred feet inland from the mouth of the river.

The limits of the Port of Sandakan is delineated as Sandakan Bay bounded by a line drawn in a west-south-west direction from the northern point of Pulau Berhala to the mainland, and by another from the same point on Pulau Berhala to the most eastern extremity of Tanjong Aru and thence to the opposite bank at the entrance Terusan Duyong, including all passages and rivers entering same for a distance of two hundred metres inland from the mouth of the river.

The limits of the Port of Kunak is a line drawn from the point (headland) on the mainland south of Wise Hill to the most easterly point of Pulau Sakar, thence in south-south-east direction to the most easterly point of Pulau Bohihan, and thence in a south-south-west direction to the peak of Mt. Cook at a point where the said line intersects the coastline, including rivers entering the same for a distance of 200 m inland from the mouth of the river.

The Port of Tawau comprises the waters of Cowie Bay and Wallace Bay and rivers entering same contained within the following geographical coordinates: Lat. 4° 13.3′ N., Lon. 118° 00.0′ E. to the National boundary mark on the eastern shore of Sebatik Island, thence to the National boundary mark on the western shore of the said island, thence on an azimuth of 270° to the shore at Burs Point thence in a 311° direction to 4° 20.0′ N 117° 25.9′ E, thence in a 070° direction to 4° 25.0′ N 117° 40.0′ E, thence in a 120° direction to the point of commencement.

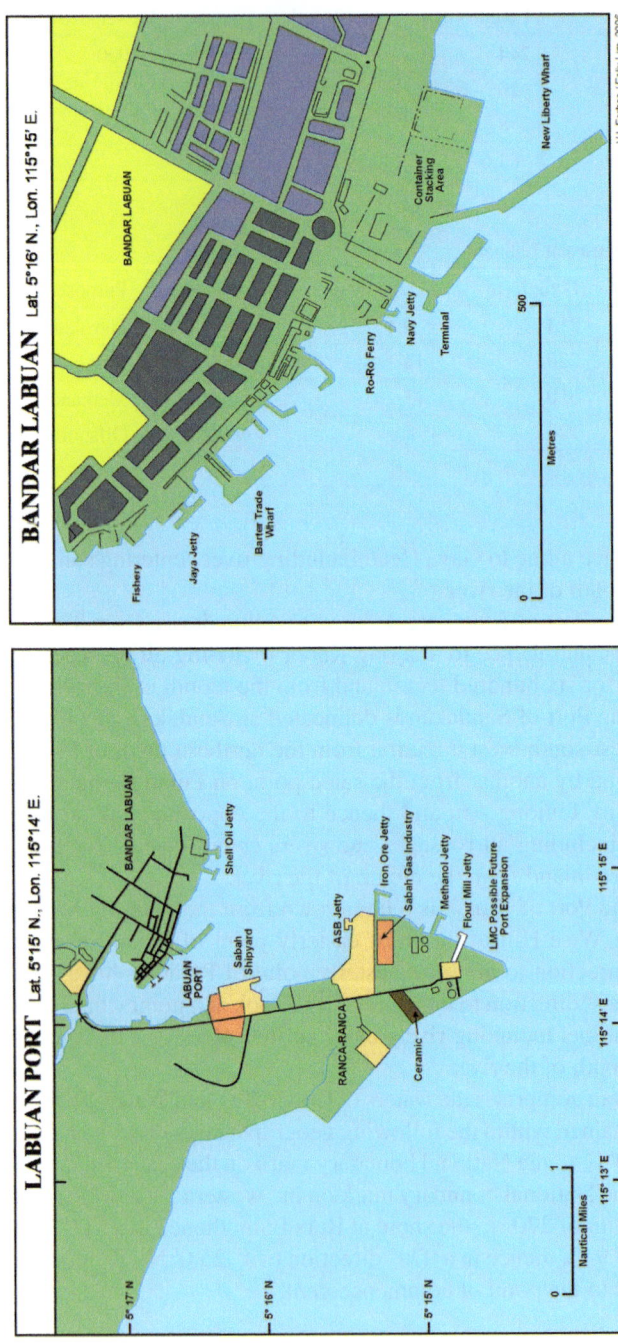

Fig. 7.21 Port of Labuan

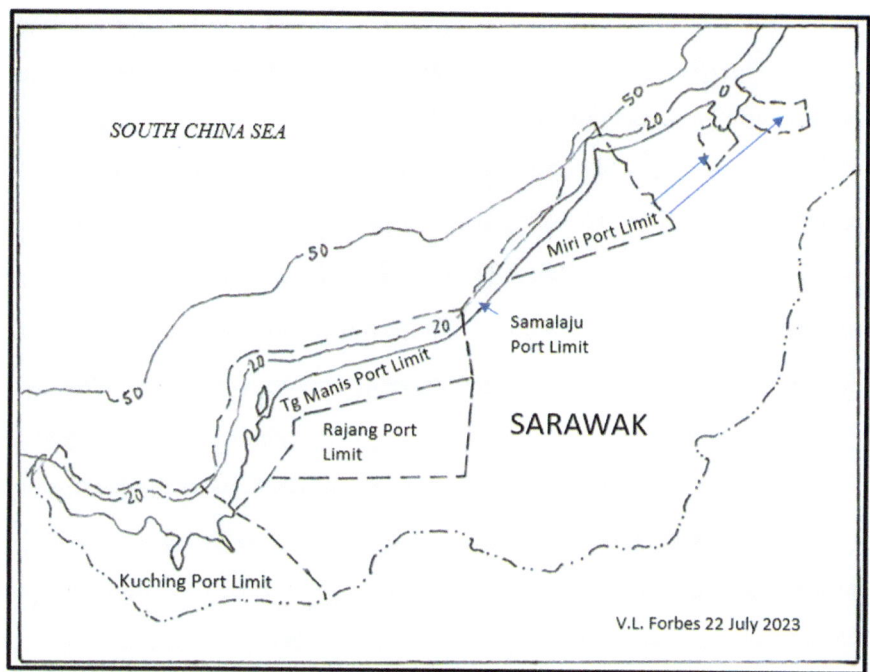

Fig. 7.22   Sarawak: Limits of Ports. *Source* Present author

7.6   Prescribed Sea Lanes to Ports of Sabah

Publication MP 1240 prescribes Guidelines for the establishment of specific corridors—sea lanes—for the protection and security of the coastal zone of Sabah. The designated routes have been prescribed for Indonesian- and Philippine-flagged cargo and passenger ships that are bound for ports in Sabah. These routes, as depicted on the maps, are mandatory as of 1 March 2002.[10]

The rationale for the creation and implementation of these routes was based on evidence that many of the foreign-flagged vessels, particularly from Indonesia and the Philippines, were in breach of Malaysian laws and regulations especially in weapons importation, landing of aliens and people-smuggling, acts of piracy and engaging in fish-bombing activities and illegally fishing in Malaysian waters. Furthermore, the incidents on Pulau Sipadan on 23rd April 2000 and events in Pandanan and Semporna during the following years in which foreign and Malaysian national tourist were kidnapped by terrorist prompted the then Prime Minister of Malaysia to propose the scheme.

Following an analysis of the proposal by the Ministry of Defence the guidelines were adopted on 27th June 2001. The corridors are designed to observe the activities of the ships *en route* to the ports and present an opportunity to safeguard national integrity and the marine environment and of course, to ensure the safety of navigation.

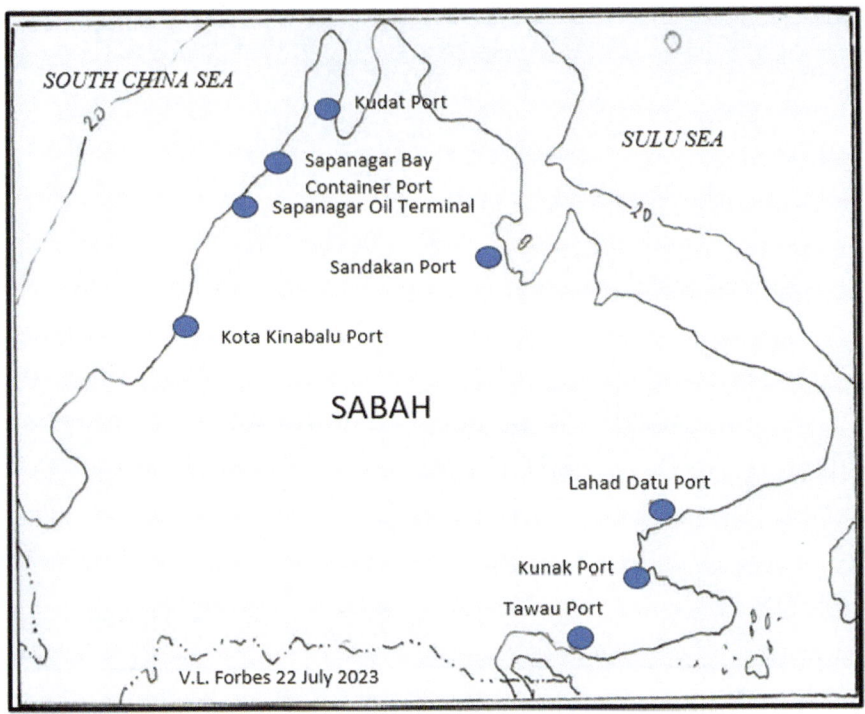

Fig. 7.23 Ports of Sabah

The designated corridors, as depicted on the map, and whose geographical coordinates are listed below, are not categorised as ships' routeing schemes, because the charted course are within national coastal waters. Hence there was no need to refer the adoption of the scheme to the International Maritime Organisation (IMO). The prescribed routes were discussed and approved at a meeting between the Prime Minister of Malaysia and the President of the Philippines on 9th August 2001.

The primary agencies responsible for the enforcement of the use of these routes are the Royal Malaysian Police, the Marine Police, and the Marine Department of Sabah. The secondary agencies are the Royal Malaysian Army, the Customs and Immigration Departments and the National Security Council, Sabah Branch. These agencies are tasked to take appropriate action if the ships are not within the designated corridors; or if the crew and passengers of these ships are carrying weapons; or the ships did not exhibit their flag of registry; and if the cargo manifest was not in order or contrary to what was being carried onboard. In any case, ships using the designated sea lanes were not exempt from being monitored, checked, or searched. Some of the tasks may be undertaken by associated agencies.

The geographical coordinates for the defined sea-lanes to the various ports of Sabah are in Table 7.11 and portrayed in Fig. 7.24.

Table 7.11 Geographical coordinates of prescribed Sea Lanes to Ports of Sabah

Port destination	Point	Latitude (North)	Longitude (East)
Tawau	1A	4° 00.0′	118° 01.0′
	1B	4° 08.2′	118° 01.0′
	1C	4° 10.9′	117° 53.0′
	1D	4° 13.8′	117° 53.0′
Semporna	2A	4° 33.0′	119° 05.1′
	2B	4° 37.0′	118° 56.0′
	2C	4° 40.5′	118° 52.4′
	2D	4° 42.0′	118° 37.0′
	2E	4° 31.0′	118° 37.1′
Lahad Datu	2A	4° 33.0′	119° 05.1′
	2B	4° 37.0′	118° 56.0′
	2C	4° 40.5′	118° 52.4′
	2D	4° 42.0′	118° 37.0′
	3A	4° 54.0′	118° 24.0′
	3B	4° 59.0′	118° 24′0′
	3C	5° 00.7′	118° 21.0′
Sandakan	4A	6° 04.8′	118° 18.0′
	4B	5° 51.4′	118° 09.1′
	4C	5° 50.2′	118° 07.8′
Labuan	5A	7° 43.9′	116°50.0′
	5B	7° 12.9′	116° 50.0′
	5C	6° 15.5′	115° 16.5′
	5D	5° 17.0′	115° 17.5′
	5E	5° 13.9′	115° 14′0.0′
Kudat	5A	7° 43.9′	116° 50.0′
	5B	7° 12.9′	116° 50.0′
	7A	6° 55.9′	116° 54.4′
	7B	6° 51.9′	116° 51.3′
Kota Kinabalu	5A	7° 43.9′	116° 50.0′
	5B	7° 12.9′	116° 50.0′
	8A	7° 43.9′	116° 01.0′
	8B	6° 50.0′	116° 01.0′
	8C	6° 02.2′	116° 04.9′

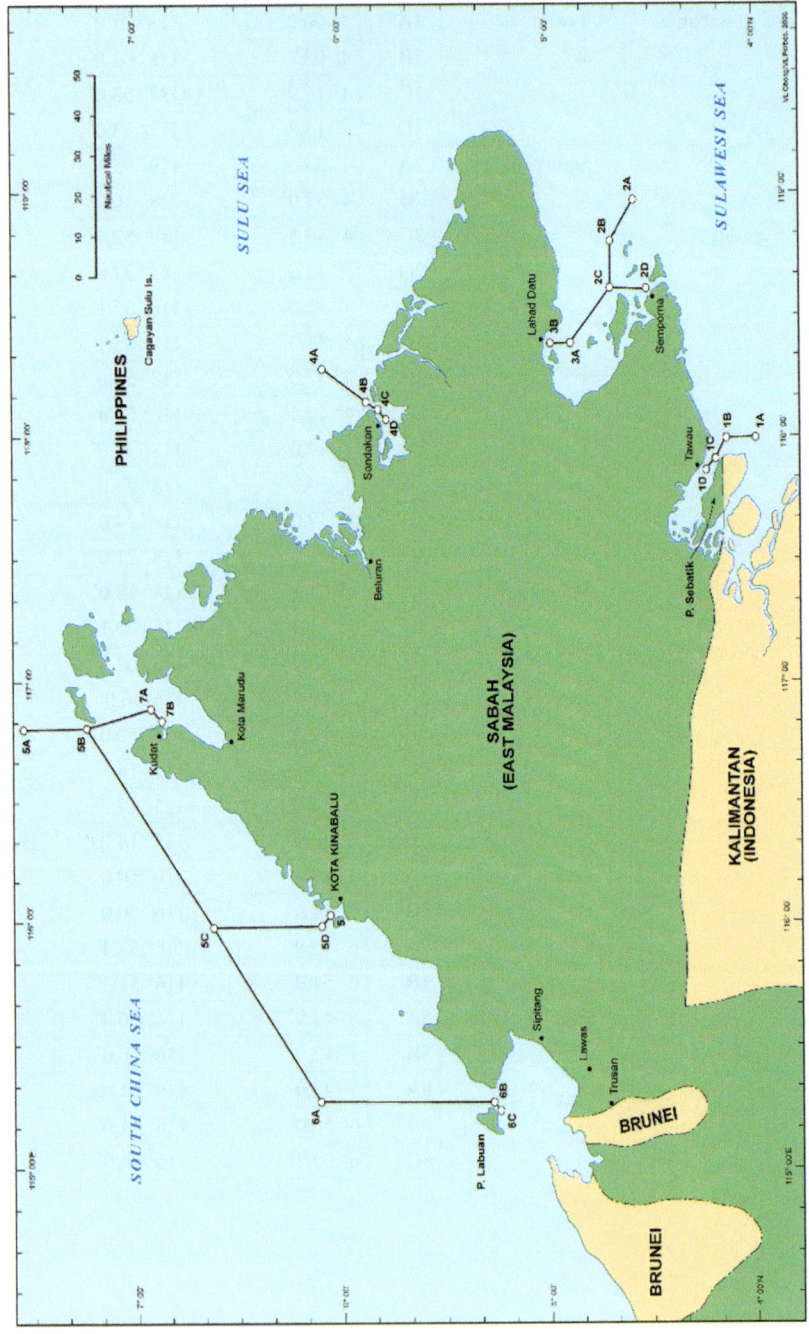

Fig. 7.24 Prescribed navigational approaches to ports of Sabah

7.7 Indonesia's Archipelagic Baseline System (Revised in 2002) and Archipelagic Sea Lanes

A revision of Indonesia's archipelagic straight baseline system was proclaimed in Act No. 38 on 28th June 2002, in accordance with obligations to the provisions of Article 47 of the 1982 Convention. The revision was discussed by Forbes, 2014 and analysed in *Limits of the Sea* No. 141, 2014.[11] The legislation, comprising 14 Articles, offered a rationale for defining these points whose geographical coordinates were listed.

The shortest are three straight base lines each of about 0.1 M in length; the two longest line segments, namely lines linking Points 44 and 45 and Points 54 and 55 are 122.75 M and 122.74 M respectively. Five-line segments, or three per cent of the total, measure between 100 and 123 M (see Table 7.3) which are in accordance with Article 47.

A point of curiosity is that of the absence of defined potential base points on the north coast of West Timor and on the south coast of Indonesian Islands facing the Ombai and Wetar Straits (Selat). These will be defined and published when officials from the Parties meet to negotiate their common maritime boundary. By August 2024, no official statement was issued.

The following tabulations present an overview of the number of line segments, total distances of the sections and other information for ease of comparison with the criteria set out in Article 47 (Tables 7.12 and 7.13).

The 2002 revision of the Indonesian archipelagic baseline system does not pose any problems to the delimited Indonesian maritime boundaries. However, the revised baseline system may impact on the negotiations between Indonesia and its neighbours Australia and Timor-Leste when the discussions of lateral maritime boundaries in the Timor Sea are raised. Its effect on the delimitation of maritime boundaries in the eastern and western approaches to the Straits of Singapore will also be viewed with great interest (Fig. 7.25).

Section Two in the Annex appended below is not relevant and should not be seen as a problem to Malaysia's ocean policy. It was necessary to include it to present the facts and to illustrate the way the system was devised.

Table 7.12 Sections, number of line segments and distances

Section	Base points	Distance	Number	Average
1	Point 1 to point 16	533.83	15	35.59
2	Point 17 to point 20	76.41	3	25.47
3	Point 20 to point 64	2015.20	44	45.80
4	Point 65 to point 101	1171.26	36	32.54
5	Point 102 to point 1	3008.27	82	36.69
	Total	6804.97	180	37.81

Table 7.13 Number of base lines categorized by length of line

Range of length (M)	Number of lines	Percentage
0–10.0	45	25.0
0.1–20.0	28	15.5
20.1–30.0	20	11.1
30.1–40.0	17	9.4
40.1–50.0	12	6.6
50.1–60.0	12	6.6
60.1–70.0	7	3.9
70.1–80.0	9	5.0
80.1–90.0	11	6.1
90.1–100.0	14	7.8
100.1–110.0	2	1.2
110.1–120.0	1	0.6
120.1–130.0	2	1.2
Total	180	100

Indonesia is to be congratulated for the promptness and its open policy in making public the document which lists the geographical coordinates of the archipelagic base points and thus permit researchers to analyse and disseminate the information.

Indonesia's Archipelagic Sea Lanes

On 24th October 1988, Indonesia asserted its rights to close international shipping lanes through parts of its archipelagic waters. A similar assertion was made in early April 1996. Since then, the debate has attracted attention and concern by Australia and the United States; and indeed, in other maritime areas of semi-enclosed seas and straits.

Since about March 1995, Indonesian military officials had refused on four separate occasions to grant requests from Australian and United States research ships to conduct marine scientific research within the South Java Current.

The issue of archipelagic sea lanes (Article 53 of the 1982 Convention) was debated at the IMO annual meetings. It was discussed at the 43rd Meeting held in London during July 1997. Indonesian Government *Regulation No. 37* of 2002 was proclaimed on 28 June 2002, and is in force since 28th December 2002.

7.8   Reported Acts of Armed Robbery and Piracy

The International Maritime Bureau (IMB) concerned at the alarming growth in the phenomenon, was prompted to create the IMB Piracy Reporting Centre in 1992. The Centre is based in Kuala Lumpur, Malaysia. It maintains a round-the-clock watch

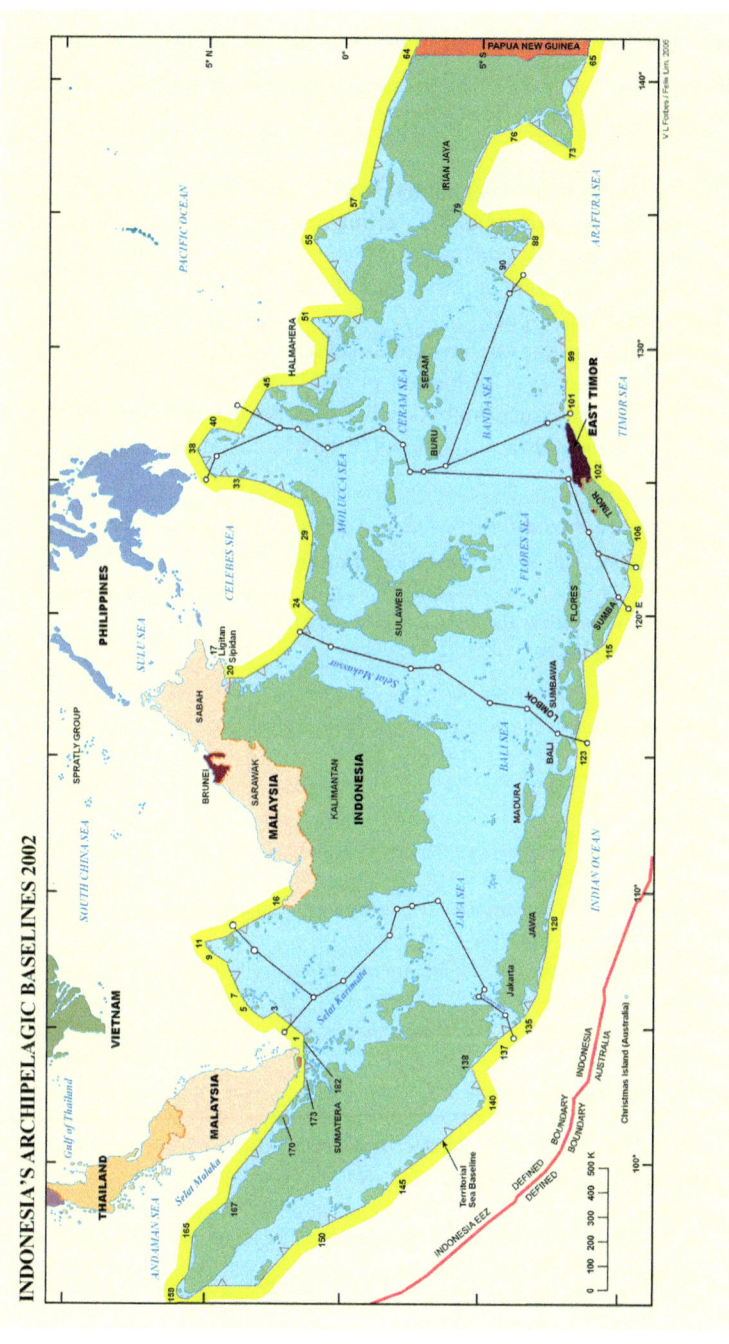

Fig. 7.25 Indonesia's Archipelagic Straight Baselines and Designated Sea Lanes

on the world's shipping lanes, reporting pirate attacks to local law enforcement, and issuing warnings about piracy hotspots to shipping.[12]

For statistical purposes, the IMB defines acts of Armed Robbery and Piracy as:

An act of boarding or attempting to board any ship with the apparent intent to commit theft or any other crime and with the apparent intent or capability to use force in the furtherance of that act.

This definition covers actual or attempted attacks whether the ship is alongside a berth, at anchor or at sea. Petty thefts are excluded unless the thieves are armed. It was adopted by the IMB as most attacks against ships take place within the authority of States and piracy as defined under provisions of the 1982 Convention (reproduced below) does not address this aspect.

At the International Maritime Organisation's (IMO) 74th Meeting of the Maritime Security Council the matter was addressed in the Draft Code of Practice for the Investigation of Crimes of Piracy and Armed Robbery against Ships (MSC/Circ. 984) Article 2.2. The Code of Practice defines 'Piracy' and 'Armed Robbery against Ships' as follows:

Piracy means unlawful acts as defined in Article 101 of the 1982 Convention.[13]

Article 100 Duty to cooperate in the repression of piracy.

All States shall cooperate to the fullest possible extent in the repression of piracy on the high seas or in any other place outside the jurisdiction of any State.

Article 101 Definition of piracy.

Piracy consists of any of the following acts:

(a) any illegal acts of violence or detention, or any act of depredation, committed for private ends by the crew or the passengers of a private ship or a private aircraft, and directed:

(i) on the high seas, against another ship or aircraft, or against persons or property on board such ship or aircraft;

(ii) against a ship, aircraft, persons, or property in a place outside the jurisdiction of any State;

(b) any act of voluntary participation in the operation of a ship or of an aircraft with knowledge of facts making it a pirate ship or aircraft;

(c) any act of inciting or of intentionally facilitating an act described in subparagraph (a) or (b).

Article 102 Piracy by a warship, government ship or government aircraft whose crew has mutinied.

The acts of piracy, as defined in article 101, committed by a warship, government ship or government aircraft whose crew has mutinied and taken control of the ship or aircraft are assimilated to acts committed by a private ship or aircraft.

Article 103 Definition of a pirate ship or aircraft.

A ship or aircraft is considered a pirate ship or aircraft if it is intended by the persons in dominant control to be used for the purpose of committing one of the acts referred to in article 101. The same applies if the ship or aircraft has been used to commit any such act, so long as it remains under the control of the persons guilty of that act.

Article 104 Retention or loss of the nationality of a pirate ship or aircraft.

A ship or aircraft may retain its nationality although it has become a pirate ship or aircraft. The retention or loss of nationality is determined by the law of the State from which such nationality was derived.

Article 105 Seizure of a pirate ship or aircraft.

On the high seas, or in any other place outside the jurisdiction of any State, every State may seize a pirate ship or aircraft, or a ship or aircraft taken by piracy and under the control of pirates, and arrest the persons and seize the property on board. The courts of the State which carried out the seizure may decide upon the penalties to be imposed, and may also determine the action to be taken with regard to the ships, aircraft, or property, subject to the rights of third parties acting in good faith.

In response to Acts of Armed Robbery and Piracy, on 9th July 2004, the then Malaysia's Deputy Prime Minister announced that the Malaysian Maritime Enforcement Agency (APMM) will enforce the country's maritime law within 50 nautical miles off its coast. The APMM commenced operations in March 2005. The Royal Malaysian Navy protects the remaining area of Malaysia's maritime space. Under the APMM Bill passed by Parliament in June 2004, the agency would operate up to 200 M of the limits of the nation's EEZ. APMM takes over responsibility of federal laws on sea piracy, marine pollution, and illegal immigrants, as well as search and rescue operations. An allocation of RM 26 million was made for APMM in 2004 and around $260 million for 2005. APMM was expected to take over maritime enforcement responsibilities which were undertaken by 11 Government Agencies, including the Royal Malaysian Marine Police and Royal Malaysian Navy.[14]

On 15th July 2004, Indonesia, Malaysia, and Singapore confirmed that each would begin co-ordinated military patrols in the Malacca Strait. The initiative fell short as the agreement did not allow patrol vessels from neighbouring States to enter each other's waters in 'hot pursuit of pirates'. The issue of sovereignty—understandably— has slowed co-operation on security in the Strait, especially after reports that the US navy was considering becoming involved in security and anti-piracy measures.

Indonesia and Malaysia ruled out a role for outside troops, despite independent groups who had stressed the limitations of the littoral states' forces. The Government of Singapore had suggested links between pirates in the Straits and regional terrorist groups such as JI. Indonesia noted that such fears were overblown.

The Governing Council of the Information Sharing Centre (ISC) of the Regional Cooperation Agreement on Combating Piracy and Armed Robbery against Ships in Asia (ReCAAP) releases regular reports of incidents reported in the regional seas. ReCAAP agreed to cooperate with the International Maritime Organization (IMO) on anti-piracy and armed robbery efforts in the Asian region, as well as to share expertise and best practices where possible.

The Information Sharing Centre (Information Fusion Centre), which was set up in Singapore in November 2006, is a central element of the ReCAAP initiative to harness the collective resources and expertise of regional governments to combat piracy and publish reports.

Reported Acts of Armed Robbery and Piracy: off Peninsula Malaysia

Opening the IMB/ICC (International Maritime Bureau/International Chamber of Commerce) Conference in Kuala Lumpur on 12th June 2007, Malaysia's Deputy Minister for Internal Security Y. B. Dato' Mohd. Johari Bin Baharum observed that the dramatic reduction in attacks in the Malacca Straits was the result of firm action taken by Malaysia and the littoral states. His address was followed by presentations outlining contemporary responses to piracy and security, and an analysis of hot spots. Tan Sri Musa Hassan, Inspector General of the Royal Malaysian Police, highlighted the importance of neighbouring law enforcement agencies of the Malacca Straits in maintaining open channels of communication. He noted that multi-jurisdictional measures, such as coordinated patrols and 'eye-in-the-sky' programmes, had resulted in a dramatic drop in piracy attacks in the region.

Reported incidents of piracy dropped significantly in South-East Asia during the first half of 2007. Indonesia recorded nine incidents, down dramatically from 19 in 2020. Two incidents were recorded in the Malacca Straits. This area now represents an excellent example of how cooperation between authorities can tackle and suppress piracy attacks. There was also notable improvement in Bangladesh with only two reported incidents compared to nine in the last quarter of 2006. Table 7.14 illustrates the trend of actual and attempted attacks in these important, indeed, vital shipping lanes.

Reported acts of armed robbery and piracy: off Sabah and Sarawak

The statistics for the region covered in this map may not appear as dramatic as that for the Malacca Strait. However, the dangers are ever so present, and mariners are advised to maintain anti-piracy watches and report all piratical attacks and suspicious movements of craft to the IMB Piracy Reporting Centre, Kuala Lumpur, Malaysia.

The number of incidents in 2007 as depicted on the map are for the first quarter of that year. Figures 7.26 and 7.27 illustrate the extent of these illegal acts in theses regional seas. Examples of what transpired in a few of those incidents:

On 29 March 2007, at about 0200 h (Local Time) —the mv *Grace Casablanca*, registered in Hong Kong, carrying general cargo was anchored off Bintulu, Sarawak in Lat. 3° 14′ N, Lon. 112° 58′ E. Five robbers armed with knives boarded the ship. The Second Officer of the ship spotted the robbers and raised the alarm; crew were alerted. The crew proceeded to the forecastle of the ship. The robbers threw shackles at the crew, but no one was injured. The robbers stole the ship's stores and escaped in their boat. Port Security was informed, and a patrol boat arrived on the scene two hours later and investigated.

Table 7.14 Reported attempted attacks from 2003 to 2007

Locality	2003	2004	2005	2006	2007
Malacca strait	15	20	8	3	2
Singapore straits	–	7	6	3	3

**REPORTED ACTS OF ARMED ROBBERY AND PIRACY
IN THE VICINITY OF PENINSULAR MALAYSIA**

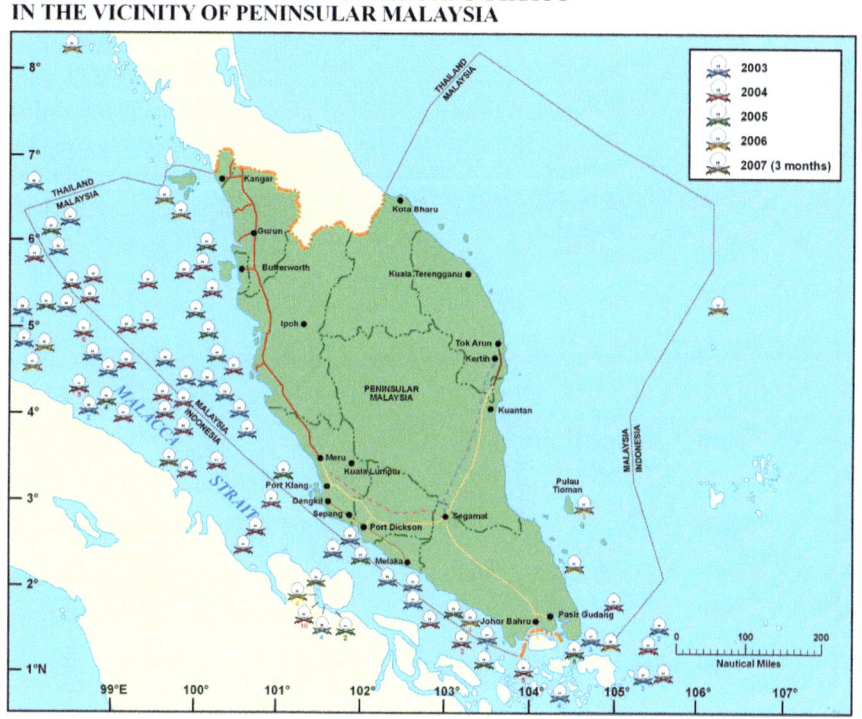

Fig. 7.26   Reported Acts of Armed Robbery and Piracy, 2003–2007, off Peninsula Malaysia

On 9 May 2007, a Malta-registered, bulk carrier, *Nin* was alongside berth number 4 at Lahad Datu Port, Sabah. Several stevedores stole ship's property at night during discharging of cargo operations. The Master of the ship reported the incident to the Agents. Police boarded the ship and allegedly arrested two stevedores (Fig. 7.28).

Acts of Armed Robbery and Hi-Jacking, 2009–2010.

The International Chamber of Commerce's (ICC) International Maritime Bureau (IMB)'s annual piracy report recorded an increase of piracy and armed robbery incidents in 2020. In that year, IMB's Piracy Reporting Centre (PRC), located in Kuala Lumpur, received 195 incidents of piracy and armed robbery against ships worldwide, in comparison to 162 in 2019. The incidents included three hijacked vessels, 11 vessels fired upon, 20 attempted attacks, and 161 vessels boarded. The rise was attributed to an increase of piracy and armed robbery reported within the Gulf of Guinea as well as increased armed robbery activity in the Singapore Straits (Fig. 7.29).

The total number of attacks in the Singapore Strait reached a seven-year high of 55 cases in 2022. The vessels attacked—actual and attempted—included bulk carriers,

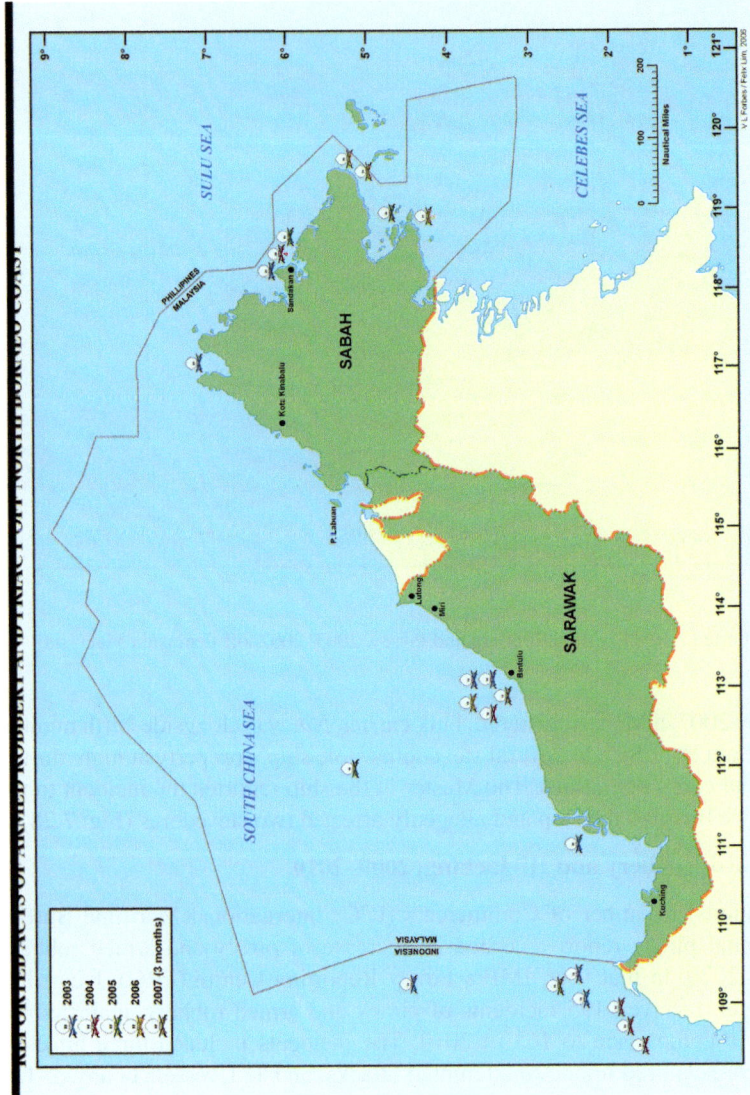

Fig. 7.27 Reported acts of Armed Robbery and Piracy, 2003–2007, off North Borneo coast

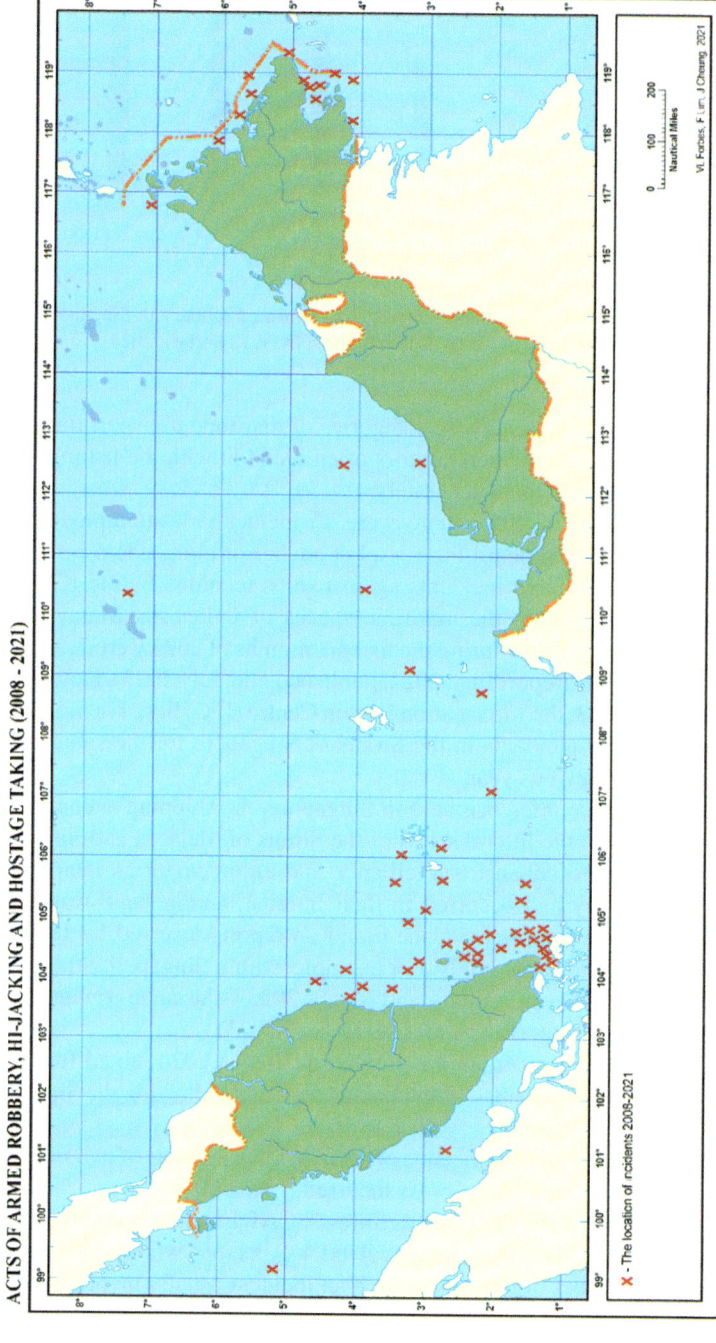

Fig. 7.28 Acts of Armed Robbery, Hijacking and Hostage Taking, 2008–2021

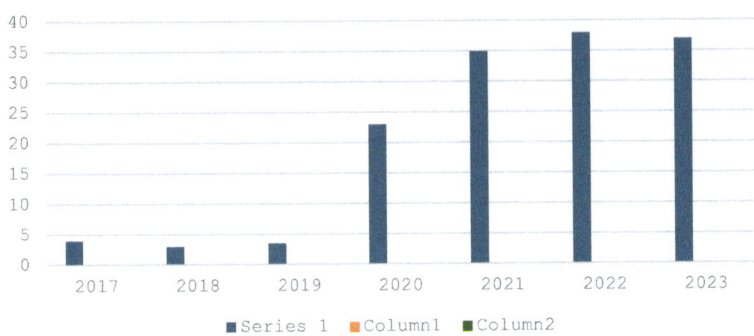

Fig. 7.29 Reported incidents of actual and attempted acts of armed robbery in Singapore Strait. *Source* Data from *Annual Report* of ICC-IMB, 2023, graphic by V. L. Forbes, 2024

tankers, and tugboats and barges. These categories of maritime transport transit the straits at a slower speed, and the perpetrators often blend in with the fishing boats, making it difficult to identify them accurately.

During the first quarter of 2023, there were 25 incidents of attempted attacks on ships. Six attacks were on board the vessel at anchor or alongside a berth. The number of actual and attempted incidents against ships recorded by the ICC-IMB Piracy Reporting Centre within the maritime spaces of Indonesia, Malaysia, the Philippines, and Singapore Strait during the first six months of 2023 were 7, 1, 5, and 20 respectively. (ICC-IMB Report June 2023, published mid-July 2023). However, in its mid-year report of 2023, the Information Fusion Centre (IFC) based in Singapore, noted that the number of incidents in the Straits of Singapore for the same period was 37 up from 27 the previous year.

In mid-July 2023, at a conference held in Singapore, the shipping agencies, and the regional authorities of the littoral states of the Straits of Malacca and Singapore were urged to continue to enhance surveillance and enforcement as incidents of armed-robbery and piracy had occurred in their internal, archipelagic waters and territorial seas. There was ample evidence based on reports received by IMB and ReCAAP ISC and reported accordingly. For example, within Singapore Strait, there were 18 piracy attacks between January and March 2023—the same as that for the corresponding period in 2022. (ReCAAP ISC Report, 2023).

The number of incidents recorded for the first quarter of 2024, noted ReCAAP, was 26 of which 21 were actual attacks and five were attempted within the Singapore Strait.

ReCAAP ISC evaluates the significance of each incident in terms of two factors—the level of *violence* and the *economic loss* incurred (Fig. 7.30).

The categorisation of the incidents are as follows: CAT 1 incidents involve a large number of perpetrators; CAT 2 incidents involved 4–9 persons who are likely to be armed; CAT 3—the number of persons in such an incident usually involved groups of between 1 and 6; and CAT 4—the perpetrators were not armed and the crew not harmed. Generally, one to three persons were involved who escaped empty-handed

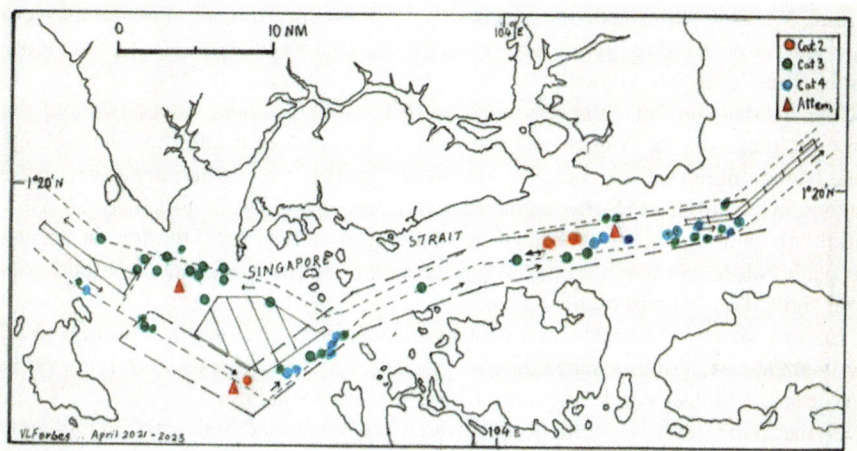

Fig. 7.30 Incidents on ships in Singapore Strait. *Sources* ReCAAP and IMB

upon being sighted by the crew of the vessel. During the 1st Quarter of 2024 there were eight incidents of Category 4; as many in Category 3; five in Category 2; and zero in Category 1. They were all armed robbery and zero record of piracy.

According to the ISC Report, in the above illustration, the red icon, represents an incident where there were four to nine armed culprits. Crew were either threatened, held hostage, or injured. Cash and equipment such as engine spare parts were stolen in that incident. In the instance of the blue icons there were one to six culprits. The ship's crew was subjected to duress but were unharmed. Nothing was stolen. The green icons represent incidents with one to three unarmed culprits. The crew was unharmed; the perpetrators escaped empty-handed after being spotted.

On average, at least 1500 ships transit the Singapore Strait daily, hence the environment for illegal activities is conducive despite the 'Eye in the Sky' concept which commenced in 2005 and the actions of the water police and other law enforcement agencies. The perpetrators are usually struggling in economic and social terms and just wish to 'make a quick dollar;' however, they will assess the situation but would not hesitate to physically strike out if challenged.

Economic uncertainty, regional geopolitical tensions, and high inflation in many parts of the world are likely to spark a rise in piracy attacks in Asian waters. But then, piracy is the second oldest profession in the world.

Increased and aggressive patrols by the littoral states' authorities since July 2005 in the Malacca Strait has ensured a drop in the number of incidents. Masters of ships are advised to continue maintaining strict anti-piracy/robbery watches whilst transiting the Straits of Malacca and Singapore. Incidents in the past have been reported by vessels anchored of Bandar Penawar, Johor, and Tanjung Piai. Off Eastern Sabah acts of kidnapping of the crew of merchant ships have ceased for several years. Due care must be taken is the advice of the authorities. Sabah *Notice to Mariners* (*NtM*) 14 of 2017 on the Ship Reporting System should be consulted.

ReCAAP-ISC informed the international community that four incidents were reported on board ships while underway in the eastbound lane of the Singapore Strait, from 7th to 9th August 2023.

The incident on 7th August occurred off Tanjung Tondong, Indonesia and the three incidents on 8th August and 9th August occurred in the Philip Channel, off Cula Island, Indonesia. All incidents occurred to bulk carriers while underway in the eastbound lane of the TSS. In one incident, the perpetrators were armed with knives, and the perpetrators were not armed in the other three incidents. In three incidents, the perpetrators were sighted in the engine room, and in one incident perpetrators were sighted on the port quarter deck.

Engine spares were stolen in all four incidents. The crew was not injured in all the incidents. With these four incidents, a total of 55 occurrences were reported in the Straits of Malacca and Singapore (SOMS) since January 2023.

Of the 55 incidents, 51 incidents occurred in the Singapore Strait and four incidents in the Malacca Strait. A total of 55 incidents were reported in SOMS in the entire year of 2022, all in the Singapore Strait. The ReCAAP-ISC urged the littoral States to increase patrols/surveillance in their respective waters, respond promptly to incidents reported by ships, strengthen coordination, and promote information sharing on incidents and criminal groups involved to arrest the perpetrators. (ReCAAP-ISC, 10th August 2023).

Within the narrow straits, tugs, barges carrying scrap metal and slow-moving bulk carriers are targeted by 'perpetrators' in stealing whatever they can for gainful purposes. The MMEA and its counterpart in Singapore are in constant watch according to local authorities and international organisations such as IMO, ICC-IMB, ReCAAP and UKMTO. The MMEA recorded 37 arrests within the Sandakan Maritime Zone in the first half of 2024 and two arrests in the vicinity of Penang Island in July 2014.[15]

The event reported by the BBC World Service on 22nd July 2024,[16] as the present author was finalising this chapter, best illustrates the rationale for this study. Malaysia's coast guard intercepted the crude oil carrying supertanker, *Ceres I*, that was involved in a collision with *Hafnia Nile* on Friday 19th July 2024. The latter, with a crew of 22, was Singapore-flagged whilst the former, with a crew of 40, was Sáo Tomé and Principe-flagged and apparently Chinese owned. The *Hafnia Nile* was transporting highly flammable naphtha cargo.

The collision, the cause of which is presently unclear, allegedly occurred about 30 M northeast of Pedra Branca Island in what is consider a busy sea-lane (Refer: Fig. 4.18), The former-named ship fled the scene of the accident, an oil spill area as a result of the collision. Forty crew members were rescued from the blazing ships, and about 22 of them remained on *Ceres I* to tackle the fire.

7.9 Summary

There is concern among the maritime industry and international community that often large, slow-moving vessels transiting the Straits of Singapore and the southern sector of the Malacca Strait are targeted and boarded by persons intent on opportunistic crimes. The significant 25% increase in reported incidents in the six-month of 2023 compared to the same period in the previous year. The Indonesia archipelagic region has shown a sustained decrease in reported incidents compared to years preceding 2020.

The context of this chapter has demonstrated the positive and progressive way the commercial ports of Malaysia have operated in the past and their prospects for the future. Collectively, the ports have contributed to Malaysia's Blue Economy. The administrators of the ports have shown resilience during the period of the COVID-19 Pandemic and are now reaping the benefits. To this end the Government of Malaysia must be congratulated.

A brief discussion on the nominated sea lanes through Indonesia's Archipelagic Waters was offered including the Special Agreement whereby fishers from Malaysia are given special status to transit Natuna Sea in the vicinity of the Natuna Archipelago as they passage between Peninsular Malaysia and Northern Borneo—Sabah and Sarawak.

Notes

1. UNCTAD Maritime Profile of Malaysia < unctadstat.unctad.org/countryprofile… >.
2. Y.S. Loong and W.A.W. Usamah 'The Malaysian Economy and Covid-19: Policies and Responses from January 2020-April 2021, UNCTAD/BRI Project/RP30.
    https://unctad.org/system/files/official-document/BRI-Project_RP30_en.pdf
3. Ministry of Foreign Affairs, Netherlands (2022) Port Development in Malaysia.
4. Robin Bush, 'BRI in Malaysia—a tool for domestic political elites', BBC, 2 Aug 2022.
5. Malaysia Marine Department https://www.marine.gov.my/jlm/en/public…
6. As above.
7. Microsoft Word—Draf Ord 1952—Port Klg Port Limits Final70610-2.doc (marine.gov.my).
8. BBC World News 'Forest City' https://www.bbc.com/news/business-67610677.
    See also: Country Garden Forest City https://www.forestcitycgpv.com
9. Limits of Port of Labuan: 20,200,911,154,041-ae3ef-labuan_17012017.pdf (marine.gov.my).
10. PortsAndHarbours(PortsHarboursAndDues)Regulations2008.pdf (sabah.gov.my).
11. Forbes, V.L. *Indonesia's Delimited Maritime Boundaries*, Springer, 2014.
    https://link.springer.com/book/10.1007/978-3-642-54395-1; see also *Limits in the Seas* No. 141 'Indonesia: Archipelagic and Other maritime Claims and Boundaries' 2014, Office of Ocean and Polar Affairs, US Department of State.
12. ICC-IMB (2023) 'Piracy and Armed Robbery Against Ships' for the period 1 January-30 June 2023. (This Report and other Quarterly and Annual Reports have been consistently consulted over many years during the compilation of the maps in this publication. ReCAAP-ISC Reports are a useful source for information relating to acts of armed robbery and piracy in the regional seas for Weekly, Quarterly, and Annual Reports. www.recaap.org/.
13. UN Law of the Sea Convention 1982 https://www.un.org/depts/los/convention….

14. APMM Official Maritime Malaysia Website (mmea.gov.my). https://www.chinadaily.com.cn/world/2007-07/06/content_911077.htm.
15. As above, Note 14.; also https://www.seafarertimes.com/2018-19/node/5675.
16. Lipika Pelham, 'Malaysia tracks down missing oil tanker which fled after collision', *BBC World Service*, 22 July 2024, 08:00 GMT. <accessed 16:00 WAST, 22 July 2024>. https://www.bbc.com/news/articles/c047jm8984mo.

Chapter 8
Conclusion

The objective of this volume has been fulfilled by demonstrating that maps and associated graphics are important documents and could potentially be deemed as legal instruments. Maps are of fundamental value to a nation's developmental aims and a valuable source in enhancing a nation's blue economy.

Malaysia's maritime jurisdiction is extensive despite its perceived 'zone-locked' status in the parlance of the 1982 UN *Convention of the Law of the Sea.* However, it is not geographically disadvantage, like its southern neighbour, Singapore, nor does it accrue greater benefits from the oceans than its archipelagic state neighbours, Indonesia, and the Philippines. Malaysia's adjacent tropical ocean waters stretch from the Malacca Strait in the West to the Sulawesi and Sulu Seas in the East and includes the southern sector of the South China Sea where sovereignty issues persist at the time of publication of this volume.

The Malaysian Government is committed to protecting its maritime areas in the South China Sea which includes its sovereignty, sovereign rights to the marine biotic and mineral resources and its interests in accordance with international law in its stated territorial claim over certain marine features, water column, seabed, and its substratum.

In standing its ground, successive Governments of Malaysia, nevertheless, have adopted a cautious diplomatic approach with its neighbours with reference to maritime boundary issues. On 8 June 2023, the leaders of Indonesia and Malaysia met in Kuala Lumpur and signed two 'Treaties' for the delimitation of Territorial Sea boundaries in the southern sector of the Malacca Strait and north-western segment of the Celebes (Sulawesi) Sea. Unfortunately, the statement relating to this event was tantalisingly vague on details of the boundary alignments and whether the Treaties could be considered as being ratified when announced. *Were they just Agreements in principle?* The official statement inferred that further negotiations would aim to resolve the maritime issues between the two nations at the earliest opportunity. Negotiations on resolving the maritime boundary alignments have been in progress for at least 18 years.

V. L. Forbes, *Malaysia's Maritime Jurisdictional Limits,*
https://doi.org/10.1007/978-3-031-78783-6_8

By late-April 2024, Malaysia had four segments of maritime boundary delimitation to resolve with Indonesia; two with Singapore; and a potential maritime boundary each with the Philippines and Vietnam and with China and Taiwan (depending upon the geopolitical inter-play between the last-two named countries) in the South China Sea. The maritime boundary with Brunei has been resolved, although the details of the Agreement are not in the public domain and hence could not be analysed in this study.

This study offered an appraisal, in commentary and maps, of seven themes: namely, mapping to enhance Malaysia's planned Blue Economy policy; geography and ocean basin environment; geographical and legal concepts with respects to maritime boundary delimitation; Malaysia's maritime jurisdictional limits; traffic monitoring and maritime safety; marine parks, protected zones and marine biotic and mineral resources; and, major commercial ports of Malaysia, sea lanes and acts of piracy and armed robbery at sea.

8.1   Mapping to Enhance the Blue Economy

In the first chapter, the narrative focused on the role of maps and other graphics in supporting a nation's claim over maritime space in its adjacent seas and the obligations and rights to enhance its blue economy. The UN infers that the Blue Economy is one that aims at the improvement of human well-being and social equity, while significantly reducing environmental risks and ecological scarcities. Realising Blue Economy benefits to a nation can lead to better and new job creations, achieve higher rates of economic growth, and secure biodiversity and sustainable development obligations. Achieving a balance between competing uses and users will be crucial for attaining a sustainable ocean economy, consistent with Malaysia's ongoing international policy commitments.

It is abundantly clear that access to marine biotic and mineral resources in a coastal state's adjacent seas could be utilized in advancing the nation's Blue Economy. Malaysia's coastal waters are home to rich coral reefs, seagrass and seaweed beds that are feeding grounds for fish. The seas surrounding Malaysia contain productive and diverse habitats with the major ecosystem being mangroves, coral reefs, and sea grasses, among others. These are productive natural ecosystems that contribute significantly to human, food, economic and environmental security. The fishery industry offers employment opportunities to thousands of workers that include fishers and others in the support businesses and manufacturing.

A vibrant Blue Economy ecosystem is predicted to potentially increase the contribution of marine and ocean resources from 21.3 to 31.5% of the GDP from 2020 to 2030 which is close to RM 1.4 trillion (or US$300,000,000) contribution to the economy of Malaysia. Ocean-based sectors contribute significantly to the economic growth of Malaysia. According to financial models and analysts' expectations the economy was expected to trend around US$380 billion in 2022, and in the following year to about US$440 billion.

The maritime sector (shipping and ports) of Malaysia contributed about 40% of the nation's gross domestic product (GDP) equating to over $24 billion in 2022 and an estimated employment for about four million persons. The fisheries' sector including aquaculture accrued about S13 billion to the economy or about 20% of GDP. The National Oil and Gas production company is a major contributor to the blue economy. The share of the offshore energy has been at the limit. A prominent issue of the blue economy is to understand and better manage the many aspects of coastal and oceanic sustainability, ranging from sustainable fisheries to ecosystem health to pollution. Marine research and education made a modest contribution of about nine per cent or about $5 billion. A substantial contributor to the nation's economy was the marine tourism and recreation bringing in about $16 billion or about 26%. This is a competitive industry in the Southeast Asian region, for example, Bali, Phuket and Singapore but to name a few destinations. A second significant issue of the Blue Economy is to realise that the sustainable management of ocean resources requires collaboration among the different stakeholders and across the public and private enterprise sectors, and on a scale that has not been previously achieved.

8.2 Geographical Setting and Ocean Basin Environment

Chapter Two of the study offered an appraisal of the geographical setting and ocean basin environment in 28 illustrations which were accompanied with narratives on topics such as the limits of the seas and straits adjacent to Malaysia; the nature of the seabed; predominant generalised wind pattern and air flow; sea-surface temperature and barometric pressure; sea-surface ocean currents; wind-roses for coastal stations; coastal geomorphology; and tidal regimes along the Malaysian littoral zone. The chapter alluded to the compound impacts of climate change, land subsidence and local human activities that could potentially lead to higher flood levels and prolonged inundation. Such problems have been experienced along the east and west coasts of Peninsular Malaysia.

With clearly defined and delimited maritime boundary as delineated on and guided by the official charts, the coastal state's fishing vessels and fishers will be able to operate and undertake livelihood activities with confidence. The MMEA and other law enforcement authorities of the state will know the maximum extent of their jurisdiction and when and whom they should prosecute.

8.3 Geographical and Legal Concepts

The geographical and legal concepts employed in the defining, delimiting, and delineating of international political maritime boundaries and in demarcating of terrestrial border/boundary markers and frontier limits (international and provincial) was an important aspect of Chapter Three of this study. During the negotiation processes of maritime boundary delimitation, the parties to an agreement will consult charts and maps of their respective countries to establish an initial alignment of the boundary that each perceives is their sovereign territory. Delegates of each party will then debate the negative and positive points of the stance they wish to adopt. They will then meet with the other party to negotiate a settlement in what they will be on favourable terms in the dispute resolution. This process may take several months or indeed, years for ratification of the Agreement and entry into force as a bilateral Treaty.

The preparation of accurate maps and nautical charts depicting coastline configuration and bathymetry of the adjacent seas and distant areas of the oceans has been an objective of maritime nations.

8.4   Malaysia's Maritime Jurisdictional Limits

In the fourth chapter of this study, 31 illustrations were presented that portray Malaysia's maritime jurisdictional limits and zones for the security of the nation and indicate the limits of the nation's sovereignty. The maps tell a story. The maritime limits are the Internal Waters, Territorial Sea, Contiguous Zone, and Exclusive Economic Zone. One map in this group depicts Malaysia's Extended Continental Shelf claim in the South China Sea, which was submitted in December 2019, to the UN Commission of the Legal Continental Shelf for the Commission's approval. Clearly defined maritime jurisdictional limits and zones are critical for decision-makers as well as for those authorities and agencies charged with implementing and enforcing the laws of the state and in meeting international and regional obligations.

8.5   Traffic Monitoring and Maritime Safety

The narrative in the fifth chapter of this study focused on marine traffic monitoring and maritime safety and air traffic systems over Malaysia's perceived airspace. The safety of mariners and other users of ocean space and the protection of the marine environment from pollution remains paramount. Ten maps showed the density and flow of aeronautical and maritime traffic over and on Malaysia's adjacent seas. One of the maps in this section portrayed the approximate route of submarine communication cables for domestic and international transmissions. In this chapter, a set of seven graphics illustrated the density of ships (maritime transport) transiting the seas adjacent to Malaysia on a specific date and time in July 2021.

The Straits of Malacca and Singapore are the vital and most utilised seaways connecting the Indian Ocean with the South China Sea and Pacific Ocean and together offer the shortest route for tankers trading between the Middle East and Far East Asian countries. Consequently, traffic transiting the region is heavy, conservatively estimated at around 100,000 vessels annually. In addition, there are a considerable number of local vessels engaged in trade across the straits; numerous fishing vessels can be encountered in most areas; naval ships; and many pleasure craft and work boats operate daily. An average of 66 containerships per day transit the Malacca Strait (or about 24,000 annually) and approximately 80% of oil by volume is carried on tankers bound for ports in Northeast Asia.

Malaysia's maritime security is challenged by non-traditional threats, such as illegal trafficking of goods and humans; illegal, unregulated, and unreported fishing (IUU fishing); armed robbery, hijacking and piracy, terrorism; threats to the marine ecosystem, like climate; as well as shipping and land-based pollution and illegal anchoring of alien ships. The Government has over the years enacted legislation and established *Orders* and *Regulations* to manage its maritime space. The plans for increased security along the coasts and borders have taken on greater urgency.

For a near flawless flow of traffic there must be clear and precise communication between agencies at local, national, and international levels; however, the onus always rests with the users—mariners, port operators and government appointed agencies and authorities. A typical example of excellent communications between agencies and stakeholders would be the regular announcements of the location of offshore drilling activities in the hydrocarbon exploration and exploitation industries to ensure safety to navigation.

Figure 8.1 portrays the density of marine traffic, on 10th April 2024, as generalisation and the scale of the graphic permits, within Malaysia's maritime jurisdictional limits. Much of this traffic will undoubtedly be beneficial to Malaysia's Blue Economy, namely generating valuable funds through the import and exports of manufactured goods and commodities and in transhipment fees and Light Dues, import Customs duty to name but a few factors.

Compare and contrast Figs. 8.2 and 8.1 with the images mentioned in Figs. 5.12 and 5.13 respectively. The latter set was captured nearly three years ago. Numerous economical and geopolitical may account for the reduction in marine traffic transiting the seas partially encompassing Malaysia during early April 2024 due to global and political issues that include the war in Ukraine and the Hamas and Israel debacle of October 2023 to July 2023. These events have attracted other actors to get involved, for example, Hezbollah, Houthis and Islamic State (IS) in attacking ships in the Red Sea, Gulf of Aden, Gulf of Oman and Arabian Sea, and by their actions disrupted the global maritime trade by causing backlog in the supply chain, the laying up of ships and the resultant illegal anchoring of ships within the Straits of Malacca and Singapore and in the vicinity of ports of Indonesia and Malaysia. It is worth bringing to the reader's attention the sheer number of ships at anchor (orange circles) within the Singapore Strait as depicted in Fig. 8.2. This body of water fits the terms 'geographic constriction', 'congested seaway' and 'maritime chokepoint' perfectly and the graphic tells a story. The graphic conveys a thousand words! Note: the number of ships (orange circle icons (that are static—stopped and possibly at anchor) awaiting further instructions (Fig. 8.3).

During 2023, Port Klang was rated the 11th busiest port in the world. Domestic and international traffic flow for container handling in the export, import and transhipment categories at Northport and Westport in the Port Klang complex recorded

Fig. 8.1 Marine traffic in the vicinity of Malaysia. Image captured at 19:10:1 on 10 April 2024 (MyST). *Source VesselFinder*

Fig. 8.2 Marine traffic within Singapore Strait. Image captured at 19:08:20 on 10 April 2024 (MyST). *Source VesselFinder*

Fig. 8.3 Marine traffic within central Malacca Strait west of Port Klang. Image captured at 19:11: 35 on 10 April 2024 (my ST). *Source VesselFinder*

a total number of TEUs of 14,061,022. On 8th December 2023, the Government of Malaysia, Port Klang Authority and Westports Malaysia Sdn. Bhd. signed the Third Supplementary Agreement of Privatisation. Westports expect to begin the expansion phase of the planned development sometime during the third quarter of 2024.

8.6 Marine Parks, Protected Zones and Marine Biotic and Mineral Resources

Malaysia's marine parks, protected zones and marine biotic and mineral resources are illustrated with maps and described in brief commentaries in Chapter Six. To mention these topics in the same breath may raise eyebrows, however, it is reasonable to discuss the issues in an academic study within one chapter. In Malaysia, Marine Parks protect and conserve various habitats and aquatic marine life and the waters around Malaysia's 42 islands that are gazetted in accord with the *Fisheries Act* 1985.

Whilst the concept of sustainable development is being promoted, and the Blue Economy and Green Technology are endorsed, there is deep concern for a warming Earth and Climate Change. In Malaysia, Green Technology master plan for 2017–2030 demonstrates the Government's aim to create a low-carbon and resource efficient economy. By the end of 2030, there is anticipation that businesses adopting the green technology will contribute to the nation's GDP.

8.7 Commercial Ports of Malaysia, Sea Lanes and Piracy

Chapter Seven presented an appraisal of the major commercial ports of Malaysia's, the importance of the sea lanes of communication and the impact of acts of armed robbery and piracy at sea. Thirty illustrations are offered in this chapter. In terms of trade movements, the adjacent seas and the ports of the country have contributed significantly to the growth of the Malaysian economy as conduits for the movement of exports to earn much-needed foreign exchange and for imports used in consumption and inputs for the industries of the country. Trade has grown significantly since the country's independence in 1957 and especially since the 1990s when the foundations of modern industrial Malaysian nation had its beginnings.

The commercial ports of Malaysia are ideally located to capture a fair share of the maritime trade that transit the seas of Southeast Asia. This fact is assisted by the indicator that Malaysia readily adopted the initiative taken by the People's Republic of China (PRC) concept of One Belt, One Road (OBOR) or the Belt and Road Initiative (BRI).

China was warmly greeted by Malaysia when signing a Memorandum of Understanding (MOU) relating to the OBOR. The Chinese-funded development projects being undertaken are not for geostrategic purposes, it has been argued, but rather non-strategic in context, for example, investments in the sectors of real estate, reclamation projects in the Johor Strait, for example Forest City, and entertainment and industry. In addition, there are port and railway commercial project developments. However, two major projects, one is the Malacca Gateway in the Malacca Strait and the other at the port of Kuantan and adjacent industrial park being upgraded to attract Chinese industrial investment. Malaysia has harnessed Chinese economic expansion.

Despite this 'cuddly diplomacy' between China and Malaysia, major issues at sea have reared their ugly heads with incidents at James and Luconia Shoals, harassment by the Chinese Coast Guard vessels to workboats operating in the offshore exploration and production platforms, and the occasional Chinese military flight incursions into Malaysian airspace and Chinese marine militia operating in Malaysian waters, however defined.

Malaysia is to be congratulated in delimitating a continental shelf boundary (1969) and territorial sea boundary (1970) with its neighbour, Indonesia. It has delimited maritime boundaries with Brunei, Singapore, and Thailand. It has established aspects of joint developments zones for the exploration and exploitation of hydrocarbon resources each with Brunei, Thailand, and Vietnam.

There remain contentious issues, by mid-July 2024, with Indonesia, the Philippines and Singapore with respect to closing existing gaps in the maritime boundaries, however, termed and settling territorial issues with the Philippines and with China in the South China Sea. Some of these issues impact on the Federal/State level which is to be expected in a democratic and multicultural society.

Maritime transport, sea lanes, illegal anchoring and operation of 'dark ships' and illegal fishing are major issues that are of concern for the law enforcement agencies of Malaysia. Secure and well-defined maritime boundaries that are delineated on accurate and maps are essential and vital to boost the Blue Economy and confidence for the regional states and the international community.

Annex I
Articles of the 1982 Un Law of the Sea Convention

Pertaining to International Jurisdictional Maritime Limits

Article 2   Legal status of the territorial sea, of the air space over the territorial sea and of its bed and subsoil

1. The sovereignty of a coastal State extends, beyond its land territory and internal waters and, in the case of an archipelagic State, its archipelagic waters, to an adjacent belt of sea, described as the territorial sea.
2. This sovereignty extends to the air space over the territorial sea as well as to its bed and subsoil.
3. The sovereignty over the territorial sea is exercised subject to this Convention and to other rules of international law.

Article 3   Breadth of the territorial sea

Every State has the right to establish the breadth of its territorial sea up to a limit not exceeding 12 nautical miles, measured from baselines determined in accordance with this Convention.

Rules for Defining the Territorial Sea Baseline System

Article 4   Outer limit of the territorial sea

The outer limit of the territorial sea is the line every point of which is at a distance from the nearest point of the baseline equal to the breadth of the territorial sea.

Article 5   Normal baseline

Except where otherwise provided in this Convention, the normal baseline for measuring the breadth of the territorial sea is the low-water line along the coast as marked on large-scale charts officially recognised by the coastal State.

Article 6   Reefs

In the case of islands situated on atolls or of islands having fringing reefs, the baseline for measuring the breadth of the territorial sea is the seaward low-water line of the reef, as shown by the appropriate symbol on charts officially recognised by the coastal State.

Article 7   Straight baselines

1. In localities where the coastline is deeply indented and cut into, or if there is a fringe of islands along the coast in its immediate vicinity, the method of straight baselines joining appropriate points may be employed in drawing the baseline from which the breadth of the territorial sea is measured.
2. Where because of the presence of a delta and other natural conditions the coastline is highly unstable, the appropriate points may be selected along the furthest seaward extent of the low-water line and notwithstanding subsequent regression of the low-water line, the straight baselines shall remain effective until changed by the coastal State in accordance with this Convention.
3. The drawing of straight baselines must not depart to any appreciable extent from the general direction of the coast, and the sea areas lying within the lines must be sufficiently closely linked to the land domain to be subject to the regime of internal waters.
4. Straight baselines shall not be drawn to and from low-tide elevations, unless lighthouses or similar installations, which are permanently above sea level, have been built on them or except in instances where the drawing of baselines to and from such elevations has received general international recognition.
5. Where the method of straight baselines is applicable under paragraph 1, account may be taken, in determining particular baselines, of economic interest's peculiar to the region concerned, the reality and the importance of which are clearly evidenced by long usage.
6. The system of straight baselines may not be applied by a State in such a manner as to cut off the territorial sea of another State from the high seas or an exclusive economic zone.

Article 8   Internal Waters

1. Except as provided in Part IV, waters on the landward side of the baseline of the territorial sea form part of the internal waters of the State.

2. Where the establishment of a straight baseline in accordance with the method set forth in Article 7 has the effect of enclosing as internal waters areas which had not previously been considered as such, a right of innocent passage as provided in this Convention shall exist in those waters.

Article 9   Mouths of rivers

If a river flows directly into the sea, the baseline shall be a straight line across the mouth of the river between points on the low-water line of its banks.

Article 10   Bays

1. This article relates only to bays the coasts of which belong to a single State.
2. For the purposes of this Convention, a bay is a well-marked indentation whose penetration is in such proportion to the width of its mouth as to contain land-locked waters and constitute more than a mere curvature of the coast. An indentation shall not, however, be regarded as a bay unless its area is as large as, or larger than, that of the semi-circle whose diameter is a line drawn across the mouth of that indentation.
3. For the purpose of measurement, the area of an indentation is that lying between the low-water mark around the shore of the indentation and a line joining the low-water mark of its natural entrance points. Where, because of the presence of islands, an indentation has more than one mouth, the semi-circle shall be drawn on a line as long as the sum total of the lengths of the lines across the different mouths. Islands within an indentation shall be included as if they were part of the water area of the indentation.
4. If the distance between the low-water marks of the natural entrance points of a bay does not exceed 24 nautical miles, a closing line may be drawn between these two low-water marks, and the waters enclosed thereby shall be considered as internal waters.
5. Where the distance between the low-water marks of the natural entrance points of a bay exceeds 24 nautical miles, a straight baseline of 24 nautical miles shall be drawn within the bay in such a manner as to enclose the maximum area of water that is possible with a line of that length.
6. The foregoing provisions do not apply to so-called "historic" bays, or in any case where the system of straight baselines provided for in article 7 is applied.

Article 11   Ports

For the purpose of delimiting the territorial sea, the outermost permanent harbour works which form an integral part of the harbour system are regarded as forming part of the coast. Off-shore installations and artificial islands shall not be considered as permanent harbour works.

Article 12   Roadsteads

Roadsteads which are normally used for the loading, unloading, and anchoring of ships, and which would otherwise be situated wholly or partly outside the outer limit of the territorial sea, are included in the territorial sea.

Article 13   Low-tide elevations

1. A low-tide elevation is a naturally formed area of land which is surrounded by and
    above water at low tide but submerged at high tide. Where a low-tide elevation is
    situated wholly or partly at a distance not exceeding the breadth of the territorial
    sea from the mainland or an island, the low-water line on that elevation may be
    used as the baseline for measuring the breadth of the territorial sea.
2. Where a low-tide elevation is wholly situated at a distance exceeding the breadth
    of the territorial sea from the mainland or an island, it has no territorial sea of its
    own.

Article 14   Combination of methods for determining baselines

The coastal State may determine baselines in turn by any of the methods provided
for in the foregoing articles to suit different conditions.

Article 15   Delimitation of the territorial sea between States

Where the coasts of two States are opposite or adjacent to each other, neither of the
two States is entitled, failing agreement between them to the contrary, to extend its
territorial sea beyond the median line every point of which is equidistant from the
nearest points on the baselines from which the breadth of the territorial seas of each
of the two States is measured. The above provision does not apply, however, where
it is necessary by reason of historic title or other special circumstances to delimit the
territorial seas of the two States in a way which is at variance therewith.

Article 16   Charts and lists of geographical coordinates

1. The baselines for measuring the breadth of the territorial sea determined in accor-
    dance with articles 7, 9 and 10, or the limits derived therefrom, and the lines of
    delimitation drawn in accordance with articles 12 and 15 shall be shown on charts
    of a scale or scales adequate for ascertaining their position. Alternatively, a list
    of geographical coordinates of points, specifying the geodetic datum, may be
    substituted.
2. The coastal State shall give due publicity to such charts or lists of geographical
    coordinates and shall deposit a copy of each such chart or list with the Secretary-
    General of the United Nations.

Article 33   Contiguous Zone

1. In a zone contiguous to its territorial sea, described as the contiguous zone, the
    coastal State may exercise the control necessary to:

    (a) prevent infringement of its customs, fiscal, immigration or sanitary laws and
        regulations within its territory or territorial sea;
    (b) punish infringement of the above laws and regulations committed within its
        territory or territorial sea.

2. The contiguous zone may not extend beyond 24 nautical miles from the baselines from which the breadth of the territorial sea is measured.

Article 55   Specific legal regime of the Exclusive Economic Zone

The exclusive economic zone is an area beyond and adjacent to the territorial sea, subject to the specific legal regime established in this Part, under which the rights and jurisdiction of the coastal State and the rights and freedoms of other States are governed by the relevant provisions of this Convention.

Article 56   Rights, jurisdiction and duties of the coastal State in the exclusive economic zone

1. In the exclusive economic zone, the coastal State has:

   (a) sovereign rights for the purpose of exploring and exploiting, conserving and managing the natural resources, whether living or non-living, of the waters superjacent to the seabed and of the seabed and its subsoil, and with regard to other activities for the economic exploitation and exploration of the zone, such as the production of energy from the water, currents and winds;

   (b) jurisdiction as provided for in the relevant provisions of this Convention with regard to:
      (i)   the establishment and use of artificial islands, installations and structures;
      (ii) marine scientific research;
      (iii) the protection and preservation of the marine environment;

   (c) other rights and duties provided for in this Convention.

2. In exercising its rights and performing its duties under this Convention in the exclusive economic zone, the coastal State shall have due regard to the rights and duties of other States and shall act in a manner compatible with the provisions of this Convention.

3. The rights set out in this article with respect to the seabed and subsoil shall be exercised in accordance with Part VI.

Article 57   Breadth of the Exclusive Economic Zone

The exclusive economic zone shall not extend beyond 200 nautical miles from the baselines from which the breadth of the territorial sea is measured.

Article 58   Rights and duties of other States in the Exclusive Economic Zone

1. In the exclusive economic zone, all States, whether coastal or land-locked, enjoy, subject to the relevant provisions of this Convention, the freedoms referred to in Article 87 of navigation and overflight and of the laying of submarine cables and pipelines, and other internationally lawful uses of the sea related to these freedoms, such as those associated with the operation of ships, aircraft and submarine cables and pipelines, and compatible with the other provisions of this Convention.

2. Articles 88–115 and other pertinent rules of international law apply to the exclusive economic zone in so far as they are not incompatible with this Part.

3. In exercising their rights and performing their duties under this Convention in the exclusive economic zone, States shall have due regard to the rights and duties of the coastal State and shall comply with the laws and regulations adopted by the coastal State in accordance with the provisions of this Convention and other rules of international law in so far as they are not incompatible with this Part.

Article 76 Definition of the Continental Shelf

1. The continental shelf of a coastal State comprises the seabed and subsoil of the submarine areas that extend beyond its territorial sea throughout the natural prolongation of its land territory to the outer edge of the continental margin, or to a distance of 200 nautical miles from the baselines from which the breadth of the territorial sea is measured where the outer edge of the continental margin does not extend up to that distance.

2. The continental shelf of a coastal State shall not extend beyond the limits provided for in paragraphs 4–6.

3. The continental margin comprises the submerged prolongation of the land mass of the coastal State and consists of the seabed and subsoil of the shelf, the slope, and the rise. It does not include the deep ocean floor with its oceanic ridges or the subsoil thereof.

4. (a) For the purposes of this Convention, the coastal State shall establish the outer edge of the continental margin wherever the margin extends beyond 200 nautical miles from the baselines from which the breadth of the territorial sea is measured, by either:

   (i) a line delineated in accordance with paragraph 7 by reference to the outer-most fixed points at each of which the thickness of sedimentary rocks is at least 1 per cent of the shortest distance from such point to the foot of the continental slope; or.

   (ii) a line delineated in accordance with paragraph 7 by reference to fixed points not more than 60 nautical miles from the foot of the continental slope.

   (b) In the absence of evidence to the contrary, the foot of the continental slope shall be determined as the point of maximum change in the gradient at its base.

5. The fixed points comprising the line of the outer limits of the continental shelf on the seabed, drawn in accordance with paragraph 4 (a) (i) and (ii), either shall not exceed 350 nautical miles from the baselines from which the breadth of the territorial sea is measured or shall not exceed 100 nautical miles from the 2500-m isobath, which is a line connecting the depth of 2500 m.

6. Notwithstanding the provisions of paragraph 5, on submarine ridges, the outer limit of the continental shelf shall not exceed 350 nautical miles from the

baselines from which the breadth of the territorial sea is measured. This paragraph does not apply to submarine elevations that are natural components of the continental margin, such as its plateaux, rises, caps, banks, and spurs.

7. The coastal State shall delineate the outer limits of its continental shelf, where that shelf extends beyond 200 nautical miles from the baselines from which the breadth of the territorial sea is measured, by straight lines not exceeding 60 nautical miles in length, connecting fixed points, defined by coordinates of latitude and longitude.

8. Information on the limits of the continental shelf beyond 200 nautical miles from the baselines from which the breadth of the territorial sea is measured shall be submitted by the coastal State to the Commission on the Limits of the Continental Shelf set up under Annex II on the basis of equitable geographical representation. The Commission shall make recommendations to coastal States on matters related to the establishment of the outer limits of their continental shelf. The limits of the shelf established by a coastal State on the basis of these recommendations shall be final and binding.

9. The coastal State shall deposit with the Secretary-General of the United Nations charts and relevant information, including geodetic data, permanently describing the outer limits of its continental shelf. The Secretary-General shall give due publicity thereto.

10. The provisions of this article are without prejudice to the question of delimitation of the continental shelf between States with opposite or adjacent coasts.

Article 77   Rights of the coastal State over the Continental Shelf

1. The coastal State exercises over the continental shelf sovereign rights for the purpose of exploring it and exploiting its natural resources.

2. The rights referred to in paragraph 1 are exclusive in the sense that if the coastal State does not explore the continental shelf or exploit its natural resources, no one may undertake these activities without the express consent of the coastal State.

3. The rights of the coastal State over the continental shelf do not depend on occupation, effective or notional, or on any express proclamation.

4. The natural resources referred to in this Part consist of the mineral and other non-living resources of the seabed and subsoil together with living organisms belonging to sedentary species, that is to say, organisms which, at the harvestable stage, either are immobile on or under the seabed or are unable to move except in constant physical contact with the seabed or the subsoil.

Article 78   Legal status of the superjacent waters and air space and the rights and freedoms of other States

1. The rights of the coastal State over the continental shelf do not affect the legal status of the superjacent waters or of the air space above those waters.

2. The exercise of the rights of the coastal State over the continental shelf must not infringe or result in any unjustifiable interference with navigation and other rights and freedoms of other States as provided for in this Convention.

Article 79   Submarine cables and pipelines on the Continental Shelf

1. All States are entitled to lay submarine cables and pipelines on the continental shelf, in accordance with the provisions of this article.
2. Subject to its right to take reasonable measures for the exploration of the continental shelf, the exploitation of its natural resources and the prevention, reduction, and control of pollution from pipelines, the coastal State may not impede the laying or maintenance of such cables or pipelines.
3. The delineation of the course for the laying of such pipelines on the continental shelf is subject to the consent of the coastal State.
4. Nothing in this Part affects the right of the coastal State to establish conditions for cables or pipelines entering its territory or territorial sea, or its jurisdiction over cables and pipelines constructed or used in connection with the exploration of its continental shelf or exploitation of its resources or the operations of artificial islands, installations, and structures under its jurisdiction.
5. When laying submarine cables or pipelines, States shall have due regard to cables or pipelines already in position. In particular, possibilities of repairing existing cables or pipelines shall not be prejudiced.

Article 80   Artificial islands, installations, and structures on the continental shelf

Article 60 applies *mutatis mutandis* to artificial islands, installations, and structures on the continental shelf.

Article 81   Drilling on the continental shelf

The coastal State shall have the exclusive right to authorise and regulate drilling on the continental shelf for all purposes.

Article 82   Payments and contributions with respect to the exploitation of the continental shelf beyond 200 nautical miles

1. The coastal State shall make payments or contributions in kind in respect of the exploitation of the non-living resources of the continental shelf beyond 200 nautical miles from the baselines from which the breadth of the territorial sea is measured.
2. The payments and contributions shall be made annually with respect to all production at a site after the first five years of production at that site. For the sixth year, the rate of payment or contribution shall be 1 per cent of the value or volume of production at the site. The rate shall increase by 1 per cent for each subsequent year until the twelfth year and shall remain at 7 per cent thereafter. Production does not include resources used in connection with exploitation.

3. A developing State which is a net importer of a mineral resource produced from its continental shelf is exempt from making such payments or contributions in respect of that mineral resource.

4. The payments or contributions shall be made through the Authority, which shall distribute them to States Parties to this Convention, on the basis of equitable sharing criteria, taking into account the interests and needs of developing States, particularly the least developed and the land-locked among them.

Article 83   Delimitation of the continental shelf between States with opposite or adjacent coasts

1. The delimitation of the continental shelf between States with opposite or adjacent coasts shall be effected by agreement on the basis of international law, as referred to in Article 38 of the Statute of the International Court of Justice, in order to achieve an equitable solution.

2. If no agreement can be reached within a reasonable period of time, the States concerned shall resort to the procedures provided for in Part XV.

3. Pending agreement as provided for in paragraph 1, the States concerned, in a spirit of understanding and cooperation, shall make every effort to enter into provisional arrangements of a practical nature and, during this transitional period, not to jeopardise or hamper the reaching of the final agreement. Such arrangements shall be without prejudice to the final delimitation.

4. Where there is an agreement in force between the States concerned, questions relating to the delimitation of the continental shelf shall be determined in accordance with the provisions of that agreement.

Article 84   Charts and lists of geographical coordinates

1. Subject to this Part, the outer limit lines of the continental shelf and the lines of delimitation drawn in accordance with article 83 shall be shown on charts of a scale or scales adequate for ascertaining their position. Where appropriate, lists of geographical coordinates of points, specifying the geodetic datum, may be substituted for such outer limit lines or lines of delimitation.

2. The coastal State shall give due publicity to such charts or lists of geographical coordinates and shall deposit a copy of each such chart or list with the Secretary-General of the United Nations and, in the case of those showing the outer limit lines of the continental shelf, with the Secretary-General of the Authority.

Article 85   Tunnelling

This Part does not prejudice the right of the coastal State to exploit the subsoil by means of tunnelling, irrespective of the depth of water above the subsoil.

Annex II
Primary Documents Relating to Maritime Boundary Delimitation

1958 No. 1517 Overseas Territories.

The North Borneo (Definition of Boundaries) Order in Council, 1958.

1958 No. 1518 Overseas Territories.

The Sarawak (Definition of Boundaries) Order in Council, 1958.

1969 Agreement between the Government of Malaysia and the Government of Indonesia on the Delimitation of the Continental Shelves between the Two Countries.

1970 Treaty between the Republic of Indonesia and Malaysia on Determination of Boundary Lines of Territorial Waters of the Two Nations at the Strait of Malacca.

Agreement between the Government of the Republic of Indonesia, the Government of Malaysia and the Government of the Kingdom of Thailand Relating to the Delimitation of the Continental Shelf Boundaries in the Northern Part of the Strait of Malacca.

Memorandum of Understanding between Malaysia and the Kingdom of Thailand on the Delimitation of the Continental Shelf Boundary between the Two Countries in the Gulf of Thailand.

1927 Agreement between the Government of Malaysia and the Government of the Republic of Singapore to Delimit Precisely the Territorial Waters Boundary in Accordance with the Straits Settlement and Johore Territorial Waters Agreement.

Memorandum of Understanding between Malaysia and the Socialist Republic of Vietnam for the Exploration and Exploitation of Petroleum in a Defined Area of the Continental Shelf Involving the two Countries. 5th June 1992.

Sovereignty over Pulau Ligitan and Pulau Sipadan (Decision of the ICJ in the case of Indonesia/Malaysia).

© The Editor(s) (if applicable) and The Author(s), under exclusive license
to Springer Nature Switzerland AG 2025
V. L. Forbes, *Malaysia's Maritime Jurisdictional Limits*,
https://doi.org/10.1007/978-3-031-78783-6

Order in the Case concerning Land Reclamation by Singapore in and around the Straits of Johor (Malaysia vs. Singapore).

Press Release 2006/38 of the International Court of Justice Sovereignty over Pedra Branca (Pulau Batu Puteh), Middle Rocks and South Ledge (Malaysia/Singapore).

1958 No. 1517

Overseas Territories

The North Borneo (Definition of Boundaries) Order in Council, 1958

Made … 11th September 1958

At the court at Balmoral, the 11th day of September 1958

Present,

The Queen's Most Excellent Majesty in Council

Her Majesty, in pursuance of the power conferred upon Her by the Colonial Boundaries Act, 1895(a), and of all other powers enabling Her in that behalf, is pleased, by and with the advice of Her Privy Council, to order, and it is hereby ordered, as follows:

1. This Order may be cited as the North Borneo (Definition of Boundaries) Order in Council, 1958.
2. (1) The boundary of the waters and continental shelf of the Colony of North Borneo in and in the neighbourhood of Brunei Bay shall follow a series of straight lines joining in sequence the following positions:

    (i)   the termination of the land boundary between the Colony of North Borneo and the Colony of Sarawak in the mouth of the Benkulit River (Latitude 4 degrees, 59 min and 10 sec North, Longitude 115 degrees, 26 min and 48 sec East, approximately;

    (ii)   a position bearing 070 degrees, distant 12.65 miles from Sapo Point light-structure on Pulau Muara;

    (iii)   a position bearing 061 3/4 degrees, distant 13.1 miles from Sapo Point light structure;

    (iv)   a position bearing 050 degrees, distant 10.5 miles from Sapo Point light-structure;

    (v)   a position bearing 045 1/2 degrees, distant 8.65 miles from Sapo Point light-structure;

    (vi)   a position bearing 350 degrees, distant 7.40 miles from Sapo Point light-structure;

    (vii)   a position bearing 009 degrees, distant 4.65 miles from Pelong Rocks light-structure;

    (viii)   a position bearing 330 1/4 degrees, distant 8.55 miles from Pelong Rocks light- structure;

    (ix) a position bearing 311 degrees, distant 17.4 miles from Pelong Rocks light-structure;

    (x) a position bearing 310 3/4 degrees, distant 20.4 miles from Pelong Rocks light- structure; and

    (xi) the intersection with the 100-fathom depth contour of a straight line drawn in a direction of 316 degrees from the position given in paragraph (x) of this subsection.

(2) In this section

    (i) references to Sapo Point light-structure and Pelong Rocks light-structure are references to those light-structures as shown on Admiralty Chart Number 1844 at the date of this Order (their respective positions being approximately Latitude 4 degrees, 59 min and 45 sec North, Longitude 115 degrees, 7 min and 45 sec East, and Latitude 5 degrees, 4 min and 45 sec North, Longitude 115 degrees, 3 min and 9 sec East);

    (ii) references to bearings are references to bearings by the true compass reckoned clockwise from 000 degrees (North) to 359 degrees; and

    (iii) references to miles are references to sea miles *(nautical miles M)*, each comprising one sixtieth of a degree of latitude.

3. Nothing in this Order shall be construed as prescribing boundaries of waters beyond the territorial waters of the Colony of North Borneo and above the continental shelf of the Colony or as affecting the character as high seas of any such waters.

W.G. Agnew.

1958 No. 1518

Overseas Territories

The Sarawak (Definition of Boundaries) Order in Council, 1958

Made ... 11th September 1958

At the Court at Balmoral, the 11th day of September 1958
    Present,

The Queen's Most Excellent Majesty in Council

Her Majesty, in pursuance of the powers conferred upon Her by the Colonial Boundaries Act, 1895(a), and of all other powers enabling Her in that behalf, is pleased, by and with the advice of Her Privy Council, to order, and it is hereby ordered, as follows:

1. This Order may be cited as the Sarawak (Definition of Boundaries) Order in Council, 1958.

2. (1) The boundary of the waters and continental shelf of the Colony of Sarawak in Brunei Bay and the approach to Sungei Brunei shall follow a series of straight lines joining in sequence the following positions:

   (i) the termination of the land boundary between the Colony of North Borneo and the Colony of Sarawak in the mouth of the Benkulit River (Latitude 4 degrees, 59 min and 10 s North, Longitude 115 degrees, 26 min and 48 s East, approximately);

   (ii) a position bearing 070 degrees, distant 12.65 miles from Sapo Point light-structure.

   (iii) a position bearing 061 3/4 degrees, distant 13.1 miles from Sapo Point light-structure.

   (iv) a position bearing 050 degrees, distant 10.5 miles from Sapo Point light-structure.

   (v) a position bearing 031 degrees, distant 8.35 miles from Sunda Spit beacon.

   (vi) a position bearing 025 3/4 degrees, distant 5.77 miles from Sunda Spit beacon.

   (vii) a position bearing 010 degrees, distant 3.47 miles from Sunda Spit beacon.

   (viii) a position bearing 357 degrees, distant 2.90 miles from Sunda Spit beacon.

   (ix) a position bearing 340 1/4 degrees, distant 2.15 miles from Sunda Spit beacon,

   (x) a position bearing 321 degrees, distant 1.67 miles from Sunda Spit beacon.

   (xi) a position bearing 287 1/2 degrees, distant 1.37 miles from Sunda Spit beacon.

   (xii) a position bearing 242 1/2 degrees, distant 2.32 miles from Sunda Spit beacon.

   (xiii) a position bearing 223 1/2 degrees, distant 3.25 miles from Sunda Spit beacon.

   (xiv) a position bearing 221 degrees, distant 3.40 miles from Sunda Spit beacon.

   (xv) a position bearing 216 1/2 degrees, distant 3.45 miles from Sunda Spit beacon.

   (xvi) a position bearing 203 1/2 degrees, distant 3.49 miles from Sunda Spit beacon.

   (xvii) a position bearing 198 degrees, distant 3.80 miles from Sunda Spit beacon.

   (xviii) the termination of the land boundary between the Colony of Sarawak and the State of Brunei in approximate position bearing 191 degrees, distant 4.05 miles from Sunda Spit beacon.

2. (2) In this section the references to Sapo Point light-structure and to Sunda Spit beacon are references respectively, to that light-structure on Pulau Muara and to

the red and white beacon on Sunda Spit as shown on Admiralty Chart Number 1844 at the date of this Order (their respective positions being approximately Latitude 4 degrees, 59 min and 45 sec North, Longitude 115 degrees, 7 min and 45 sec East, and Latitude 4 degrees, 58 min and 48 sec North, Longitude 115 degrees and 10 min East).

3. (1) In this section "the beacon" means the northern of the western pair of beacons (Latitude 4 degrees, 51 min and 24 sec North, Longitude 115 degrees, 2 min and 54 sec East, approximately) shown on Admiralty Chart Number 1844 at the date of this Order and marking the approach channel to Batang Limbang.

(2) The boundary of the waters and continental shelf of the Colony of Sarawak in the approach to Batang Limbang shall follow a series of straight lines joining in sequence the following positions:

   (i) the termination of the land boundary between the Colony of Sarawak and the State of Brunei at the northern extremity of the drying mud flat extending from Mau Siarau in approximate position bearing 195 degrees, distant 0.75 miles from the beacon.

   (ii) a position bearing 114 degrees, distant 0.35 miles from the beacon.

   (iii) a position bearing 094 degrees, distant 0.44 miles from the beacon.

   (iv) a position bearing 077 degrees, distant 0.72 miles from the beacon.

   (v) a position bearing 052 1/2 degrees, distant 1.40 miles from the beacon.

   (vi) a position bearing 035 1/2 degrees, distant 1.20 miles from the beacon.

   (vii) a position bearing 018 1/2 degrees, distant 1.13 miles from the beacon.

   (viii) a position bearing 001 degrees, distant 1.20 miles from the beacon.

   (ix) a position bearing 353 degrees, distant 1.25 miles from the beacon; and

   (x) the termination of the land boundary between the Colony of Sarawak and the State of Brunei in approximate position bearing 344 1/2 degrees, distant 1.48 miles from the beacon.

Definition of boundary in the neighbourhood of Tanjong Baram

4. (1) In this section "the light-structure" means the light-structure at Tanjong Baram (Latitude 4 degrees, 35 min and 45 sec North, Longitude 113 degrees, 58 min and 30 sec East, approximately) shown on Admiralty Chart Number 2109 at the date of this Order.

(2) The boundary of the waters and continental shelf of the Colony of Sarawak in the neighbourhood of Tanjong Baram shall be as follows:

   (i) from the termination at the coast of the land boundary between the Colony of Sarawak and the State of Brunei in approximate position bearing 090 degrees, distant 5.9 miles from the light-structure in a straight line to a position bearing 058 degrees, distant 6.8 miles from the light-structure.

(ii) thence in a straight line to a position bearing 055 degrees, distant 7.1 miles from the light-structure.

(iii) thence along the arc of a circle, radius 4.6 miles, centred upon a point bearing 016 degrees, distant 4.8 miles from the light-structure to a position bearing 030 degrees, distant 9.1 miles from the light-structure; and

(iv) thence in a straight line in a direction of 314 degrees to the intersection of this line with the 100-fathom depth contour.

Interpretation

5. In this Order

(i) references to bearings are references to bearings by the true compass reckoned clockwise from 000 degrees (North) to 359 degrees; and

(ii) references to miles are references to sea miles, each comprising one-sixtieth of a degree of Latitude.

Order not to affect the high seas.

6. Nothing in this Order shall be construed as prescribing boundaries of waters beyond the territorial waters of the Colony of Sarawak and above the continental shelf of the Colony or as affecting the character as high seas of any such waters.

WG. Agnew.

Agreement between the Government of Malaysia and the Government of Indonesia on the Delimitation of the Continental Shelves between the Two Countries. Done at Kuala Lumpur on 27 October 1969

Article I

(1) The boundaries of the Malaysian and the Indonesian continental shelves in the Straits of Malacca and the South China Sea are the straight lines connecting the points specified in column 1 below whose coordinates are specified opposite those points in columns 2 and 3 below:

A. *In the Straits of Malacca:*

(1)	(2)	(3)
Point	Longitude E	Latitude N
1	98° 17.5'	05° 27.0'
2	98° 41.5'	04° 55.7'
3	99° 43.6'	03° 59.6'
4	99° 55.0'	03° 47.4'
5	101° 12.1'	02° 41.5'
6	101° 46.5'	02° 15.4'
7	102° 13.4'	01° 55.2'

(continued)

(continued)

(1)	(2)	(3)
Point	Longitude E	Latitude N
8	102° 35.0′	01° 41.2′
9	103° 03.9′	01° 19.5′
10	103° 22.8′	01° 15.0′

B. *In the South China Sea (Western Side—Off the East Coast of West Malaysia):*

(1)	(2)	(3)
Point	Longitude E	Latitude N
11	104° 29.5′	01° 23.9′
12	104° 53.0′	01° 38.0′
13	105° 05.2′	01° 54.4′
14	105° 01.2′	02° 22.5′
15	104° 51.5′	02° 55.2′
16	104° 46.5′	03° 50.1′
17	104° 51.9′	04° 03.0′
18	105° 28.8′	05° 04.7′
19	105° 47.1′	05° 40.6′
20	105° 49.2′	06° 05.8′

C. *In the South China Sea (Eastern Side—Off the Coast of Sarawak):*

(1)	(2)	(3)
Point	Longitude E	Latitude N
21	109° 38.8′	02° 05.0′
22	109° 54.5′	03° 00.0′
23	110° 02.0′	04° 40.0′
24	109° 59.0′	05° 31.2′
25	109° 38.6′	06° 18.2′

(2) The coordinates of the points specified in Paragraph (1) are geographical coordinates and the straight lines connecting them are indicated on the chart attached as Annexure "A" to this Agreement.

(3) The actual location of the above-mentioned points at sea shall be determined by a method to be mutually agreed upon by the competent authorities of the two Governments.

(4) For the purposes of paragraph (3) "competent authorities" in relation to Malaysia means the Pengarah, Pemetaan Negara, Malaysia and includes any person authorized by him and in relation to the Republic of Indonesia, the Direktur, Direktorat Hidrografi Angkatan Laut, Republik Indonesia and includes any person authorized by him.

Article II

Each Government hereby undertakes to ensure that all the necessary steps shall be taken at the domestic level to comply with the terms of this Agreement.

Article III

This Agreement shall not in any way affect any future agreement which may be entered into between the two Governments relating to the delimitation of the territorial sea boundaries between the two Countries.

Article IV

If any single geological petroleum or natural gas structure extends across the straight lines referred to in Article I and the part of such structure which is situated on one side of the said lines is exploitable, wholly or in part, from the other side of the said lines, the two Governments will seek to reach agreement as to the manner in which the structure shall be most effectively exploited.

Article V

Any dispute between the two Governments arising out of the interpretation or implementation of this Agreement shall be settled peacefully by consultation or negotiation.

Article VI

This Agreement shall be ratified in accordance with the constitutional requirements of the two Countries.

Article VII

This Agreement shall enter into force on the date of the exchange of the instruments of ratification.

Treaty between the Republic of Indonesia and Malaysia on Determination of Boundary Lines of Territorial Waters of the Two Nations at the Strait of Malacca

The Republic of Indonesia and Malaysia,

OBSERVING that coastlines of the two countries confront each other at the Strait of Malacca and the width of the territorial waters of the respective countries is 12 nautical miles.

DESIRIOUS to fortify the friendly tie which has long bound the two countries.

DESIRIOUS ALSO to determine the boundary lines of territorial waters of the two countries at the narrow part of the Strait of Malacca, bounded:

a. In the North by the line which connects Tandjung Tbu, Latitude 02° 51.1′ N, Longitude 101° 16.9′ E to Point 1, Lat. 02° 51.6′ N, Long. 101° 00.2′ E to Batu Mandi Isle Lat. 02° 52.2′ N, Long. 100° 41.0′ E and

b. In the South by the line which connects Tandjung Piai, Lat. 01° 16.2′ N, Long. 103° 30.5′ E to Point No. 8, lat. 01° 15.0′ N, Lat. 103° 22.8′ E to Iju Ketjil Isle Lat. 01° 11.2′ N, Long. 103° 21.0′ E and Tandjung Kedabu, Lat. 01° 05.9′ N, Long. 102° 58.5′ E.

HAVE APPROVED AS FOLLOWS:

Article I

(1) Without curtailment of provision in Section (2) of this Article, boundary lines of territorial waters of Indonesia and Malaysia at the Strait of Malacca in areas as stated in the preamble of this Treaty shall be the line at the centre drawn from base lines of the respective parties in said areas.

(2) (a) Except that which is stated in sub (b), Section (2) of the Article, co-ordinates of points of said boundary lines shall be as follows:

Point 1	101° 00.2′ E	02° 51.6′ N
Point 2	101° 12.1′ E	02° 41.5′ N
Point 3	101° 46.5′ E	02° 15.4′ N
Point 4	102° 13.4′ E	01° 55.2′ N
Point 5	102° 35.0′ E	01° 41.2′ N
Point 6	103° 02.1′ E	01° 19.1′ N
Point 7	103° 03.9′ E	01° 19.6′ N
Point 8	103° 22.8′ E	01° 16.0′ N

   (b) Point 6 shall not apply to Malaysia.

(3) Co-ordinates of points stipulated in Sect. (2) shall be geographical co-ordinates and boundary lines which connect them as shown on the map attached to this Treaty as Attachment 'A'.

(4) Actual sites of points stated above shall be determined through means jointly approved by authorized officials of both parties.

(5) What are referred to by "authorized officials" stated in Section (4) shall be for Indonesia the Director of Naval Hydrography of the Republic of Indonesia, including every person so authorized, and for Malaysia, Director of Mapping of the State of Malaysia including every person so authorized.

Article II

The respective parties herewith shall promise assurances that every necessary measure shall be taken in their countries to comply to provisions inserted in this Treaty.

Article III

Any dispute which may raise (*sic*) between the two parties from interpretation or implementation of this Treaty shall be settled amicably via consultation or negotiation.

Article IV

This Treaty shall be legitimate in accordance with the constitutional procedure of the respective countries.

Article V

This Treaty shall be effective as of the date of exchange of Charters of Legalization.

DONE IN DUPLICATES in Kuala Lumpur on March 17, 1970, in Indonesian, Malaysian and English languages. In case of differences of interpretations between the drafts, the English draft shall be decisive.

FOR THE REPUBLIC OF INDONESIA

ADAM MALIK, FOREIGN MINISTER

FOR MALAYSIA: HAJI ABDUL RAZAK BIN DATO HUSSEIN

VICE-PRIME MINISTER

Agreement between the Government of the Republic of Indonesia, the Government of Malaysia and the Government of the Kingdom of Thailand Relating to the Delimitation of the Continental Shelf Boundaries in the Northern Part of the Strait of Malacca

THE GOVERNMENT OF THE REPUBLIC OF INDONESIA, THE GOVERNMENT OF MALAYSIA AND THE GOVERNMENT OF THE KINGDOM OF THAILAND,

DESIRING to strengthen the existing historical bonds of friendship between the three Countries,

AND DESIRING to establish the boundaries of the continental shelves of the three Countries in the northern part of the Strait of Malacca. HAVE AGREED as follows:

Article I

The boundaries of the continental shelves of Indonesia, Malaysia and Thailand in the northern part of the Strait of Malacca shall start from a point whose co-ordinates are Latitude 5° 57.0′ N Longitude 98° 01.5′ E (hereinafter referred to as 'the Common Point').

The boundary of the continental shelves of Indonesia and Thailand shall be formed by the straight lines drawn from the Common Point in a north-westerly direction to a point whose co-ordinates are Latitude 6° 21.8′ N Longitude 97° 54.0′ E and from there in a westerly direction to a point whose coordinates are Latitude 7° 05.8′ N

Longitude 96° 36.5′ E as specified in the Agreement between the Government of the Republic of Indonesia and the Government of the Kingdom of Thailand relating to the delimitation of a continental shelf boundary between the two Countries in the northern part of the Strait of Malacca and in the Andaman Sea, signed at Bangkok on the 17th day of December, 1971.

The boundary of the continental shelves of Indonesia and Malaysia shall be formed by the straight line drawn from the Common Point in a southward direction to Point 1 specified in the Agreement signed at Kuala Lumpur on the 27th day of October, 1969 between the Government of the Republic of Indonesia and the Government of Malaysia relating to the delimitation of the continental shelves between the two Countries whose co-ordinates are Latitude 5° 27.0′ N Longitude 98° 17.5′ E.

The boundary of the continental shelves of Malaysia and Thailand shall be formed by the straight lines drawn from the Common Point in a northeasterly direction to a point whose co-ordinates are Latitude 6° 18.0′ N Longitude 99° 06.7′ E and from there in a south-easterly direction to a point whose co-ordinates are Latitude 6° 16.3′ N Longitude 99° 19.3′ E and from there in a north-easterly direction to a point whose co-ordinates are Latitude 6° 18.4′ N Longitude 99° 27.5′ E.

The co-ordinates of the points specified above are geographical coordinates derived from the British Admiralty Charts No. 793 and No. 830 and the straight lines connecting them are indicated on the chart attached as Annexure 'A' to this Agreement.

The actual location of the above-mentioned points at sea shall be determined by a method to be mutually agreed upon by the competent authorities of the respective Governments concerned.

For the purposes of paragraph (6), 'competent authorities' in relation to the Republic of Indonesia means the Chief of the Co-ordinating Body for National Survey and Mapping, Republic of Indonesia, and includes any person authorised by him; in relation to Malaysia the Director of National Mapping, Malaysia, and includes any person authorised by him; and in relation to the Kingdom of Thailand the Director of the Hydrographic Department, Thailand, and includes any person authorised by him.

Article II

Each Government hereby undertakes to ensure that all the necessary steps shall be taken at the domestic level to comply with the terms of this Agreement.

Article III

If any single geological petroleum or natural gas structure extends across the boundary line or lines referred to in Article I and the part of such structure which is situated on one side of the said line or lines is exploitable, wholly or in part from the other side or sides of the said line or lines, the Governments concerned shall seek to reach agreement as to the manner in which the structure will be most effectively exploited.

Article IV

Any dispute between the three Governments arising out of the interpretation or implementation of this Agreement shall be settled peacefully by consultation or negotiation.

Article V

This Agreement shall be clarified in accordance with the legal requirements of the three Countries.

Article VI

This Agreement shall enter into force on the date of the exchange of the Instruments of Ratification.

IN WITNESS WHEREOF the undersigned being duly authorised thereto by their respective Governments, have signed this Agreement.

DONE IN TRIPLICATE at Kuala Lumpur the 21st day of December 1971 in Indonesian, Malaysia, Thai and English languages. In the event of any conflict between the texts, the English text shall prevail.

For the Government of the Republic of Indonesia	For the Government of Malaysia	For the Government of the Kingdom of Thailand
(signed) Prof	*(signed)* Tan Sri Haji	*(signed)* Mr. Vija Sethaput,
Dr. Soemantri Brodjonegoro, Minister of Mines, Republic of Indonesia	Abdul Kadir bin Yusof, Attorney General Malaysia	Under-Secretary of State for National Development, in charge of the Ministry of National Development

Memorandum of Understanding between Malaysia and the Kingdom of Thailand on the Delimitation of the Continental Shelf Boundary

between the Two Countries in the Gulf of Thailand

MALAYSIA AND THE KINGDOM OF THAILAND,

DESIRING to strengthen the existing historical bonds of friendship between the two Countries,

AND DESIRING to establish the continental shelf boundary of the two Countries in the Gulf of Thailand,

HAVE AGREED AS FOLLOWS:

Article I

(1) The boundary of the continental shelf in the Gulf of Thailand between Malaysia and the Kingdom of Thailand shall consist of straight lines joining in the order specified below the points whose co-ordinates are:

    (i) Latitude 6° 27'.5 N; Longitude 102° 10'.0 E
    (ii) Latitude 6° 27'.8 N; Longitude 102° 09'.6 E
    (iii) Latitude 6° 50'.0 N; Longitude 102° 21'.2 E

(2) The co-ordinates of point (ii) above have been determined by reference to a point whose co-ordinates are Latitude 6° 16'.6 N Longitude 102° 03'.8 E, this point being the former position of Kuala Tabar under the Boundary Protocol annexed to the Treaty between Siam and Great Britain signed at Bangkok on the 10th March 1909.

Article II

(1) The co-ordinates of the points specified in Article 1 above are geographical co-ordinates derived from the British Admiralty Chart No. 3961 and the boundary lines connecting them are indicated on the chart attached as an Annexure to this Memorandum.
(2) The actual location of these points at sea and of the lines connecting them shall be determined by a method to be mutually agreed upon by the competent authorities of the two Countries.
(3) For the purpose of paragraph (2) of this Article, the term "competent authorities" in relation to Malaysia shall mean the Director of National Mapping and include any person authorised by him, and in relation to the Kingdom of Thailand the Director of the Hydrographic Department and include any person authorised by him.

Article III

The Governments of the two Countries shall continue negotiations to complete the delimitation of the continental shelf boundary of the two Countries in the Gulf of Thailand.

Article IV

If any single geological petroleum or natural gas structure or field, or any mineral deposit of whatever character, extends across the boundary lines referred to in Article I, the two Governments shall communicate to each other all information in this regard and shall seek to reach agreement as to the manner in which the structure, field or deposit will be most effectively exploited; and all expenses incurred and benefits derived therefrom shall be equitably shared.

Article V

Any difference or dispute arising out of the interpretation or implementation of the provisions of this Memorandum shall be settled peacefully by consultation or negotiation between the Parties.

Article VI

This Memorandum shall be ratified in accordance with the constitutional requirements of each Country. It shall enter into force on the date of the exchange of the Instruments of Ratification.

DONE IN DUPLICATE at Kuala Lumpur, the Twenty-fourth Day of October, One Thousand Nine Hundred and Seventy-nine in the Malaysian, Thai and English languages. In the event of any conflict between the texts, the English text shall prevail.

FOR MALAYSIA: Datuk Hussein Onn, Prime Minister

FOR THE KINGDOM OF THAILAND: General Tun Kriangsak Chomanan: Prime Minister

Memorandum of Understanding between the Kingdom of Thailand and Malaysia on the Establishment of a Joint Authority for the Exploitation of the Resources of the Sea-Bed in a Defined Area of the Continental Shelf of the Two Countries in the Gulf of Thailand

The Kingdom of Thailand and Malaysia

DESIRING to strengthen further the existing bonds of traditional friendship between the two countries;

RECOGNIZING that, as a result of overlapping claims made by the two countries regarding the boundary line of their continental shelves in the Gulf of Thailand, there exists an overlapping area on their adjacent continental shelves;

NOTING that the existing negotiations between the two countries on the delimitation of the boundary of the continental shelf in the Gulf of Thailand may continue for some time;

CONSIDERING that it is in the best interests of the two countries to exploit the resources of the sea-bed in the overlapping area as soon as possible; and

CONVINCED that such activities can be carried out jointly through mutual cooperation.

HAVE AGREED AS FOLLOWS:

Article I

Both Parties agree that as a result of overlapping claims made by the two countries regarding the boundary line of their continental shelves in the Gulf of Thailand, there exists an overlapping area, which is defined as that area bounded by straight lines joining the following coordinated points:

(A)	N 6° 50'.0	E 102° 21'.2
(B)	N 7° 10'.25	E 102° 29'.0
(C)	N 7° 49'.0	E 103° 02'.5
(D)	N 7° 22'.0	E 103° 42'.5
(E)	N 7° 20'.0	E 103° 39'.0
(F)	N 7° 03'.0	E 103° 06'.0
(G)	N 6° 53'.0	E 102° 34'.0

and shown in the relevant part of the British Admiralty Chart No. 2414, Edition 1967, annexed hereto.

Article II

Both Parties agree to continue to resolve the problem of the delimitation of the boundary of the continental shelf in the Gulf of Thailand between the two countries by negotiations or such other peaceful means as agreed to by both Parties, in accordance with the principles of international law and practice especially those agreed to in the Agreed Minutes of the Malaysia-Thailand Officials' Meeting on Delimitation of the Continental Shelf Boundary Between Malaysia and Thailand in the Gulf of Thailand and in the South China Sea, 27 February—1 March 1978, and in the spirit of friendship and in the interest of mutual security.

Article III

(1) There shall be established a Joint Authority to be known as "Malaysia-Thailand Joint Authority" (hereinafter referred to as "the Joint Authority") for the purpose of the exploration and exploitation of the non-living natural resources of the sea-bed and subsoil in the overlapping area for a period of fifty years commencing from the date this Memorandum comes into force.

(2) The Joint Authority shall assume all rights and responsibilities on behalf of both Parties for the exploration and exploitation of the non-living natural resources of the sea-bed and subsoil in the overlapping area (hereinafter referred to as the joint development area) and also for the development, control and administration of the joint development area. The assumption of such rights and responsibilities by the Joint Authority shall in no way affect or curtail the validity of concessions or licences hitherto issued or agreements or arrangements hitherto made by either Party.

(3) The Joint Authority shall consist of:

    (a) two joint-chairmen, one from each country, and
    (b) an equal number of members from each country.

(4) Subject to the provisions of this Memorandum, the Joint Authority shall exercise on behalf of both Parties all the powers necessary for, incidental to or connected with the discharge of its functions relating to the exploration and exploitation of the non-living natural resources of the sea-bed and subsoil in the joint development area.

(5) All costs incurred, and benefits derived by the Joint Authority from activities carried out in the joint development area shall be equally borne and shared by both Parties.

(6) If any single geological petroleum or natural gas structure or field, or other mineral deposits of whatever character, extends beyond the limit of the joint development area defined in Article I, the Joint Authority and the Party or Parties concerned shall communicate to each other all information in this regard and shall seek to reach agreement as to the manner in which the structure, field or deposit will be most effectively exploited; and all expenses incurred and benefits derived therefrom shall be equitably shared.

Article IV

(1) The rights conferred or exercised by the national authority of either Party in matters of fishing, navigation, hydrographic and oceanographic surveys, the prevention and control of marine pollution and other similar matters (including all powers of enforcement in relation thereto) shall extend to the joint development area and such rights shall be recognised and respected by the Joint Authority.

(2) Both Parties shall have a combined and coordinated security arrangement in the joint development area.

Article V

The criminal jurisdiction of Malaysia in the joint development area shall extend over that area bounded by straight lines joining the following coordinated points:

A	N 6° 50′.0	E 102° 21′.2
X	N 7° 35′.0	E 103° 23′.0
D	N 7° 22′.0	E 103° 42′.5
E	N 7° 20′.0	E 103° 39′.0
F	N 7° 03′.0	E 103° 06′.0
G	N 6° 53′.0	E 102° 34′.0

The criminal jurisdiction of the Kingdom of Thailand in the joint development area shall extend over that area bounded by straight lines joining the following coordinated points:

A	N 6° 50′.00	E 102° 21′.2
B	N 7° 10′.25	E 102° 29′.0

(continued)

(continued)

C	N 7° 49'.00	E 103° 02'.5
X	N 7° 35'.00	E 103° 23'.0

The areas of criminal jurisdiction of both Parties defined under this Article shall not in any way be construed as indicating the boundary line of the continental shelf between the two countries in the joint development area, which boundary is to be determined as provided for by Article II, nor shall such definition in any way prejudice the sovereign rights of either Party in the joint development area.

Article VI

(1) Notwithstanding Article III, if both parties arrive at a satisfactory solution on the problem of the delimitation of the boundary of the continental shelf before the expiry of the said fifty-year period, the Joint Authority shall be wound up and all assets administered, and liabilities incurred by it shall be equally shared and borne by both Parties. A new arrangement may, however, be concluded if both Parties so decide.

(2) If no satisfactory solution is found on the problem of the delimitation of the boundary of the Continental Shelf within the said fifty-year period, the existing arrangement shall continue after the expiry of the said period.

Article VII

Any difference or dispute arising out of the interpretation or implementation of the provisions of this Memorandum shall be settled peacefully by consultation or negotiation between the Parties.

Article VIII

This Memorandum shall come into force on the date of exchange of instruments of ratification.

DONE in duplicate at Chiang Mai, the Twenty-first Day of February in the year One thousand Nine hundred and Seventy-nine, in the Thai, Malay and English Languages.

In the event of any conflict among the texts, the English text shall prevail.

For The Kingdom of Thailand:

General Kriangsak Chomanan, Prime Minister

For Malaysia: Datuk Hussein Onn, Prime Minister

(Unofficial Translation)

Proclamation Establishing the Exclusive Economic Zone of the Kingdom of Thailand in the Gulf of Thailand Adjacent to the Exclusive Economic Zone of Malaysia

16 February 1988

By Royal Command of His Majesty the King, it is hereby proclaimed that:

Whereas on the 23rd day of February, B.E. 2514 (1981 A.D.) the Kingdom of Thailand issued a Proclamation Establishing the Exclusive Economic Zone of the Kingdom of Thailand, stipulating that the exclusive economic zone of the Kingdom of Thailand is an area beyond and adjacent to two hundred nautical miles measured from the baselines used for measuring the breadth of the territorial sea;

It is now deemed appropriate to issue a further Proclamation, pursuant to the generally accepted principles of international law, that outer limit of the exclusive economic zone of the Kingdom of Thailand in the Gulf of Thailand adjacent to the exclusive economic zone of Malaysia is formed by the lines connecting each geographical co-ordinate as follows:

No	Latitude	Longitude
1	6° 14'.5 N	102° 05'.6 E
2	6° 27'.5 N	102° 10'.0 E
3	6° 27'.8 N	102° 09'.6 E
4	6° 50'.0 N	102° 21'.2 E
5	6° 53'.0 N	102° 34'.0 E
6	7° 03'.0 N	103° 06'.0 E
7	7° 20'.0 N	103° 39'.0 E
8	7° 22'.0 N	103° 42'.5 E

The outer limit of the exclusive economic zone as mentioned above is shown on the annexed map.

Proclaimed on the 16th of February, B.E. 2531 (1988 A.D.), being the forty-third year of the present Reign.

Counter-signed by: General Prem Tinsulanonda, Prime Minister

Agreement between the Government of Malaysia and the Government of the Kingdom of Thailand on the Constitution and Other Matters Relating to the Establishment of the Malaysia-Thailand Joint Authority

THE GOVERNMENT OF MALAYSIA AND THE GOVERNMENT OF THE KINGDOM OF THAILAND, hereinafter referred to as 'the Governments,'

DESIRING to implement the Memorandum of Understanding between Malaysia and the Kingdom of Thailand on the Establishment of a Joint Authority for the Exploitation of the Resources of the Sea-bed in a Defined Area of the Continental Shelf of the Two Countries in the Gulf of Thailand dated 21 February 1979, hereinafter referred to as the Memorandum of Understanding, 1979.

HAVE HEREBY AGREED on the establishment of the Malaysia-Thailand Joint Authority, hereinafter referred to as 'the Joint Authority,' which shall operate in accordance with the following provisions:

CHAPTER I LEGAL STATUS AND ORGANIZATION

Article 1 Juristic Personality and Capacity

(1) The Joint Authority shall have a juristic personality and such capacities as shall be provided for in the Acts of Parliament to be enacted by the Government of Malaysia and the Government of the Kingdom of Thailand, respectively, for the establishment of the Joint Authority, hereinafter referred to as 'the Acts.'

(2) The drafts of the Acts, referred to in paragraph (1) and attached hereto as Appendix A and Appendix B respectively, shall form an integral part of this Agreement.

Article 2 Purpose

(1) The purpose of the Joint Authority shall be the exploration and exploitation of the non-living natural resources of the sea-bed and subsoil, in particular petroleum, in the Joint Development Area as defined in the Memorandum of Understanding, 1979 for the period of validity of the Memorandum of Understanding, 1979.

(2) In this Agreement, 'petroleum' means any mineral oil or relative hydro-carbon and natural gas existing in its natural condition and casing head petroleum spirit, including bituminous shales and other stratified deposits from which oil can be extracted.

Article 3 Membership

(1) The Joint Authority shall comprise

    (a) two Co-Chairmen, one each to be appointed by the respective Governments; and

    (b) an equal number of members to be appointed by each Government, provided that the initial number of members, excluding the Co-Chairmen, to be appointed by each Government shall be six.

(2) The word 'member' shall, for the purposes of this Agreement, and unless the context otherwise requires, include a Co-Chairman.

Article 4 **Procedure**

(1) The Co-Chairmen shall alternate to perform the functions of a Chairman at meetings of the Joint Authority. In the absence of a Co-Chairman during his chairmanship, the members of the Joint Authority shall elect a Chairman from amongst the members representing the same Government as the corresponding Co-Chairman. When so elected, he shall have all the powers of the Chairman.

(2) The quorum for any meeting of the Joint Authority shall not be less than ten. Decisions shall be taken jointly at a meeting by the Co-Chairmen.

Article 5 **Personal Liability**

No member of the Joint Author←ity shall incur any personal liability for any loss or damage caused by any act or omission in the administration of the affairs of the Joint Authority unless such loss or damage is occasioned by a wrongful act, gross negligence or omission on his part.

Article 6 **Emoluments**

The members of the Joint Authority shall be paid such emoluments and other allowances as the Joint Authority may determine with the approval of the Governments.

CHAPTER II POWERS AND FUNCTIONS

Article 7 **Powers and Functions**

(3) The Joint Authority shall control all exploration and exploitation of the non-living natural resources in the Joint Development Area and shall be responsible for the formulation of policies for the same.

(4) Without prejudice to the generality of the foregoing, the powers and functions of the Joint Authority shall include the following:

  (a) to decide, with the approval of the Governments, on the organisational structure of the Joint Authority;

  (b) subject to subparagraph (a), to appoint the chief executive officer and other officers of the Joint Authority, provided that the appointment of the chief executive officer and the deputy chief executive officer shall require the approval of the Governments;

  (c) to decide on the terms and conditions of service of the chief executive officer and other officers of the Joint Authority;

  (d) to decide on the plan of operation and the working programme for the administration of the Joint Development Area;

  (e) to permit operations and to conclude transactions or contracts for or relating to the exploration and exploitation of the non-living natural resources in the Joint Development Area subject to the approval of the Governments;

(f) in respect of any contract referred to in subparagraph (e) for the exploration and exploitation of petroleum
   (i) to approve and extend the period of exploration and exploitation;
   (ii) to approve the work programmes and budgets of the contractor; and
   (iii) to approve the production programmes of the contractor, including the production costs, conditions and schedules of the production;
(g) in respect of an operator of any contract referred to in subparagraph (f)
   (i) to inspect and audit the operator's books and accounts relating to its operations in the Joint Development Area;
   (ii) to take periodic inventories of the properties and assets procured by the operator for petroleum operations; and
   (iii) to receive, collate and store all data supplied by the operator relating to its operations in the Joint Development Area;
(h) to approve and award tenders and contracts relating to goods and services required in carrying out petroleum operations in the Joint Development Area;
(i) to appoint committees, sub-committees or independent experts and consultants where necessary for the administration of the Joint Authority;
(j) to regulate any meeting of the Joint Authority, any committee and sub-committee thereof; and
(k) to do any other thing incidental to or necessary for the performance of any of its functions.

Article 8 Production Sharing

(1) For the purpose of paragraph (2)(f) of Article 7, any contract awarded to any person for the exploration and exploitation of petroleum in the Joint Development Area shall, in accordance with subsection (3) of Section 14 of the Acts, be a production sharing contract.

(2) A production sharing contract referred to in paragraph (1) shall include, without prejudice to paragraph (3), the terms and conditions specified in sub-Section (3) of Section 14 of the Acts as follows:

(a) the contract shall be valid for a period not exceeding thirty-five years but shall not exceed the period of validity of this Agreement;
(b) payment in the amount of ten per centum of gross production of petroleum by the contractor to the Joint Authority as royalty in the manner and at such times as may be specified in the contract.
(c) fifty per centum of gross production of petroleum shall be applied by the contractor for the purpose of recovery of costs for petroleum operations;
(d) the remaining portion of gross production of petroleum, after deductions for the purpose of subparagraphs (b) and (c), shall be deemed to be profit petroleum and be divided equally between the Joint Authority and the contractor;
(e) all costs of petroleum operations shall be borne by the contractor and shall, subject to subparagraph (c), be recoverable from production;

(f) a minimum amount that the contractor shall extend on petroleum operations under the contract as a minimum commitment as may be agreed to by the Joint Authority and the contractor;

(g) payment of a research access by the contractor to the Joint Authority in the amount of one half of one per centum of the aggregate of that portion of gross production which is applied for the purpose of recovery of costs under subparagraph (c) and the contractor's share of profit petroleum under subparagraph (d) in the manner and at such times as may be determined by the Joint Authority, provided that such payment shall not be recoverable from production; and

(h) any disputes or differences arising out of or in connection with the contract which cannot be amicably settled shall be referred to arbitration before a panel consisting of three arbitrators, one arbitrator to be appointed by each party, and a third to be jointly appointed by both parties. If the parties are unable to concur on the choice of a third arbitrator within a specified period, the third arbitrator shall be appointed upon application to the United Nations Commission of International Trade Law (UNCITRAL). The arbitration proceedings shall be conducted in accordance with the rules of UNCITRAL. The venue of arbitration shall be either Bangkok or Kuala Lumpur, or any other place as may be agreed to by the parties.

(3) In addition to the matters specified in paragraph (2), production sharing contract may, at the option of the Joint Authority, include the following:

(a) the period referred to under subparagraph (a) of paragraph (2) shall be applied as follows:

(i) in respect of a contract for petroleum (other than gas), the periods for the purposes of exploration, development and production shall not exceed five years, five years and twenty-five years, respectively; and

(ii) in respect of a contract for gas, the periods for the purposes of exploration, the identification and nomination of the gas market, development and production shall not exceed, in respect of the first three periods, five years each, and in respect of the fourth period, twenty years;

Provided that any period referred to in subparagraphs (a)(i) and (a)(ii) may be varied by the Joint Authority from time to time as may be necessary on condition that where any such variation affects a subsisting contract it shall only be made with the agreement of the contractor;

Provided further that where the first commercial production occurs, in the case of

(A) petroleum (other than gas), before the expiry of the development period, the balance of that development period shall be added to the production period; and

(B) gas, before the expiry of the period for the identification and nomination of the gas market or the development period, the balance of either of those periods shall be added to the production period;

(b) title to any equipment or assets purchased or acquired by the contractor for the purpose of petroleum operations shall pass to the Joint Authority upon such purchase or acquisition;

(c) the Joint Authority shall have title to all original data (raw, processed or interpreted) resulting from petroleum operations, including but not limited to geological, geophysical, core samples, petro-physical, well completion reports, engineering and other data reports and actual samples as the contractor may collect and compile, and

(d) the contractor shall purchase or acquire equipment, facilities, goods, materials, supplies, including services and research facilities, professional or otherwise, from sources in Malaysia or the Kingdom of Thailand where technically and economically feasible.

(4) The Joint Authority may vary any of the amounts referred to in sub-paragraphs (c), (d) and (g) of paragraph (2) in respect of any contract with the approval of the Governments:

Provided that there shall be no variation of any of these amounts in respect of a subsisting contract without the agreement of the contractor.

(5) For the purpose of this Article:

(a) 'first commercial production' in relation to petroleum (other than gas) means the date that production has continued for a period of twenty-four hours following completion of testing from the first production well, and, in relation to gas, means the date within the first sixty days on which a cumulative 10^6 Giga Joule (approximately 947 billion BTU) of gas is first sold or the sixtieth day after the gas is first sold if the cumulative sale within the first sixty days does not exceed 10^6 Giga Joule; and

(b) 'gross production' with reference to gas means gross proceeds of sale of gas.

(6) The contractor shall pay export duty on its share of profit oil and petroleum income tax in accordance with Article 16 and Article 17 of this Agreement, respectively.

CHAPTER III FINANCIAL PROVISIONS

Article 9 Finance

(1) All costs incurred, and benefits derived by the Joint Authority from activities carried out in the Joint Development Area shall be equally borne and shared by the Governments.

(2) Until such time as the Joint Authority shall have sufficient income to finance its annual operational expenditure, the Governments shall annually provide the Joint Authority with the agreed amounts of money in equal shares to be paid into the Malaysia-Thailand Joint Authority Fund, hereinafter referred to as 'the Fund.'

(3) Thereupon, as specified in paragraph (2), and unless otherwise decided by the Governments, all such annual government contributions shall cease.

Article 10 Accounts and Records

(1) The Joint Authority shall cause proper accounts and other records of its transactions and affairs to be kept in accordance with generally accepted accounting principles and shall do all things necessary to ensure that all incomes are properly accounted for and that all expenditures out of the Fund including payments of salaries, remuneration and other monetary benefits to members of the Joint Authority and its employees, are properly authorised and that adequate control is maintained over the assets of or in the custody of the Joint Authority.
(2) Either Government may at any time direct any accounts or records to be made available to it and the Joint Authority shall comply with such direction.

Article 11 Budget

The annual budget of the Joint Authority shall be submitted to the Governments well in advance before the financial year of the respective Governments for their approval.

Article 12 Audit

(1) The Joint Authority shall have a financial year beginning on the first day of January.
(2) The accounts of the Joint Authority shall be audited annually by an auditor appointed by the Joint Authority with the approval of the Governments.
(3) The Joint Authority shall, within six months after the end of each financial year, have its accounts audited and transmitted to both Governments together with a copy of any observations made by the auditor on any statement or on the accounts of the Joint Authority and a copy of the annual report dealing with the activities of the Joint Authority in the preceding year.

CHAPTER IV REGULATIONS AND RELATIONS WITH OTHER ORGANISATIONS

Article 13 Regulations

The Joint Authority may, in accordance with and for the purpose of Sect. 15 of the Acts, submit recommendations on regulations in respect of any matter falling thereunder to the Governments for consideration.

Article 14 Relations with other Organisations

In order to fulfil its purpose, the Joint Authority may cooperate with any government or organisation, and, to this end, may, subject to the approval of the Governments, conclude any agreement or arrangement with such government or organisation.

CHAPTER V   THE ACTS

Article 15   Amendment of the Acts

In order to facilitate the efficient management and operation of this Agreement, the Governments agree that the Acts shall not be amended without prior agreement between the Governments.

CHAPTER VI   CUSTOMS AND EXCISE, AND TAXATION

Article 16   Customs Matters

(1) For the purpose of Part X of the Acts

(a) the rate of export duty payable by the contractor in respect of the contractor's share of profit oil sold outside Malaysia and the Kingdom of Thailand shall be ten per centum subject to the provision of subparagraph (b)(ii);

(b) the Customs and Excise Authorities shall continue to exercise all powers in relation to all matters relating to the regulation of the movement of goods imported into or exported from the Joint Development Area in accordance with the existing legislation of Malaysia or the Kingdom of Thailand, as the case may be, subject to the following:

(i) customs approved goods, equipment and materials for use in the Joint Development Area shall be accorded duty exemption if they are imported by the Joint Authority or any person authorised by it;

Provided that where one of the Governments proposes to impose any duties or taxes on any such customs approved goods, equipment and materials, it may impose such duties or taxes after consultation with the other Government;

(ii) Malaysia and the Kingdom of Thailand shall collect their respective duties and taxes collectible under their respective legislation but shall reduce the applicable rates by fifty per centum;

(iii) any goods entering the Joint Development Area from:

(A) a third country, any licensed warehouse or bonded area of either Malaysia or the Kingdom of Thailand shall be deemed an import; and

(B) Malaysia or the Kingdom of Thailand, shall be deemed an internal movement, provided they are customs approved goods, equipment and materials for use in the Joint Development Area;

(iv) any goods produced in the Joint Development Area entering Malaysia or the Kingdom of Thailand or a third country shall be deemed an export;

(v) any goods which has entered the Joint Development Area under the situation described in subparagraph (b)(iii)(B) and is to be moved into Malaysia or the Kingdom of Thailand shall be subject to the laws of Malaysia or the Kingdom of Thailand, as the case may be;

(vi) any goods falling within the category of goods appearing in both the lists of prohibited goods made in accordance with the laws of Malaysia and the Kingdom of Thailand, respectively, shall not be permitted to be brought into the Joint Development Area:

Provided that where an exception is required in respect of any specific importation, such an exception may be made with the agreement of the competent authorities of the other Country;

(vii) proceeds from any sale of forfeited goods which are the produce of the Joint Development Area shall be equally shared by Malaysia and the Kingdom of Thailand;

(c) the Customs and Excise Authorities shall use common customs forms for the purpose of import, export and internal movement of goods in the Joint Development Area as specified in subparagraphs (b)(iii) and (iv); and

(d) the Country where the headquarters of the Joint Authority is located shall empower the officers of the Customs and Excise Authority of the other Country to exercise their authority with regard to customs clearance, including the collection of duties and taxes, within the premises of the Joint Customs Office.

(2) Notwithstanding Sect. 18 of the Acts, and insofar as it applies to customs and excise matters, the following arrangements shall apply:

(a) where an act is committed in the Joint Development Area and that act is an offence under the laws of one of the Countries only, such Country whose laws are alleged to have been breached may assume jurisdiction over such alleged offence;

(b) where the act referred to in subparagraph (a) is an offence under the laws of the Countries, the Country which may assume jurisdiction over the act shall be that whose officer first makes an arrest or seizure in respect of the alleged offence; and

(c) where the act referred to in subparagraph (a) is an offence under the laws of the Countries and in respect of which there are simultaneous arrests or seizures by the Customs and Excise Authorities, the jurisdiction over the alleged offence shall be determined through consultation between such Authorities.

(3) For the purpose of this Article:

(a) 'Countries' means Malaysia and the Kingdom of Thailand, and when used in the singular means Malaysia or the Kingdom of Thailand, as the context requires;

(b) 'customs approved goods' means goods in respect of which customs duties are exempted under both the laws of Malaysia and the Kingdom of Thailand relating to customs;

(c) 'Customs and Excise Authority' in relation to Malaysia means the Royal Customs and Excise, Malaysia and in relation to the Kingdom of Thailand means the Customs Department of Thailand, and when used in the plural means both such Authorities;

(d) 'Joint Customs Committee' means the committee comprising officers of the Customs and Excise Authorities established for the purpose of the coordination of the administration of customs and excise laws in the Joint Development Area; and

(e) 'Joint Customs Office' means the office of the Joint Customs Committee established in the headquarters of the Joint Authority for the purpose of the coordination of the administration of the customs and excise laws in the Joint Development Area.

Article 17 **Taxation**

(1) For the purpose of Part X of the Acts, the Revenue Authorities of the Governments shall, subject to the following, continue to impose and collect taxes in respect of income from the Joint Development Area in accordance with the existing tax legislation of Malaysia and the Kingdom of Thailand, as the case may be:

(a) the taxation of such income of any person who holds the right to explore and exploit any petroleum in the Joint Development Area under a contract awarded by the Joint Authority shall be in accordance with the following rates:

First 8 years of production 0% of taxable income
Next 7 years   10% of taxable income
Subsequent years   20% of taxable income:

Provided that the tax chargeable by each of the Governments shall be reduced by fifty per centum of the amount so chargeable;

Provided further that where the tax chargeable for each year by one of the Governments exceeds that chargeable by the other Government, such excess shall be shared equally by the two Governments and shall be effected by the Joint Authority through appropriate adjustments of payments made under paragraph (d) of Sect. 10 of the Act.

(b) the taxation of such income of a person who is a Malaysian or Thai national exercising employment in the Joint Development Area or with the Joint Authority shall be based on his residence; and

(c) the taxation of such income of any person other than a person mentioned in subparagraphs (a) and (b) shall be in accordance with the laws and regulations of Malaysia and the Kingdom of Thailand, provided that where the same income is subject to tax in both countries, the tax chargeable in each country shall be reduced by fifty per centum of the amount so chargeable.

(2) The Governments agree that any law for taxation which is in the nature of general sales tax, including any tax imposed on the provision of goods and services in the Joint Development Area, shall not be applicable to the Joint Development Area.

(3) The Joint Authority shall be exempt from taxation in Malaysia and the Kingdom of Thailand.

(4) The Revenue Authorities of the Governments shall continue to communicate with and consult each other in respect of the implementation and administration of any tax law in the Joint Development Are

CHAPTER VII MISCELLANEOUS PROVISIONS

Article 18 Entry into Force and Termination

(1) This Agreement shall enter into force upon the exchange of instruments of ratification, and, unless otherwise agreed to by the Governments, shall remain in force for the period of validity of the Memorandum of Understanding, 1979.

(2) Upon termination of this Agreement, the Joint Authority shall be wound up in accordance with a liquidation procedure as may be approved by the Governments.

Article 19 Application

The application and interpretation of the provisions of this Agreement shall be consistent with the spirit and provisions of the Memorandum of Understanding, 1979.

Article 20 Amendment

This Agreement may be amended by a joint decision of the Co-Chairmen with the approval of the Governments.

Article 21 Settlement of Disputes

Any differences or disputes arising out of the interpretation or application of the provisions of this Agreement shall be settled peacefully by consultation or negotiation between the Governments. In the event that no settlement is reached within a period of three months either Government may refer the matter to the Prime Minister of Malaysia and the Prime Minister of the Kingdom of Thailand who shall jointly decide on the mode of settlement for the purpose of the particular matter referred to them.

Article 22

This Agreement is made in duplicate at Kuala Lumpur, Malaysia, on the Thirtieth day of May, in the year One thousand Nine hundred and Ninety, in the English language.

For The Government of Malaysia: Dato' Haji Abu Hassan bin Haji Omar, Minister of Foreign Affairs

For The Government of the Kingdom of Thailand: Air Chief Marshal Siddhi Savetsila, Minister of Foreign Affairs.

Agreement Between the Government of Malaysia and the Government of the Republic of Singapore to Delimit Precisely the Territorial Waters Boundary in Accordance with the Straits Settlement and Johore Territorial Waters Agreement 1927

Whereas by the Straits Settlement and Johore Territorial Waters Agreement dated 19 October 1927, hereinafter referred to as 'the 1927 Agreement', made between His Excellency Sir Hugh Charles Clifford, Governor and Commander-in-Chief of the Colony of the Straits Settlements, on behalf of His Britannic Majesty and His Highness Ibrahim bin Almarhum Sultan Abu Bakar, Sultan of the State and Territory of Johore, the boundary between the territorial waters of the Settlement of Singapore and the State and Territory of Johore was agreed upon;

And whereas the State and Territory of Johore has been succeeded to by Malaysia and is a State within Malaysia and the Settlement of Singapore has been succeeded to by the Republic of Singapore;

And whereas the Government of Malaysia and the Government of the Republic of Singapore, hereinafter referred to as 'the Contracting Parties', recognizing the need to precisely delimit the territorial waters boundary in accordance with the 1927 Agreement, agreed to conduct a joint hydrographic survey based on the Memorandum of Procedure relating to the said survey as agreed upon on 29 January 1980;

And whereas upon the successful completion of the joint hydrographic survey on 12 May 1982 and the adoption of its report by the Contracting Parties on 16 April 1985, the Contracting Parties are desirous of entering into an agreement to precisely delimit the territorial waters boundary between Malaysia and the Republic of Singapore in the areas described in Article 1 of the 1927 Agreement;

Now, therefore it is agreed and declared as follows:

Article 1 **The Boundary**

1. The territorial waters boundary between Malaysia and the Republic of Singapore in the areas described in Article 1 of the 1927 Agreement is defined by straight lines joining the points, the geographical coordinates of which are specified in Annex I.
2. The latitude and longitude of the geographical coordinates specified in Annex I have been determined on the Revised Kertau Datum, Everest Spheroid (Malaya), Malaysian Rectified Skew Orthomorphic Projection (Projection Tables published by Directorate of Military Survey, Ministry of Defence, United Kingdom-March 1965). Chart Datums used are as described in the Joint Hydrographic Survey Fair Sheets 1980/1982 listed in Annex II.
3. As an illustration, the territorial waters boundary referred to in paragraph (1) is shown in red on the map attached hereto as Annex III.
4. Where the actual location of the points specified by the geographical coordinates in Annex 1 or any other points along the boundary is required to be determined, it shall be determined jointly by the competent authorities of the Contracting Parties.
5. For the purpose of paragraph 4 of this Article the term 'competent authorities', in relation to Malaysia shall mean the Director General of Survey and Mapping, Malaysia and any person authorized by him, and in relation to the Republic of

Singa- pore shall mean the Head of the Mapping Unit, Ministry of Defence, Singapore and any person authorized by him.

Article 2 Finality of Boundary

There shall be no alteration to the territorial waters boundary as defined in Article 1.

Article 3 Settlement of Disputes

Any dispute between the Contracting Parties arising out of the interpretation or implementation of this Agreement shall be settled by consultation or negotiation.

Article 4 Relationship with 1927Agreement

In the event of any inconsistency between Article 1 of this Agreement and Article I of the 1927 Agreement, Article 1 of this Agreement shall prevail.

Article 5 Ratification

This Agreement shall be subject to ratification by the Contracting Parties.

Article 6 Entry into Force

This Agreement shall enter into force on the date of exchange of the instruments of ratification by the Contracting Parties.

In witness whereof the undersigned, being duly authorized by their respective Governments, have signed this Agreement.

Done at Singapore on this seventh day of August one thousand nine hundred and ninety-five in four original texts, two each in Malay and English languages, all texts being equally authentic. In case of any divergence, the English text shall prevail.

For the Government of Malaysia:	Datuk Abdullah Ahmad Badawi, Minister of Foreign Affairs
For the Government of the Republic of Singapore:	Professor S. Jayakumar, Minister of Foreign Affairs

Annex I

Geographical Coordinates

1. *East of Johor Causeway*

Designating Points	Latitude North	Longitude East
E1	01° 27′ 10.0″	103° 46′ 16.0″
E2	01° 27′ 54.5″	103° 47′ 25.7″
E3	01° 28′ 35.4″	103° 48′13.2″

(continued)

(continued)

Designating Points	Latitude North	Longitude East
E4	01° 28′ 42.5″	103° 48′ 45.6″
E5	01° 28′ 36.1″	103° 49′ 19.8″
E6	01° 28′ 22.8″	103° 50′ 03.0″
E7	01° 27′ 58.2″	103° 51′ 07.2″
E8	01° 27′ 46.6″	103° 51′ 31.2″
E9	01° 27′ 31.9″	103° 51′ 53.9″
E10	01° 27′ 23.5″	103° 52′ 05.4″
E11	01° 26′ 56.3″	103° 52′ 30.1″
E12	01° 26′ 06.5″	103° 53′ 10.1″
E13	01° 25′ 40.6″	103° 53′ 52.3″
E14	01° 25′ 39.1″	103° 54′ 45.9″
E15	01° 25′ 36.0″	103° 55′ 00.6″
E16	01° 25′ 41.7″	103° 55′ 24.0″
E17	01° 25′ 49.5″	103° 56′ 00.3″
E18	01° 25′ 49.7″	103° 56′ 15.7″
E19	01° 25′ 40.2″	103° 56′ 33.1″
E20	01° 25′ 31.3″	103° 57′ 09.1″
E21	01° 25′ 27.9″	103° 57′ 27.2″
E22	01° 25′ 29.1″	103° 57′ 38.4″
E23	01° 25′ 19.8″	103° 58′ 00.5″
E24	01° 25′ 19.0″	103° 58′ 20.7″
E25	01° 25′ 27.9″	103° 58′ 47.7″
E26	01° 25′ 27.4″	103° 59′ 00.9″
E27	01° 25′ 29.7″	103° 59′ 10.2″
E28	01° 25′ 29.2″	103° 59′ 20.5″
E29	01° 25′ 30.0″	103° 59′ 34.5″
E30	01° 25′ 25.3″	103° 59′ 42.9″
E31	01° 25′ 14.2″	104° 00′ 10.3″
E32	01° 26′ 20.9″	104° 01′ 23.9″
E33	01° 26′ 38.0″	104° 02′ 27.0″
E34	01° 26′ 23.5″	104° 03′ 26.9″
E35	01° 26′ 04.7″	104° 04′ 16.3″
E36	01° 25′ 51.3″	104° 04′ 35.3″
E37	01° 25′ 03.3″	104° 05′ 18.5″
E38	01° 24′ 55.8″	104° 05′ 22.6″
E39	01° 24′ 44.8 "	104° 05′ 26.7″
E40	01° 24′ 21.4″	104° 05′ 33.6″

(continued)

(continued)

Designating Points	Latitude North	Longitude East
E41	01° 23′ 59.3″	104° 05′ 34.9″
E42	01° 23′ 39.3″	104° 05′ 32.9″
E43	01° 23′ 04.9″	104° 05′ 22.4″
E44	01° 22′ 07.5″	104° 05′ 00.9″
E45	01° 21′ 27.0″	104° 04′ 47.0″
E46	01° 20′ 48.0″	104° 05′ 07.0″
E47	01° 17′ 21.3″	104° 07′ 34.0″

2. West of Johor Causeway

Designating Points	Latitude North	Longitude East
W1	01° 27′ 09.8″	103° 46′ 15.7″
W2	01° 25′ 54.2″	103° 45′ 38.5″
W3	01° 27′ 01.4″	103° 44′ 48.4″
W4	01° 27′ 16.6″	103° 44′ 23.3″
W5	01° 27′ 36.5″	103° 43′ 42.0″
W6	01° 27′ 26.9″	103° 42′ 50.8″
W7	01° 27′ 02.8″	103° 42′ 13.5″
W8	01° 26′ 35.9″	103° 41′ 55.9″
W9	01° 26′ 23.6″	103° 41′ 38.6″
W10	01° 26′ 14.1″	103° 41′ 00.0″
W11	01° 25′ 41.3″	103° 40′ 26.0″
W12	01° 24′ 56.7″	103° 40′ 10.0″
W13	01° 24′ 37.7″	103° 39′ 50.1″
W14	01° 24′ 01.5″	103° 39′ 25.8″
W15	01° 23′ 28.6″	103° 39′ 12.6″
W16	01° 23′ 13.5″	103° 39′ 10.7″
W17	01° 22′ 47.7″	103° 38′ 57.1″
W18	01° 21′ 46.7″	103° 38′ 27.2″
W19	01° 21′ 26.6″	103° 38′ 15.5″
W20	01° 21′ 07.3″	103° 38′ 08.0″
W21	01° 20′ 27.8″	103° 37′ 48.2″
W22	01° 19′ 17.8″	103° 37′ 04.2″
W23	01° 18′ 55.5″	103° 37′ 01.5″
W24	01° 18′ 51.5″	103° 36′ 58.2″
W25	01° 15′ 51.0″	103° 36′ 10.3″

ANNEX II

Joint Hydrographic Survey Fair Sheets, 1980/1982

Sheet 1 of 21	(JS/5/IIa-1)
Sheet 2 of 21	(JS/5/IIa-2)
Sheet 3 of 21	(JS/5/IIb-1)
Sheet 4 of 21	(JS/5/IIb-2)
Sheet 5 of 21	(JS/5/Ia)
Sheet 6 of 21	(JS/5/Ib)
Sheet 7 of 21	(JS/5/IIIa-1)
Sheet 8 of 21	(JS/5/IIIa-2)
Sheet 9 of 21	(JS/5/IIIb-1)
Sheet 10 of 21	(JS/5/IIIb-2)
Sheet 11of 21	(JS/5/IVa)
Sheet 12of 21	(JS/5/IVb-1)
Sheet 13 of 21	(JS/5/IVb-2)
Sheet 14 of 21	(JS/5/Va-1)
Sheet 15 of 21	(JS/5/Va-2)
Sheet 16 of 21	(JS/5/Vb-1)
Sheet 17 of 21	(JS/5/Vb-2)
Sheet 18 of 21	(JS/5/VIa-1)
Sheet 19 of 21	(JS/5VIa-2)
Sheet 20 of 21	(JS/5/VIb-1)
Sheet 21 of 21	(JS/5/VIb-2)

Memorandum of Understanding between Malaysia and the Socialist Republic of Vietnam for the Exploration and Exploitation of Petroleum in a Defined Area of the Continental Shelf Involving the two Countries Malaysia and the Socialist Republic of Vietnam

Desiring to further strengthen the cooperation between the two countries;

Recognizing that as a result of overlapping claims made by the two countries regarding the boundary lines of their continental shelves located off the northeast coast of West Malaysia and off the southwest coast of Vietnam, there exists an overlapping area of their continental shelves;

Consistent with the agreement reached by the leaders of the two countries to cooperate in that part of the overlapping area involving the two countries only;

Mindful of the decision by the leaders of the two countries to resolve peacefully the question of all overlapping claims involving multiple parties with the parties concerned at an appropriate time;

Considering that it is in the best interests of both countries, pending delimitation of their continental shelves located off the northeast coast of West Malaysia and off the southwest coast of Vietnam, to enter into an interim arrangement for the purpose of exploring and exploiting petroleum in the seabed in the overlapping area;

Convinced that such activities can be carried out through mutual cooperation;

Have agreed as follows:

Article 1

(1) Both parties agree that as a result of overlapping claims made by the two countries regarding the boundary lines of their continental shelves located off the north- east coast of West Malaysia and off the southwest coast of Vietnam, there exists an overlapping area (the Defined Area), being that area bounded by straight lines joining the following coordinated points:

A	N 7° 22.0'	E 103° 42.5'
B	N 7° 20.0'	E 103° 39.0'
C	N 7° 18.31'	E 103° 35.71'
D	N 7° 03.0'	E 103° 52.0'
E	N 6° 05.8'	E 103° 49.2'
F	N 6° 48.25'	E 103° 30.0'
A	N 7° 22.0'	E 103° 42.5'

and shown in the relevant part of the British Admiralty Chart No: 2414, Edition 1967, annexed hereto. (*Reference to the Agreement document.*)

(2) The actual location at sea of the points referred to in Clause (1) of this Article shall be determined by a method to be mutually agreed upon by the competent authorities of both parties. The competent authorities in relation to Malaysia means the Directorate of National Mapping Malaysia and includes any person authorized by it and in relation to the Socialist Republic of Vietnam means the Department of Geo-Cartography and the Navy Geo-Cartography Section and includes any person authorized by them.

Article 2

(1) Both parties agree, pending final delimitation of the boundary lines of their continental shelves pertaining to the Defined Area, through mutual cooperation, to explore and exploit petroleum in that area in accordance with the terms of, and for a period of the validity of this Memorandum of Understanding.

(2) Where a petroleum field is located partly in the Defined Area and partly outside that area in the continental shelf of Malaysia or the Socialist Republic of Vietnam, as the case may be, both parties shall arrive at mutually acceptable terms for the exploration and exploitation of petroleum therein.

(3) All costs incurred, and benefits derived from the exploration and exploitation of petroleum in the Defined Area shall be borne and shared equally by both parties.

Article 3

For the purpose of this Memorandum of Understanding-

(a) Malaysia and the Socialist Republic of Vietnam agree to nominate PETRONAS and PETROVIETNAM, respectively, to undertake, on their respective behalf, the exploration and exploitation of petroleum in the Defined Area;

(b) Malaysia and the Socialist Republic of Vietnam shall cause PETRONAS and PETROVIETNAM, respectively, to enter into a commercial arrangement as between them for the exploration and exploitation of petroleum in the Defined Area provided that the terms and conditions of that arrange- merit shall be subject to the approval of the Government of Malaysia and the Government of the Socialist Republic of Vietnam;

(c) both parties agree, taking into account the significant expenditures already incurred in the Defined Area, that every effort shall be made to ensure continued early exploration of petroleum in the Defined Area.

Article 4

Nothing in this Memorandum of Understanding shall be interpreted so as to in any way-

(a) prejudice the position and claims of either party in relation to and over the Defined Area; and

(b) without prejudice to the provisions of Article III, confer any rights, interests or privileges to any person not being a party hereto in respect of any petroleum resources in the Defined Area.

Article 5

This Memorandum of Understanding shall continue for a period to be specified by an exchange of Diplomatic Note between the two parties.

Article 6

Any dispute arising out of the interpretation or implementation of the provisions of this Memorandum of Understanding shall be settled peacefully by consultation or negotiation between both parties.

Article 7

This Memorandum of Understanding shall come into force on the date to be specified by an exchange of Diplomatic Notes between the two parties.

Article 8

For the purpose of this Memorandum of Understanding-

(a) 'Defined Area' means the area referred to in Article I(1) of this Memorandum of Understanding;

(b) 'petroleum' means any mineral oil or relative hydrocarbon and natural gas existing in its natural condition and casing head petroleum spirit, including bituminous shales and other stratified deposits from which oil can be extracted;

(c) 'petroleum field' means an area consisting of a single reservoir or multiple reservoirs all grouped on, or related to, the same individual geological structural feature, or stratigraphic conditions from which petroleum may be produced commercially;

(d) 'PETRONAS' is the short form of Petroliam Nasional Berhad, a company incorporated under the Malaysia Companies Act 1965; and

(e) 'PETROVIETNAM' is the short form of Vietnam National Oil and Gas Company established by the Decree of No. 250/HDBT of 6 July 1990.

Done in duplicate at Kuala Lumpur the 5th day of June in the year One Thousand Nine Hundred and Ninety-Two in the English language.

For Malaysia:

H.E. Datuk Ahmad Kamil Jaafar, Minister of Foreign Affairs

H.E. Vu Khoan, Secretary-General, Ministry of Vice Minister of Foreign Affairs

Socialist Republic of Vietnam

Sovereignty over Pulau Ligitan and Pulau Sipadan (Indonesia/Malaysia)

The Court finds [on this day] that sovereignty over the islands of Ligitan and Sipadan belongs to Malaysia

At THE HAGUE, 17 December 2002. The International Court of Justice (ICJ), principal judicial organ of the United Nations, has today given Judgment in the case concerning Sovereignty over Pulau Ligitan and Pulau Sipadan (Indonesia/Malaysia).

In its Judgment, which is final, without appeal and binding for the Parties, the Court finds, by sixteen votes to one, that "sovereignty over Pulau Ligitan and Pulau Sipadan belongs to Malaysia". Ligitan and Sipadan are two very small islands located in the Celebes Sea, off the north-east coast of the island of Borneo.

Reasoning of the Court

The Court begins by recalling the complex historical background of the dispute between the Parties. It then examines the titles invoked by them. Indonesia's claim to sovereignty over the islands is based primarily on a conventional title, the 1891 Convention between Great Britain and the Netherlands. Indonesia thus maintains that that Convention established the 4° 10' north parallel of latitude as the dividing line between the British and Dutch possessions in the area where Ligitan and Sipadan are situated. As the disputed islands lie to the south of that parallel, "[i]t therefore follows that under the Convention title to those islands vested in The Netherlands, and now vests in Indonesia". Malaysia, for its part, asserts that the

1891 Convention, when seen as a whole, clearly shows that Great Britain and the Netherlands sought by the Convention solely to clarify the boundary between their respective land possessions on the islands of Borneo and Sebatik, since the line of delimitation stops at the easternmost point of the latter island.

After examining the 1891 Convention, the Court finds that the Convention, when read in context and in the light of its object and purpose, cannot be interpreted as establishing an allocation line determining sovereignty over the islands out to sea, to the east of the island of Sebatik, and as a result the Convention does not constitute a title on which Indonesia can found its claim to Ligitan and Sipadan. The Court states that this conclusion is confirmed both by the *travaux préparatoires* and by the subsequent conduct of the parties to the Convention. The Court further considers that the cartographic material submitted by the Parties in the case does not contradict that conclusion.

Having rejected this argument by Indonesia, the Court turns to consideration of the other titles on which Indonesia and Malaysia claim to find their sovereignty over the islands of Ligitan and Sipadan. The Court determines whether Indonesia or Malaysia obtained a title to the islands by succession. The Court begins in this connection by observing that, while the Parties both maintain that the islands of Ligitan and Sipadan were not *terrae nullius* during the period in question in the present case, they do so on the basis of diametrically opposed reasoning, each of them claiming to hold title to those islands. The Court does not accept Indonesia's contention that it retained title to the islands as successor to the Netherlands, which allegedly acquired it through contracts concluded with the Sultan of Bulungan, the original titleholder. Nor does the Court accept Malaysia's contention that it acquired sovereignty over the islands of Ligitan and Sipadan further to a series of alleged transfers of the title originally held by the former sovereign, the Sultan of Sulu, that title having allegedly passed in turn to Spain, the United States, Great Britain on behalf of the State of North Borneo, the United Kingdom of Great Britain and Northern Ireland and finally to Malaysia.

Having found that neither of the Parties has a treaty-based title to Ligitan and Sipadan, the Court next considers the question whether Indonesia or Malaysia could hold title to the disputed islands by virtue of the *effectivités* cited by them. In this regard, the Court determines whether the Parties' claims to sovereignty are based on activities evidencing an actual, continued exercise of authority over the islands, i.e., the intention and will to act as sovereign.

Indonesia cites in this regard a continuous presence of the Dutch and Indonesian navies in the vicinity of Ligitan and Sipadan. It adds that the waters around the islands have traditionally been used by Indonesian fishermen. In respect of the first of these arguments, it is the opinion of the Court that "it cannot be deduced [from the facts relied upon in the present proceedings] that the naval authorities concerned considered Ligitan and Sipadan and the surrounding waters to be under the sovereignty of the Netherlands or Indonesia". As for the second argument, the Court considers that "activities by private persons cannot be seen as *effectivités* if they do not take place on the basis of official regulations or under governmental authority".

Having rejected Indonesia's arguments based on its *effectivités*, the Court turns to consideration of the *effectivités* relied on by Malaysia. As evidence of its effective administration of the islands, Malaysia cites *inter alia* the measures taken by the North Borneo authorities to regulate and control the collecting of turtle eggs on Ligitan and Sipadan, an activity of some economic significance in the area at the time. It relies on the Turtle Preservation Ordinance of 1917 and maintains that the Ordinance "was applied until the 1950s at least" in the area of the two disputed islands. It further invokes the fact that the authorities of the colony of North Borneo constructed a lighthouse on Sipadan in 1962 and another on Ligitan in 1963, that those lighthouses exist to this day and that they have been maintained by Malaysian authorities since its independence. The Court notes that "the activities relied upon by Malaysia... are modest in number but... they are diverse in character and include legislative, administrative and quasi-judicial acts. They cover a considerable period of time and show a pattern revealing an intention to exercise State functions in respect of the two islands in the context of the administration of a wider range of islands". The Court further states that "at the time when these activities were carried out, neither Indonesia nor its predecessor, the Netherlands, ever expressed its disagreement or protest".

The Court concludes, on the basis of the *effectivités* referred to above, that "sovereignty over Pulau Ligitan and Pulau Sipadan belongs to Malaysia".

Composition of the Court

The Court was composed as follows: President Guillaume; Vice-President Shi; Judges Oda, Ranjeva, Herczegh, Fleischhauer, Koroma, Vereshchetin, Higgins, Parra-Aranguren, Kooijmans, Rezek, Al-Khasawneh, Buergenthal and Elaraby; Judges ad hoc Weeramantry and Franck; Registrar Couvreur.

Judge Oda appends a declaration to the Judgment of the Court; Judge ad hoc Franck appends a dissenting opinion to the Judgment of the Court.

A fuller summary of the Judgment was subsequently given in Press Communiqué No. 2002/39bis. The full text of the Judgment, Judge Oda's declaration and Judge ad hoc Franck's opinion, together with the Press Communiqués, is available on the Court's Internet site (www.icj-cij.org).

INTERNATIONAL TRIBUNAL FOR THE LAW OF THE SEA TRIBUNAL INTERNATIONAL DU DROIT DE LA MER

Press Release

ORDER IN THE *CASE CONCERNING LAND RECLAMATION BY SINGAPORE IN AND AROUND THE STRAITS OF JOHOR (MALAYSIA vs. SINGAPORE)*

Judge Dolliver Nelson, President of the Tribunal, today read the Order of the International Tribunal for the Law of the Sea in the *Case concerning Land Reclamation by*

Singapore in and around the Straits of Johor (Malaysia vs. *Singapore), Provisional Measures.*

THE DISPUTE

On 5 September 2003, Malaysia submitted a request for the prescription of provisional measures under article 290, paragraph 5, of the Convention.

The dispute concerns land reclamation activities carried out by Singapore which allegedly impinge upon Malaysia's rights in and around the Straits of Johor, which separate the island of Singapore from Malaysia.

Pursuant to article 290 of the Convention, the Tribunal, pending the constitution of the arbitral tribunal, may prescribe provisional measures if it considers provisional. Measures appropriate to "preserve the respective rights of the parties to the dispute or to prevent serious harm to the marine environment" and if it considers that *prima facie* the arbitral tribunal which is to be constituted would have jurisdiction and that the urgency of the situation so requires.

THE ORDER OF 8th OCTOBER 2003

In its Order, the Tribunal first addresses the issue of whether the Annex VII arbitral. Tribunal would *prima facie* have jurisdiction over the dispute. With respect to the obligation to exchange views set out in article 283 of the Convention, the Tribunal considers that Malaysia was not obliged to continue with an exchange of views when it concluded that this exchange could not yield a positive result. Turning to Singapore's contention that the parties by agreeing to meet on 13th and 14th August had embarked on a negotiation process, the Tribunal noted that the meeting took place after the institution of arbitral proceedings and that Malaysia had expressly stated that such meetings would be without prejudice to its right to proceed with the arbitration pursuant to Annex VII to the Convention or to request this Tribunal to prescribe provisional measures. Therefore, the Tribunal finds that the Annex VII arbitral tribunal would *prima facie* have jurisdiction over the dispute. The Tribunal also finds that the case is admissible.

The Tribunal then proceeds to analyse the contention of Singapore that, as the Annex VII arbitral tribunal is to be constituted not later than 9 October 2003, there is no need to prescribe provisional measures given the short period of time remaining before that date.

The Tribunal notes that, under article 290, paragraph 5, of the Convention, the Tribunal is competent to prescribe provisional measures prior to the constitution of the Annex VII arbitral tribunal, and that there is nothing in article 290 of the Convention to suggest that the measures prescribed by the Tribunal must be confined to that period.

With respect to the land reclamation works in the sector of Tuas, the Tribunal finds. That Malaysia has not shown that there is a situation of urgency or that there is a risk that its rights with respect to an area of its territorial sea would suffer irreversible damage pending consideration of the merits of the case by the arbitral tribunal. Therefore, the Tribunal does not consider it appropriate to prescribe provisional measures with respect to the land reclamation by Singapore in the sector of Tuas.

The Tribunal notes that during the oral proceedings Singapore, in response to the measures requested by Malaysia, reiterated its offer to share the information requested by Malaysia with respect to the reclamation works, stated that it would provide Malaysia with a full opportunity to comment on the reclamation works and their potential impacts, and declared that it was ready and willing to enter into negotiations. The Tribunal places on record these assurances given by Singapore.

With respect to the infilling work in Area D at Pulau Tekong, which is of primary. Concern to Malaysia, the Tribunal notes the commitment made by Singapore at the hearing not to undertake any irreversible action to construct the stone revetment around 'Area D' pending the completion of a study, jointly sponsored and funded by both States, to be undertaken by independent experts.

The Tribunal considers that, in the particular circumstances of this case, the land.

Reclamation works may have adverse effects on the marine environment in and around the Straits of Johor. For that reason, the Tribunal considers that prudence and caution require Malaysia and Singapore to establish mechanisms for exchanging information on and assessing the effects of the land reclamation work. For these reasons, the Tribunal, unanimously, prescribes the following provisional. Measures, pending a decision by the Annex VII arbitral tribunal: "Malaysia and Singapore shall cooperate and shall, for this purpose, enter into consultations forthwith in order to:

(a) establish promptly a group of independent experts with the mandate

    (i) to conduct a study, on terms of reference to be agreed by Malaysia and Singapore, to determine, within a period not exceeding one year from the date of this Order, the effects of Singapore's land reclamation and to propose, as appropriate, measures to deal with any adverse effects of such land reclamation;

    (ii) to prepare, as soon as possible, an interim report on the subject of infilling works in 'Area D' at Pulau Tekong;

(b) exchange, on a regular basis, information on, and assess risks or effects of, Singapore's land reclamation works;

(c) implement the commitments noted in this Order and avoid any action incompatible with their effective implementation, and, without prejudice to their positions on any issue before the Annex VII arbitral tribunal, consult with a view to reaching a prompt agreement on such temporary measures with respect to Area D at Pulau Tekong, including suspension or adjustment, as may be found necessary to ensure that the infilling operations pending completion of the study referred to in subparagraph (a)(i) with respect to that area do not prejudice Singapore's ability to implement the commitments referred to in paragraphs 85–87.

2. Unanimously,

Directs Singapore not to conduct its land reclamation in ways that might cause irreparable prejudice to the rights of Malaysia or serious harm to the marine environment, taking especially into account the reports of the group of independent experts.

3. Unanimously,

Decides that Malaysia and Singapore shall each submit the initial report referred to in article 95, paragraph 1, of the Rules, not later than 9 January 2004 to this Tribunal and to the Annex VII arbitral tribunal, unless the arbitral tribunal decides otherwise.

4. Unanimously, *Decides* that each party shall bear its own costs."

President Nelson and Judge Anderson appended a declaration to the Order.

Judges ad hoc Hossain and Oxman appended a joint declaration to the Order.

Judges Chandrasekhara Rao, Ndiaye, Jesus, Cot and Lucky appended separate opinions to the Order.

The text of the Order and the opinions appended thereto is available on the website of the Tribunal at www.itlos.org and www.tidm.org.

The Press Releases of the Tribunal, documents and other information are available on the Tribunal's websites: http://www.itlos.org and http://www.tidm.org and from the Registry of the Tribunal. Please contact Ms. Julia Pope at Am Internationalen Seegerichtshof 1, 22609

Hamburg, Germany, Tel.: + 49 (40) 35607-227, Fax: + 49 (40) 35607-245;

E-mail: press@itlos.org

16 November 2006

Year 2006

Press Release 2006/38 of the International Court of Justice

Sovereignty over Pedra Branca/Pulau Batu Puteh, Middle Rocks and South Ledge (Malaysia/Singapore)

Public hearings on the merits of the dispute to open on Tuesday 6 November 2007

THE HAGUE, 16 November 2006. The public hearings in the case concerning Sovereignty over Pedra Branca/Pulau Batu Puteh, Middle Rocks and South Ledge (Malaysia/Singapore) will open on Tuesday 6 November 2007 before the International Court of Justice (ICJ), principal judicial organ of the United Nations.

The detailed schedule for the hearings, which will be concerned with the merits of the dispute, will be published at a later date.

History of the proceedings

On 24 July 2003, Malaysia and Singapore jointly submitted to the Court a dispute concerning sovereignty over Pedra Branca/Pulau Batu Puteh, Middle Rocks and South Ledge. They did so by notifying the Court of a Special Agreement between them which was signed on 6 February 2003 at Putrajaya and entered into force on 9 May 2003.

In the Special Agreement, the Parties request the Court "to determine whether sovereignty over: (a) Pedra Branca/Pulau Batu Puteh; (b) Middle Rocks; (c) South Ledge, belongs to Malaysia or the Republic of Singapore". They agree in advance "to accept the Judgment of the Court... as final and binding upon them".

The Parties filed their Memorials and Counter-Memorials, as well as their Replies, within the time-limits fixed by the Court, taking into account the provisions of the Special Agreement, in the Orders dated 1 September 2003 and 1 February 2005, respectively.

In a joint letter dated 23 January 2006, the Parties informed the Court that they had agreed not to submit Rejoinders. The Court having decided that no further written pleadings were necessary, the case became ready for hearing.

Websites as Reference sources

International Court of Justice: http://www.icj-cij.org

UN Law of the Sea: http://www.un.org/Depts/los/index.htm

International Tribunal for the law of the Sea: http://www.itlos.org/

Florida State University, Law School Library: http://www.itlos.org/
THE DECISION
Press Release Unofficial
No. 2008/10 23 May 2008
Sovereignty over Pedra Branca/Pulau Batu Puteh, Middle Rocks and South Ledge (Malaysia/Singapore)

The Court finds that Singapore has sovereignty over Pedra Branca/Pulau Batu Puteh; that Malaysia has sovereignty over Middle Rocks; and that sovereignty over South Ledge belongs to the State in the territorial waters of which it is located.
THE HAGUE, 23 May 2008.

The International Court of Justice (ICJ), principal judicial organ of the United Nations, today rendered its Judgment in the case concerning Sovereignty over Pedra Branca/Pulau Batu Puteh, Middle Rocks and South Ledge (Malaysia/Singapore).

In its Judgment, which is final, binding and without appeal, the Court

- finds by twelve votes to four that sovereignty over Pedra Branca/Pulau Batu Puteh belongs to the Republic of Singapore;

- finds by fifteen votes to one that sovereignty over Middle Rocks belongs to Malaysia;
- finds by fifteen votes to one that sovereignty over South Ledge belongs to the State in the territorial waters of which it is located.

Reasoning of the Court

The Court first explains that the dispute between Malaysia and Singapore concerns sovereignty over three maritime features in the Straits of Singapore: Pedra Branca/ Pulau Batu Puteh (a granite island on which Horsburgh lighthouse stands), Middle Rocks (consisting of some rocks that are permanently above water) and South Ledge (a low-tide elevation).

Having described the historical background of the case, the Court notes that the dispute as to sovereignty over Pedra Branca/Pulau Batu Puteh crystallized on 14 February 1980, when Singapore protested against the publication in 1979 by Malaysia of a map depicting the island as lying within Malaysia's territorial waters. It further observes that the dispute as to sovereignty over Middle Rocks and South Ledge crystallized on 6 February 1993, when Singapore referred to the two features in the context of its claim to Pedra Branca/Pulau Batu Puteh during bilateral negotiations.

- *Sovereignty over Pedra Branca/Pulau Batu Puteh*

Malaysia contends that it has an original title to Pedra Branca/Pulau Batu Puteh (dating back from the time of its predecessor, the Sultanate of Johor) and that it continues to hold this title, while Singapore claims that the island was terra nullius in the mid-1800s when the United Kingdom (its predecessor) took lawful possession of the island in order to construct a lighthouse.

Having reviewed the evidence submitted by the Parties, the Court finds that the territorial domain of the Sultanate of Johor did cover in principle all the islands and islets within the Straits of Singapore and did thus include Pedra Branca/Pulau Batu Puteh. It establishes that this possession of the islands by the Sultanate was never challenged by any other Power in the region; and that it therefore satisfies the condition of "continuous and peaceful display of territorial sovereignty". The Court thus concludes that the Sultanate of Johor had original title to Pedra Branca/Pulau Batu Puteh. It adds that this ancient title is confirmed by the nature and degree of the Sultan of Johor's authority exercised over the Orang Laut ("the people of the sea", who inhabited or visited the islands in the Straits of Singapore, including Pedra Branca/Pulau Batu Puteh and made this maritime area their habitat).

The Court then looks at whether this title was affected by developments in the period between 1824 and the 1840s. In March 1824, the colonial Powers in the region, the United Kingdom and the Netherlands, signed a Treaty which had the practical effect of broadly establishing the spheres of influence of the two Powers in the East Indies. As a consequence, one part of the Sultanate of Johor (under Sultan Hussein) fell within the British sphere of influence while the other (under Sultan Abdul Rahman,

Sultan Hussein's brother) fell within a Dutch sphere of influence. In August 1824, Sultan Hussein ceded the island of Singapore, together with its adjacent seas, straits, and islets to the extent of 10 geographical miles from the coast of Singapore to the English East India Company in the so-called Crawfurd Treaty. Finally, in a letter of 25 June 1825, Sultan Abdul Rahman "donated" certain territories, which were already within the British sphere of influence, to his brother, thereby confirming the division of the "old" Sultanate of Johor. After careful consideration of the legal effects of these developments, the Court finds that none of them brought any change to the original title.

The Court turns next to the legal status of Pedra Branca/Pulau Batu Puteh after the 1840s to determine whether Malaysia and its predecessor retained sovereignty over the island. It observes that in order to do so, it needs to assess the relevant facts, consisting mainly of the conduct of the Parties (and of their predecessors) during the period under review.

The Court examines the events surrounding the selection process of the site of the lighthouse and the construction of the latter, as well as the conduct of the Parties' predecessors between 1852 and 1952 (in particular with respect to the British and Singapore legislation relating to Horsburgh lighthouse and in the context of the Straits lights system; constitutional developments of Singapore and Malaysia; and Johor regulation of fisheries in the 1860s), but is unable to draw any conclusions for the purposes of the case.

The Court notes that in a letter written on 12 June 1953 to the British Adviser to the Sultan of Johor, the Colonial Secretary of Singapore asked for information about the status of Pedra Branca/Pulau Batu Puteh in the context of determining the boundaries of the "Colony's territorial waters". In a letter dated 21 September 1953, the Acting State Secretary of Johor replied that the "Johore Government [did] not claim ownership" of the island.** The Court considers that this correspondence and its interpretation are of central importance "for determining the developing under-standing of the two Parties about sovereignty over Pedra Branca/Pulau Batu Puteh" and finds that the Johor's reply shows that as of 1953 Johor understood that it did not have sovereignty over Pedra Branca/Pulau Batu Puteh.

The Court finally examines the conduct of the Parties after 1953 with respect to the island. Having reviewed all arguments submitted to it, it finds that certain acts, inter alia the investigation of shipwrecks by Singapore within the island's territorial waters and the permission granted or not granted by Singapore to Malaysian officials to survey the waters surrounding the island, may be seen as conduct *à titre de souverain*. The Court also considers that some weight can be given to the conduct of the Parties in support of Singapore's claim (i.e., the absence of reaction from Malaysia to the flying of the Singapore ensign on the island, the installation by Singapore of military. communications equipment on the island in 1977, and the proposed reclamation plans by Singapore to extend the island, as well as a few specific publications and maps).

The Court concludes, especially by reference to the conduct of Singapore and its predecessors *à titre de souverain*, taken together with the conduct of Malaysia and its predecessors including their failure to respond to the conduct of Singapore and its predecessors, that by 1980 (when the dispute crystallized) sovereignty over Pedra Branca/Pulau Batu Puteh had passed to Singapore. The Court thus concludes that sovereignty over Pedra Branca/Pulau Batu Puteh belongs to Singapore.

- *Sovereignty over Middle Rocks and South Ledge*

Malaysia claims that the two maritime features have always been under Johor/Malaysian sovereignty while Singapore's position is that sovereignty over the features goes together with sovereignty over Pedra Branca/Pulau Batu Puteh.

With respect to Middle Rocks, the Court observes that the particular circumstances which led it to find that sovereignty over Pedra Branca/Pulau Batu Puteh rests with Singapore clearly do not apply to Middle Rocks. It therefore finds that original title to Middle Rocks should remain with Malaysia as the successor to the Sultanate of Johor.

As for South Ledge, the Court notes that this low-tide elevation falls within the apparently overlapping territorial waters generated by Pedra Branca/Pulau Batu Puteh and by Middle Rocks. Recalling that it has not been mandated by the Parties to draw the line of delimitation with respect to their territorial waters in the area, the Court concludes that sovereignty over South Ledge belongs to the State in the territorial waters of which it is located.

Composition of the Court

The Court was composed as follows: Vice-President Al-Khasawneh, Acting President in the case; Judges Ranjeva, Shi, Koroma, Parra-Aranguren, Buergenthal, Owada, Simma, Tomka, Abraham, Keith, Sepúlveda-Amor, Bennouna, Skotnikov; Judges ad hoc Dugard, Sreenivasa Rao; Registrar Couvreur.

Judge Ranjeva appends a declaration to the Judgment of the Court; Judge Parra-Aranguren appends a separate opinion to the Judgment of the Court; Judges Simma and Abraham append a joint dissenting opinion to the Judgment of the Court; Judge Bennouna appends a declaration to the Judgment of the Court; Judge ad hoc Dugard appends a dissenting opinion to the Judgment of the Court; Judge ad hoc Sreenivasa Rao appends a separate opinion to the Judgment of the Court.

A summary of the Judgment appears in the document "Summary No. 2008/1", to which summaries of the declarations and opinions are annexed. In addition, this press release, the summary, and the full text of the Judgment can be found on the Court's website (www.icj-cij.org) under "Press Room" and "Cases".

** Cited in Singapore's Memorial: Letter of 21 September 1953 is appended immediately below.

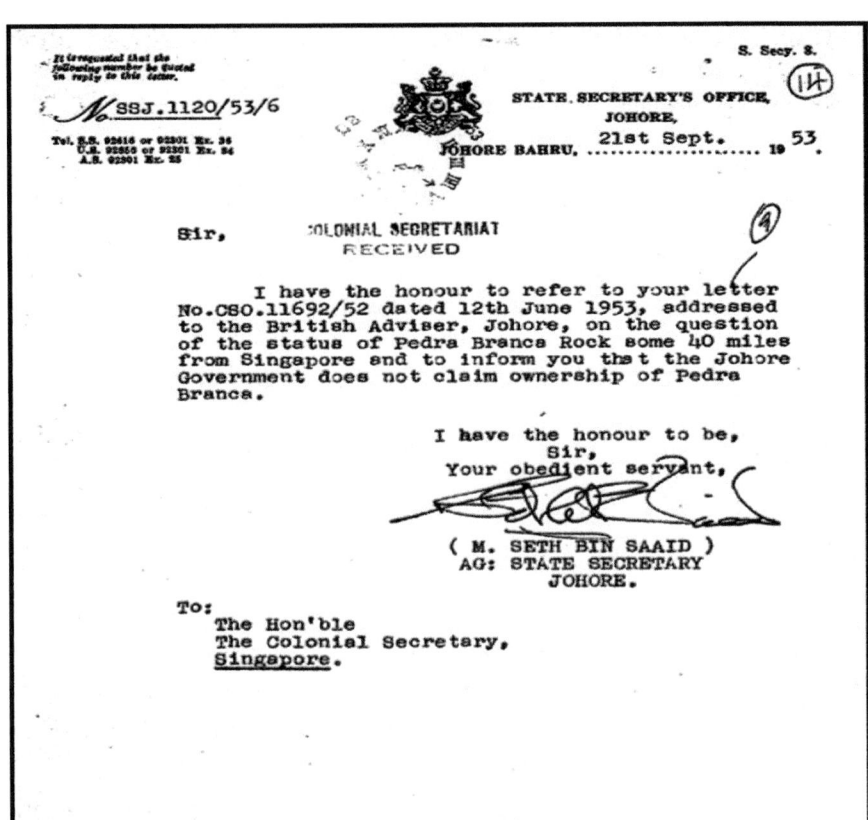

Annex III
Maritime Heritage and Cession of Territory

Introduction

An appreciation and understanding of the political territorial changes that evolved in Malaysia's history can be derived through the study of a vast array of Agreements, Treaties and pieces of legislation—primary documents—that may be cited in published works and housed in archives, public and private libraries within Malaysia and in other countries, for example, the Netherlands, Portugal, Singapore and the United Kingdom.

Within the prose of the Treaties cited in this Appendix the spelling of places and titles names may differ to those used in the body of this document. This is intentional. To that end, some local words are used, and their meanings are given below. Some clarification is therefore needed here.

For example, the term *'chop'* is a Chinese word locally used to mean 'seal' as in an official seal for an important, legal document. The expression *Iang de per Tuan*, he who ruleth (corruptly contracted into Iam Tuan or Yam Tuan) is a title of Malay sovereigns of the highest rank. An *orlong* is equal to 6,400 square yards or about one acre and a third, and, as a lineal measure is consequently 80 yards, the square root of 6,400. The word *campong* (kampong) is used as a Quarter or District of a Town. *Rooma Bechara* is a Council House or courthouse.

Agreements and Treaties Relating to Cession of Land and Boundary Determination

The Colony of the Straits Settlements

The Colony of the Straits Settlements comprised the Islands of Singapore, the Town and Province of Malacca, the Territory and Islands of the Dindings, The Island of Penang, Province of Wellesley, Christmas Island, the Cocos Islands, and the Island of Labuan, and their dependencies. For Administrative purposes the Colony was divided into four Settlements:

© The Editor(s) (if applicable) and The Author(s), under exclusive license to Springer Nature Switzerland AG 2025
V. L. Forbes, *Malaysia's Maritime Jurisdictional Limits*,
https://doi.org/10.1007/978-3-031-78783-6

The Settlement of Singapore, which also included Christmas Island and the Cocos Islands.

The Settlement of Penang, which also included Province of Wellesley and the Territory and Islands of the Dindings.

The Settlement of Malacca; and

The Settlement of Labuan.

This Briefing traces the history of acquisition of the Settlements and places and presents a list of Agreements and Treaties that support these actions.

European Connections

British connection with Malaya began by individual and trading ventures from 1576 to 1684. Sir Francis Drake came to Malaya in 1578 during his famous voyage round the world; Ralph Fitch in 1583; Sir Thomas Cavendish arrived in 1588; Lancaster in 1592 and Wood in 1596 in the course of their first voyages East. From 1594 to 1598 Houtman's Dutch expedition visited and traded in the various coastal settlements of Malaya. The importance of these early trading voyages is that vast financial gains and profits accrued from them that the English became convinced of the great value of Malayan trade with the direct result that in 1600 the East India Company (EIC) was formed and received a Royal Charter for 15 years.

The principal objective of the EIC was trade with Malaya. Thus, it may be fairly said that the early Malayan trade was the framework from which the British Indian Empire arose, and as will be seen Malaya, so far as it was administered by the British, was directly connected with India until 1684, from which date until 1762 it was a mixed commercial and political connection.

In 1684, the EIC's Government in Madras established a fort and factory at Indrapoer, and on 25 June 1685, Fort York at Bencoolen, from the establishment of which fort may be dated the dawn of British Administration in Malaya.

In 1763, the Fort and establishment at Bencoolen were formed into a separate Presidency. With a footing on Sumatra, the British desired a trading post on the Malay Peninsula aware that the Dutch, their great rivals, were established at Malacca and had been for over one hundred years.

The British accordingly judged it necessary to establish a commercial port in the Strait of Malacca and at first, Acheen was considered the proper place, but negotiations by the end of 1784 proved fruitless. Then Captain Light proposed the Island of Penang and in 1786 negotiations were opened with the King of Kedah for the cession of the Island: these proved successful (Agreement of 1786), and Captain Light and some Marines landed at Penang on 15 July 1786. On 11 August 1786, the British flag was hoisted. It was on the eve of the birthday of the Prince of Wales in whose honour the Island was re-named Prince of Wales' Island.

Agreement with the King of Quedah, for the Cession of Prince of Wales' Island, 1786

Article 1

Condition

That the Honourable Company shall be guardian of the seas; and whatever enemy may come to attack the King, shall be an enemy to the Honourable Company, and the expense shall be borne by the Honourable Company.

Reply

This Government will always keep an armed vessel stationed to guard the Island of Penang, and the coast adjacent, belonging to the King of Quedah.

Article 2

Condition

All vessels, junks, prows, small and large, coming from either east or west, and bound to the port of Quedah, shall not be stopped or hindered by the Honourable Company's Agent, but left to their own wills, either to buy and sell with us, or with the Company at Pulo Penang, as they shall think proper.

Reply

(Basically, stating that any person acting for, or on behalf of the Company will be left entirely of their own free will either to trade with the King or with the Agent of the Company).

Article 3

Condition

The articles opium, tin and rattans, being part of our revenue, are prohibited; and Qualla Mooda, Pyre, and Krean, places where these articles are produced, being so near to Penang, that when the Hon. Company's Resident remains there, this prohibition will be constantly broken through; therefore it should end, and the Governor-General allow us our profits on these articles, namely, 30,000 Spanish dollars every year.

Reply

(The G-G will take care to ensure the King will not be a sufferer by an English settlement being formed on the Island).

Article 4

(Relates to credits given to the King's relations, etc.)

Article 5

Any man in this country, without exception, be it son or brother, who shall become an enemy to us shall then become an enemy to the Honourable Company; nor shall the Hon. Company's Agent protect them, without breach of this Treaty, which is to remain while Sun and Moon endure. [Emphasis added]

Article 6

Condition

If any enemy come to attack us by land, and we require assistance from the Hon. Company, of men, arms or ammunition, the Hon. Company will supply us at our expense.

Reply

This Article will be referred for the orders of the English East India Company, together with such parts of the King of Quedah's requests as cannot be compiled with previous to their consent being obtained.

Treaty with the King of Quedah, 1791

Preamble

In the Hegira of our Prophet, 1205, year Dalakir, on the 16th Moon Saban, on the day Ahat. (Corresponding to 20 April 1971)

Whereas, on this date, this writing showeth that the Governor of Pulo Penang, vakeel of the English Company, concluded peace and friendship with his Highness, Iang de per Tuan of Quedah, and all his great officers and ryots of the two countries, to live in peace by sea and land, to continue as long as the Sun and Moon give light: the Articles of Agreement are:

Article 1

The English Company will give to His Highness … six thousand Spanish Dollars every year, for as long as the English shall continue in possession of Pulo Penang.

Article 2

His Highness … agrees that all provisions, wanted for Pulo Penang, the ships of war, and the Company's ships, may be bought at Quedah, without impediment, or being subject to any duty.

Article 3

All slaves running from Quedah to Pulo Penang, or from Pulo Penang to Quedah, shall be returned to their owners.

Article 4

[Persons in debts running from their creditors will be returned to their creditors]

Article 5

The Ruler of Quedah will not allow Europeans of any other nation to settle in any part of his country.

Article 6

[Company will not receive any person committing high treason or rebellion against the Ruler].

Article 8

All persons stealing chops (forgery) to be given up likewise.

Article 9

All persons, enemies to the English Company, and Iang de per Tuan shall not supply them with provisions.

These nine articles are settled and concluded, and peace is made between the Ruler of Quedah and the English Company; Quedah and Pulo Penang shall be as one Country.

This done and completed by
…In this Agreement, whoever departs from any part herein written, God will punish and destroy; to him there shall be no health.

The seals of …are put to this writing, with each person's handwriting.

Signed, sealed and executed, in Fort Cornwallis, on Prince of Wales' Island, this 1st day of May, in the year of our Lord 1791.

(Signed) F. Light.

Treaty with Kedah, 1791

On 1 May1791, a Treaty was concluded between the King of Kedah and Captain Light which provided for the mutual surrender of runaway slaves, debtors, forgers, and murderers: for the necessary supply of provisions, duty free, from the mainland to the residents on the Island and shipping in the harbour, and for the annual payment to the King of 6,000 Spanish dollars.

Treaty with the King of Quedah, 1800

Preamble

In the year …(corresponding to 6 June 1800) … agreed on and concluded a Treaty of Friendship and alliance …of the two countries … to continue on sea and land, as long as the Sun and Moon retain their motion and splendour: the Articles of which Treaty are as follows:

Article 1

The English Company are to pay annually to His Highness ... of Purlies and Quedah, ten thousand Dollars, as long as the English shall continue in possession of Pulo Penang, and the country on the opposite coast hereinafter mentioned.

Article 2

His Highness ... agrees to give to the English Company, for ever, all that part of the sea-coast that is between Qualla Krean and the River side of Qualla Mooda, and measuring inland from the sea side sixty Orlangs; the whole length above mentioned to be measured by people appointed by the Ruler and the Company's people. The English Company are to protect this coast from all enemies, robbers, and pirates that may attack it by sea, from north and south. [Emphasis added]

Article 3

[Provides for all kinds of provisions and ships of war and the Company's ships may be purchased at Purlies and Quedah ... without impediment or being subject to any Duty or Custom; ... and all boats going from Pulo Penang to Purlies and Quedah for such purposes will be provided proper passports, to prevent impositions].

Article 4

[Provisions for slaves, etc.]

Article 5

[Relates to debitors and creditors]

Article 6

[Europeans of other nations not permitted to settle in any part of the Ruler's dominions].

Articles 7–12, inclusive

Relate to criminals, rebels, enemies, and general irresponsible actions against the Ruler will be apprehended and returned in bonds; persons bringing produce down the river are not to be molested or impeded by the Company's people; goods required from Pulo Penang by the Ruler are to be procured by the Company's Agents, and the amount deducted from the gratuity.

Article 13

Arrears of gratuity ... to be paid off

Article 14

Ratification of the Treaty

Approved and confirmed by the Governor-General in Council, November 1802.

The King was also bound not to allow Europeans of other nations to settle in this country. The Treaty was expressed to continue "as long as the Sun and Moon give light", and it appeared to have been negotiated under the impression that the King of Kedah was an independent sovereign, whereas he was in reality a tributary to Siam. The British Government over Penang was, however, expressly acknowledged by the Siamese under the *Treaty of Bangkok* of 20 June 1826.

On 6 June 1800, Sir George Leith concluded a new Treaty with the King of Kedah which annulled the Agreement of 1786 and the Treaty of 1791. By this new Treaty, the British obtained the cession of the district known as the Province of Wellesley, and it became and remained part of the Settlement of Penang. This Treaty was not approved and confirmed by the Governor-General in Council until November 1802.

Preliminary Agreement with the Dato Temengong of Johore, 1819

[Agreement made by…] the Dato Tummungung Sree Maharajah, Ruler of Singapore, who governs the country of Singapore and all the islands which are under the government of Singapore in his own name and in the name of Sree Sultan Hussein Mahummud Shah, Rajah of Johore, with Sir Thomas Stamford Raffles, Lieutenant Governor of Bencoolen and its dependencies on behalf of the Most Noble the Governor of Bengal.

On account of the long existing friendship and commercial relations between the English Company and the countries under the authority of Singapore and Johore it is well to arrange these matters on a better footing never to be broken.

Article 1

The English Company can establish a factory (*logi*) situated at Singapore or other place in the Government of Singapore-Johore.

Article 2

On account of that the English Company agree to protect the Dato Tummungung Sree Maharajah.

Article 3

On account of the English Company having the ground on which to make a factory they will give each year to the Dato … three thousand Dollars.

Article 4

The Dato … agrees that as long as the English Company remain and afford protection according to this Agreement he will not enter into any relations with or let any other nation into his country other than the English.

Article 5

Ref: Signing and sealing of the document.

4th day of Rabil Ahkir in the year 1234 (corresponding to 30 January 1819)

(Signed) T.S. Raffles

Johore, 1819

Treaty of Friendship and Alliance

Preamble

Article 1

The Preliminary Articles of the Agreement entered into on 30 January 1819, … are hereby entirely approved, ratified, and confirmed by His Highness.

Article 2

Ref: The East India Company agree and engage to pay to His Highness the sum of Spanish Dollars five thousand annually, for and during the time that the said Company … maintain a factory or factories on any part of His Highness' hereditary dominions … The English Company in no way bound to interfere with the internal politics of the State, or engage to assert or maintain the authority of His Highness by force of arms.

Article 3

His Highness … granted his full permission to the EIC to establish a factory or factories …in recompense and in return for the said grant, settled …the yearly sum of Spanish Dollars three thousand and having received His Highness into their alliance and protection …is hereby confirmed.

Articles 4 and 5

Relate to His Highness agreeing to aid and assist the EIC against all enemies; and in turn, the EIC will support His Highness and will not enter into any treaty with any other nation, and will not admit or consent to the settlement in any part of their dominions of any other power, European or American.

Article 6

All persons belonging to the English factory or factories, or who shall hereafter desire to place themselves under the protection of its flag, shall be duly registered and considered as subject to the British Authority.

Article 7

[Mode of administering justice]

Article 8

The Port of Singapore is to be considered under the immediate protection and subject to the regulation of the British authorities.

Article 9

With regard to the Duties which it may hereafter be deemed necessary to levy on goods, merchandise, boats or vessels, His Highness …is to be entitled to a moiety or full half of all the amount collected from native vessels.

The expenses of the port and of the collection of Duties are to be defrayed by the British Government.

Signed: 6 February 1819 T.S. Raffles

Johore 1819

Arrangements Made for the Government of Singapore, in June 1819

Preamble

Article 1

The boundaries of the islands under the control of the English are as follows: from Tanjong Malang on the west, to Tanjong Katang on the east, and on the land side, as far as the range of cannon shot, all round the factory. As many persons as reside within the aforesaid boundary, and not within the *campongs* of the Sultan and Tumungong, are all to be under the control of the Resident, and with respect to the gardens and plantations that now are, or may hereafter be, made, they are to be at the disposal of the Tumungong, as heretofore, but it is understood, that he will always acquaint the Resident of the same.

Article 2

It is directed that all the Chinese move over to the other side of the river, forming a *campong* from the site of the large bridge down the river towards the mouth, and all Malays, people belonging to the Tumungong and others, are also to remove to the other side of the river, forming their *campong* from the site of the large bridge up to the river towards the source.

Articles 3–6 (inclusive)

Relate to justice; meetings of Heads and local council officials, airing of grievance or complaints.

Article 7

No Duties or Customs can be exacted, or farms established in this Settlement without the consent of the Sultan, the Tumungong, and Major William Farquhar, and without the consent of these three nothing can be arranged.

Signed: 26 June 1819 W. Farquhar, *Late Resident*

Johore 1823

Memorandum by Sir Stamford Raffles

The gist of this Memorandum is that it lays down the rules by the Lt. Governor and concurred in by their Highnesses, to form the basis of the good understanding to be maintained in the future:

First: In order to contribute to the personal comfort and respectability of their Highnesses, and at the same time to afford them an ample and liberal compensation … on account of port duties, tributes, or profits on monopolies, …their Highnesses are, in the first instant, to receive a monthly payment, His Highness the Sultan of 1500 dollars, and His Highness the Tumungong 800 dollars per month, on the following conditions;

Second: Their Highnesses to forego all right and claim to the monopoly of Kranjee and Baloo wood within Singapore, and the islets immediately adjacent, as well as all claims to presents and customs upon Chinese junks and Chinese generally coming and going.

Third: With the exception of the land appropriated to their Highnesses for their respective establishments, all land within the island of Singapore, and islands immediately adjacent, to be at the entire disposal of the British Government.

Fourth: [relates to further advances of sums of money to the Highnesses].

Fifth: [their Highnesses relieved of further personal attendances at the Court every Monday…]

Sixth: [… the laws and customs of the Malays respected, where they shall not be contrary to reason, justice, or humanity. In all other cases the laws of the British authority will be enforced …]

Seventh: The British Government do not interfere at present in the local arrangement of the countries and islands subject to their Highnesses' authority, beyond Singapore and its adjacent islands, further than to afford them general protection as heretofore.

Signed: (early June 1823) T.S. Raffles

Johore 1824

A Treaty of Friendship and Alliance between the English East India Company and their Highnesses …

Article 1

Mutual respect

Article 2

Their Highnesses …hereby cede in full sovereignty and property to the Hon. EEIC, their heirs and successors for ever, the Island of Singapore, situated in the Straits of Malacca, together with the adjacent seas, straits, and islets to the extent of ten geographical miles, from the coast of the said main Island of Singapore.

Article 3

The EEIC to pay His Highness the Sultan the sum of 33,200 Spanish dollars, ... to His Highness the Dato Tumungong the sum of 26,800 Spanish dollars.

Article 4

[Further elaboration of payment]

Article 5

[Their Highnesses to receive honours, respect, and courtesy ... when they reside at, or visit the Island of Singapore.]

Articles 6 and 7

[Further payments if their Highnesses were to remove from Singapore]

These Articles were annulled as far as they relate to the Temengong by the Treaty of Johore of 1862, which provided for the Temengong's property in the Island of Singapore.

Article 8

[Their Highnesses agreed that as long as they receive their respective monthly stipends they will not enter into an alliance ... with any foreign power or potentiate ...

Articles 9–13 (inclusive)

[These articles make provision for: Ensuring personal protection of their Highnesses; suppression of robbery and piracy within the Straits of Malacca; free and unshackled trade everywhere within their Dominions; and retainers or followers of their Highnesses who desert from their actual service not permitted to dwell or remain on the Island.]

Article 14

[All former Conventions, Treaties, or Agreements between the parties ... are hereby abrogated and annulled by the Present Treaty ... have conferred on the Hon. EEIC the right of occupation and possession of the Island of Singapore.

(Signed) 19 November 1824 Geo. Swinton,

Secretary to Government

ACQUISITION OF MALACCA AND SINGAPORE

The next events in the history of the Colony, were the acquisitions of Malacca and Singapore, which lead to the granting of the Second Charter,

In 1511, the Portuguese under Albuquerque drove the Malays out of Malacca and were in turn expelled by the Dutch in 1641. The Dutch made Malacca the headquarters of their Malayan enterprise until 1795 when the British took it and retained it until 1801 when they restored it to the Dutch. In 1807, the British again occupied Malacca and retained it until 1818 when under the Treaty of Vienna of 1814, it was restored to the Dutch once more.

Although Malacca was not actually restored until 1818 the Treaty of Vienna was dated 13 August 1814. The British were therefore bound in 1814 to restore Malacca to the Dutch and it accordingly became necessary to obtain an outpost which would command the Malacca Strait if Britain were not to lose its trade.

Singapore was the place eventually selected, and a preliminary Agreement was entered into on 30 January 1819, with the Dato' Temenggong of Johore, the Chieftain of the Island of Singapore. On 6 February 1819, a formal Treaty was entered into by Sir Stamford Raffles for the East India Company and the Sultan Husain of Johore and Dato' Temenggong.

By this Treaty, the preliminary Agreement was ratified, the Company agreed to pay the Sultan 5,000 Spanish dollars annually as long as they maintained a factory on any part of the Sultan's hereditary dominions, and the Company further agreed to afford its protection to the Sultan, who on his part agreed to enter into Treaties with no other nation.

On 26 June 1819, further Arrangements were made between the Sultan, the Temenggong, Sir Stamford Raffles and Major Farquhar whereby the boundaries of the lands under the control of the English were settled and other provisions made

... for the better guidance of the people of this Settlement, pointing out where all the different castes were severally to reside with their families and captains, or heads of their campongs.

ABSOLUTE CESSION OF SINGAPORE

When Singapore was first occupied by the British, there were only a few Malay fishermen on the island under the Dato' Temenggong who exercised a jurisdiction over them in accordance with Malayan custom. The EIC, by the 1819 Treaty only purchased the right and authority to establish a factory and consequent necessary settlement upon the island. The success of the venture was rapid and startling and was administered in so admirable a way as to attract rapidly many settlers, so that in 1824 it was deemed desirable to obtain absolute cession of the Island in full sovereignty.

With this in mind, a new Treaty was entered into with the Sultan and Dato Temenggong under the provisions of which the Island of Singapore with the Seas, Straits, and Islets within 10 Geographical Miles from its coasts became British property. The Company engaged to pay to the Sultan 33,200 Spanish Dollars and a stipend during his natural life of 1,300 Spanish Dollars [see note on Spanish dollar, below] *per mensem*, and to the Dato' Temenggong 26,800 Spanish Dollars and a monthly stipend of 700 Spanish Dollars during his natural life.

In 1823, Singapore was placed directly under administration of Fort William, under which Bencoolen also was.

By the *Treaty of London* of 17 March 1824 between the British and the Dutch, the occupation of Singapore was confirmed by the Dutch and Malacca was finally restored to the British. On 24 June 1824, the Statute 5 Geo. IV c.108 was passed whereby Malacca and Singapore were transferred to the EIC and thereby by virtue of the Statute 39 and 40 Geo. III c.79 became subordinate to Fort William and subject to the jurisdiction of the Supreme Court of Judicature in Fort William, Calcutta.

Treaty of London, 1824

Treaty between Great Britain and The Netherlands, respecting Territory and Commerce in the East Indies, signed at London, 17 March 1824.

Preamble

Pleasantries

Who, after having mutually communicated their Full Powers, found in good and due form, have agreed on the following Articles:

Article I

The High Contracting Parties engage to admit the Subjects of each other to trade with Their Possessions in the Eastern Archipelago, and on the continent of India, and in Ceylon, upon the footing of the most favoured Nation; Their respective Subjects conforming themselves to the local Regulations of each Settlement.

Article II

The Subjects and Vessels of one Nation shall not pay, upon importation or exportation, at the Ports of the other in the Eastern Seas, any Duty at a rate beyond the double of that at which the Subjects and Vessels of the Nation to which the Port belongs, are charged.

[More on duty on exports and imports]

Article III

[Treaty made with any Native Power in the Eastern Seas *vis- a- vis* unequal Duties on goods]

Article IV

(The Contracting Parties) engage to give strict Orders, as well to their Civil and Military Authorities, as to their Ships of War, to respect the freedom of Trade, established in Articles I, II, and III; and, in no case, to impede a free communication of the Natives in the Eastern Archipelago …

Article V

(The Contracting Parties), in like manner, engage to concur effectually in repressing Piracy in those Seas; They will not grant either asylum or protection to Vessels engaged in Piracy, and They will, in no case, permit the Ships or merchandise captured by such Vessels, to be introduced, deposited, or sold, in any of their Possessions.

Article VI

It is agreed that Orders shall be given by the Two Governments to their Officers and Agents in the East, not to form any new Settlement on any of the Islands in the Eastern Seas, without previous Authority from their respective Governments in Europe.

Article VII

The Molucca Islands, and especially Amboyna, Banda, Ternate, and their immediate Dependencies, are excepted from the operation of Articles I to IV until the Netherland Government shall think fit to abandon the monopoly of Spices; but if the said Government shall, at any time previous to such abandonment of the monopoly, allow the Subjects of any Power, other than a native Asiatic Power, to carry on any Commercial Intercourse with the said Islands, the Subjects of His Britannick Majesty shall be admitted to such Intercourse, upon a footing precisely similar.

Article VIII

His Netherland Majesty cedes to His Britannick Majesty all His Establishments on the Continent of India; and renounces all privileges and exemptions enjoyed or claimed in virtue of those Establishments.

Article IX

The Factory at Fort Marlborough and all the English Possessions on the Island of Sumatra, are hereby ceded to His Netherland Majesty; and his Britannick majesty further engages that no British Settlement shall be formed on that Island, nor any Treaty concluded by British Authority, with any Native Prince, Chief, or State therein.

Article X

The Town and Fort of Malacca, and its dependencies, are hereby ceded to His Britannick Majesty; and His Netherland Majesty engages, for Himself and His Subjects, never to form any Establishment on any part of the Peninsula of Malacca, or to conclude any Treaty with any Native Prince, Chief or State therein.

Article XI

His Britannic Majesty withdraws the objections which have been made to the occupation of the Island of Billiton and its dependencies, by the Agents of the Netherland Government.

Article XII

His Netherland Majesty withdraws the objections which have been made to the occupation of the Island of Singapore, by the Subjects of His Britannic Majesty.

His Britannick Majesty, however, engages, that no British Establishment shall be made on the Carimon Isles, or on the Islands of Battam, Bintang, Lingin, or on any of the other Islands South of the Straits of Singapore, nor any Treaty concluded by British Authority with the Chiefs of those Islands.

Article XIII

All the Colonies, Possessions and Establishments which are ceded by the preceding Articles shall be delivered up to the Officers of the respective Sovereigns on 1 March 1825. [Provision made herein for the Fortifications to remain in the state in which notification of this Treaty in India; no claim made shall be made by either side for ordinance, or stores etc.

Article XIV

All the Inhabitants of the Territories hereby ceded enjoy for a period of six years from the date of Ratification of the Treaty the liberty of disposing, as they please, of their property, and of transporting themselves, without let or hindrance, to any country to which they wish to remove.

Article XV

The High Contracting Parties agree that none of the Territories or Establishments mentioned in the preceding Articles, shall at any time, be transferred to any other Power. In case of any of the said Possessions being abandoned by one of the present Contracting Parties, the right of occupation thereof shall immediately pass to the other.

Article XVI

It is agreed that all accounts and reclamation arising out of the restorations of Java, and other Possessions, to the Officers of His Netherland Majesty in the East Indies,— as well those which were the subject of a Convention made at Java on 24 June 1817, between the Commissioners of the Two nations, as all others shall be finally and completely closed and satisfied, on the payment of the sum of one hundred thousand pounds, sterling money, to be made in London on the part of the Netherlands, before the expiration of the Year 1825.

Article XVII

The present Treaty shall be ratified, and the Ratifications exchanged at London, within Three Months from the date hereof, or sooner if possible.

Done at London, 17 March 1824

Signed by: George Canning, Charles W. W. Wynn, H. Fagel, and A.R. Falck

Other provisions were made the better to carry into effect the cession, and all prior treaties and arrangements were abrogated. Singapore was held by virtue of this Treaty which was dated 2 August 1824.

On 5 July 1825, the Statute 6 Geo. IV c.85 was passed whereby the King was declared to have the power to make provision for the administration of justice in Singapore and Malacca and power was given to the Directors of the EIC to declare the two places to be annexed to Penang and to be part of that Settlement or to make them separate Settlements.

Treaty with Perak, 1826

Engagement of … Supreme Ruler over the Perak Country, made and delivered to Captain James Low … and which is to be everlasting, as the revolutions and endurance of the Sun and Moon. [Emphasis added]

The Sultan, who governs the whole of the Perak Country and its dependencies, has this day … given over and ceded to the Hon. EIC of England, to be under its Government henceforward and for ever, the Pulo Dinding and the Islands of Pankgor, together with all and every one of the Islands which belonged of old and until this period to the Kings of Perak, and which have been hitherto included within the Perak State, because the said Islands afford safe abodes to the pirates and robbers, who plunder and molest traders on the coast and the inhabitants of the main land, and effectually deprive them of the means of seeking subsistence, and as the King of Perak has not the power or means singly to drive out those pirates.

For these reasons, the King of Perak has, of his own free will and pleasure, ceded and given over as aforesaid, the Islands as aforesaid, to the Hon. EIC, to be kept and governed by them, and to be placed under any one of their governments, as they think fit. To this deed, as tokens of its validity, have this day been put the great seal or chop of the Ruler of Perak Country …

This Deed is made and written this 16 day of Rabi-al-awal, Wednesday, 1242 or 18 October 1826.

THE DINDINGS ACQUIRED

On 18 October 1826, a treaty was signed with the Sultan of Perak whereby he ceded to the EIC the Island of Pangkor and the Sembilan Islands nominally in order to bring about the suppression of piracy of which the islands formed the headquarters. The Islands were as a fact never occupied until 1874 when by the *Treaty of Pangkor* the British obtained confirmation of the cession and the addition of that piece of territory known as the Dindings. The Dindings were at first administered by a British official from Perak but shortly after became a part of the Settlement of Penang for administrative purposes.

The Second Charter

By 27 November 1826, the date of the Second Charter that was granted upon the petition of the EIC, an extension of the Court's jurisdiction to cover Singapore and Malacca was established. The Court of Judicature of Prince of Wales' Island, Singapore and Malacca, and its constitution remained the same as before with Penang as its headquarters.

In 1832, the headquarters of Government was removed from Penang to Singapore. On 1 August 1851, the Straits Settlements by Proclamation, ceased to be subordinate to Fort William in Bengal and were placed in direct correspondence with the Government of India.

The Transfer of 1867

The *Government of the Straits Settlements Act* 29 and 30 Vic. C 115 was passed on 10 August 1866. It provided for the separation of the Straits Settlements from the Government of India, by Order in Council on 28 December 1866, and 1 April 1867, was the day fixed for enactment of the Statute.

The Statute was passed in consequence of the growing importance of the Settlements and the hampering effect of the connection with India. It gave the Crown power to establish laws, institutions, and ordinances, and to make provision for the Courts and administration of justice and generally for the good government of the Straits Settlements. Power was given to Her Majesty to delegate by *Letters Patent* her powers and authorities as to the Straits Settlements to resident officers, and also powers to and authorities as to Labuan.

By *Letters Patent* of 4 February 1867, Her Majesty delegated to the Legislative Council of the Straits Settlements legislative authority within the Colony and other powers to the Governor and the Executive Council. By *Straits Settlements Act I* of 1867 all appointments under the Indian Government were avoided save as to officers holding office under the Charter of 1855.

By *Straits Settlements Act III* (later termed Ordinances) of 1867 the Governor of the Straits Settlements ceased to be a Judge of the Court. In 1867 by Statute 30 and 31 Vic. c.45 the Admiralty jurisdiction of the Court was extended and assimilated to similar Courts existing in other British Colonies under the designation of the Vice-Admiralty of the Straits Settlements, which continued until 1890.

Engagement Entered into by the Chiefs of Perak at Pulo Pangkor

Dated 20 January 1874

Whereas, a state of anarchy exists in the Kingdom of Perak owing to the want of a settled government in the Country, and no efficient power exists for the protection of the people and for securing to them the fruits of their industry, and,

Whereas, large numbers of Chinese are employed and large sums of money invested in Tin mining in Perak by British subjects and others residing in Her Majesty's

Possessions, and the said mines and property are not adequately protected, and piracy, murder and arson are rife in the said country, whereby British trade and interests greatly suffer, and the peace and good order of the neighbouring British Settlements are sometimes menaced, and,

Whereas, Certain Chiefs ...have stated their inability to cope with their present difficulties ... and have requested assistance, and

Whereas, HM Government ... to assist its rulers, now,...

His Excellency, Governor of the Colony ... proposed the Following Articles of arrangement as mutually beneficial to the Independent Rulers of Perak ...

     I   Recognition of Sultan of Perak
    II   Retain pension and small Territory assigned to him
   III   Nomination of Great officers
    IV   Confirmation of power over Larut
     V   Revenues collected and appointments made in the name of the Sultan
    VI   Provision of Residence
   VII   Governor of Larut will have an Assistant Resident
  VIII   Cost of Resident to be a first charge to the Revenues of Perak
    IX   Civil list income also charged to Revenue of Perak
     X   Collection and control of Revenue regulated under the advice of the Resident
    XI   That the Treaty under which the Pulo Dinding and the Islands of Pangkor were ceded to Great Britain having been misunderstood and it being desirable to re-adjust the same, so as to carry into effect the intention of the Framers thereof, it is hereby declared that the Boundaries of the said Territory so ceded shall be rectified as follows, that is to say:

From Bukit Sigari, as laid down in Chart Sheet No. 1 Straits of Malacca a tracing of which is annexed, marked A, in a straight line to the sea, thence along the sea coast to the South, to Pulo Katta a line running North East about five miles, and thence North to Bukit Sigari.

   XII   That the Southern watershed of the Krean River, that is to say, the portion of land draining into that River from the South be declared British Territory, as a rectification of the Southern Boundary of Province Wellesley. Such Boundary to be marked out by Commissioners; one named by the Government of the Straits Settlements, and the other by the Sultan of Perak.
  XIII   On the cessation of disturbances in Perak, restoration of infrastructure etc.; payment of damages etc. will be made.
  XIV   Acknowledgment of debt due by the Mantri of Larut to the Government of theStraits Settlements.

This done and concluded at Pulo Pangkor, in the British Possessions, this 20th day of January 1874

Executed before me,

ANDREW CLARKE,

Governor, Commander-in-Chief, and Vice-Admiral of the Straits Settlements.

ANNEXATION OF THE COCOS TO THE STRAITS SETTLEMENTS, 1886

In 1857 the Cocos or Keeling Islands, which were discovered by Captain Keeling of the EIC, in 1609, were annexed in 1857 by Great Britain and in 1878 placed under the Government of Ceylon. On 1 February 1866, the Cocos or Keeling Islands were annexed to the Straits Settlements, by an Order of Her Majesty in Council and placed under the Government of the Colony.

Christmas Island Annexed to the Colony, 1900

Christmas Island was annexed by Great Britain in 1888 and a settlement was established there by a party of twenty persons from the Cocos Islands. By Letters Patent of 8 January 1889, the Governor of the Straits Settlements was made Governor of the Island. By a Proclamation of 23 May 1900, Christmas Island was annexed to the Colony of the Straits Settlements and by Ordinance XIV of 1900 provision was made for administrative purpose part of the Settlement of Singapore and the Supreme Court was given jurisdiction over the Island.

Labuan connection

The first connection of the British with Labuan was the occasion in 1775 when the British were expelled by the Sulus from Balambangan and took refuge on the island. The British Colony was obtained by cession as a *quid pro quo* for assistance in suppressing piracy. The then Sultan of Borneo made the offer in conjunction with the Raja Muda Hussin in a document addressed to Queen Victoria in 1844. In November 1846, possession was taken of Labuan and a Treaty was signed with the Sultan on 24 May 1847. Labuan became a Crown Colony, and it was given a Governor.

On 30 October 1906, by Letters Patent it was ordained that Labuan should become part of the Straits Settlements. On 1 January 1907, Labuan with its dependencies was made part of the Settlement of Singapore for administrative purposes and the boundaries of the Colony were extended to include it.

Labuan remained a part of the Settlement of Singapore until 1 December 1912, when by an Order of the Governor in Council, Ordinance III of 1911 came into operation repealing Ordinance I of 1907. This Ordinance made Labuan a separate settlement and certain alterations were made in the laws applying to it.

In 1910 the Merchant Shipping Ordinance XXXII of that year was passed, and it formed a code of maritime law for the Colony.

SUMMARY OF THE LAW PREVAILING IN THE COLONY OF THE STRAITS SETTLEMENTS

To sum up the law prevailing in the Colony of the Straits Settlement was as follows:

1. The common law, equity, civil and statute law prevailing in England on the 26 November 1826, so far as they were applicable to the circumstances of the Colony, and modified in their application to these circumstances and so far as they had not been altered by:

    (a) Statutes passed prior to 1 April 1867, extending to India, or passed after that extending to the Colony;

    (b) Indian Acts passed prior to 1 April 1867, and applying to the Straits Settlements;

    (c) Orders of the Crown in Council made under the *Government of the Straits Settlements Act*, 1867, 29 and 30 Vic. C. 115; or

    (d) Ordinances of the Colonial legislature.

2. The Statute law extending to India passed prior to 1 April 1867, and passed after that extending to the Colony;

3. Such Indian Acts passed prior to 1 April 1867, as applied to the Straits Settlements and had not been repealed;

4. Orders of the Crown in Council under the *Government of the Straits Settlements Act*, 1866, 29 and 30 Vic. c. 115;

5. Ordinances of the Colonial legislature; and

6. Such Mercantile law as was introduced by Sect. 6 of the Civil law Ordinance 1909.

Agreements and Treaties Relating to Cession of Land and Boundary Determination

The primary documents re-produced, in part, below, offer an insight into a brief period in Malaysia's complex geo-political and rich historical background when large tracts of land and islands were ceded to the English East India Company (EEIC) and as a consequence to British administration via the Government of India and later through Westminster on the demise of the EIC. These archival materials assist in an understanding of the issues that manifest themselves in present-day international relations in the context of boundary alignments.

NORTH BORNEO

TREATY OF FRIENDSHIP AND COMMERCE BETWEEN HER MAJESTY AND THE SULTAN OF BORNEO

Signed in English and the Malay languages, at Brunei, 27 May 1847

Preamble

Article I

Peace, friendship, and good understanding …for ever subsist between … [Parties to this Treaty]

Article II

The subjects of Her Britannic Majesty shall have full liberty to enter into, reside in, and trade with …

Article III

British subjects shall be permitted to purchase, rent, or occupy, or in any legal way to acquire …

Article IV

No article whatsoever shall be prohibited from the territories … trade will be perfectly free…

Article V

No duty exceeding one dollar per registered ton shall be levied on British vessels entering the ports …

Article VI

No export duty levied on goods ex Borneo

Article VII

Ships of war of HM and those ships of the EIC free to enter ports etc.…

Article VIII

British-flagged ships will be rendered assistance if wrecked upon the coast of Borneo

Article IX

[The Parties] hereby engage to use every means in their power for the suppression of Piracy within the seas, straits and rivers subject to their respective control or influence; … not to grant either asylum or protection to any persons or vessels engaged in piratical pursuits; and in no case will he permit ships, slaves, or merchandise, captured by pirates, to be introduced into his dominions, or to be exposed therein for sale …

Article X

It is desirable that British subjects should have some port … His Highness the Sultan hereby confirms the cession already spontaneously made by him in 1845, of the Island of Labuan … together with the adjacent Islets of Kuraman, Little Rusakan, Greater Rusakan, da-at, and Malankasan, and all the straits, islets, and seas, situated halfway between the fore-mentioned islets and the mainland of Borneo.

Likewise the distance of 10 geographical miles from the Island of Labuan to the westward and northward, and from the nearest point half way between the Islet of Malankasan and the Main land of Borneo, in a line running north till it intersects a line extended from west to east from a point 10 miles to the northward of the extremity of the Island of Labuan, to be possessed in perpetuity and in full sovereignty by Her Britannic Majesty and her successors; and in order to avoid occasions of difference which might otherwise arise, His Highness the Sultan engages not to make any similar cession, either of an Island, or of any settlement on the main land, in any part of his dominions, to any other nation, or to the subjects or citizens thereof, without the consent of Her Britannic Majesty.

Article XI

Her Britannic Majesty being greatly desirous of effecting the total abolition of the Trade in Slaves ... engages to suppress all such traffic ...and that all subjects of His Highness the Sultan who may be engaged in the Slave Trade, may, together with their vessels, be dealt with by cruisers of Her Majesty, as if such persons and their vessels had been engaged in piratical undertaking.

Article XII

This Treaty shall be ratified, and the ratifications thereof shall be exchanged at Bruni, within 12 months after this date. (27 May 1847)

Signed: James Brooke   His Highness, The Sultan of Borneo

SABAH DOCUMENT of 20 June 1891

CONVENTION BETWEEN GREAT BRITAIN AND THE NETHERLANDS DEFINING THE BOUNDARIES IN BORNEO

Signed at London

Her Majesty the Queen of the United Kingdom of Great Britain and Ireland ...Her Majesty the Queen-Dowager, Regent of the Netherlands ... being desirous of defining

The boundaries between the Netherlands Possessions in the Island of Borneo, and the States in that Island which are under British Protection, have resolved to conclude a Convention to that effect, and have appointed as their Plenipotentiaries for that purpose (that is to say):-

[Names of persons], ...who, having produced their full powers, found in good and due form, have agreed upon the following Articles:-

Article I

The boundary between the Netherlands Possessions in Borneo and those of the British –protected States in the same island shall start from 4° 10′ North latitude on the east coast of Borneo.

Article II

The boundary line shall be continued westward from 4° 10′ north Latitude and follow in a west-north-west direction, between the Rivers Simengaris and Soedang, up to the point where the meridian 117° East Longitude crosses the parallel 4° 20′ N. Latitude, with a view of including the Simengaris River within Dutch territory. The boundary line shall then follow westward the parallel 4° 20′ N Latitude until it reaches the summit of the range of mountains which forms on that parallel of the watershed between the rivers running to the north-west coast and those running to the east coast of Borneo, it being understood that, in the event of the Simengaris River or any other river flowing into the sea below 4° 10', being found on survey to cross the proposed boundary-line within a radius of 5 geographical miles, the line shall be diverted so as to include such small portions or bends of rivers within Dutch Territory; a similar concession being made by the Netherland Government with regard to any river debouching above 4° 10' on the Territory of British North Borneo, but turning southwards.

Article III

From the summit of the range of mountains mentioned in Article II, to Tandjong-Datoe on the west coast of Borneo, the boundary-line shall follow the watershed of the rivers running to the north-west and west coasts, north of Tandjong-Datoe, and to those running to the west coast south of Tandjong-Datoe, the south coast, and the east coast south of 4° 10 north Latitude.

Article IV

From 4° 10' north latitude on the *east coast* the boundary-line shall be continued eastward along that parallel, across the Island of Sebittik; that portion of the island situated to the north of the parallel shall belong unreservedly to the British North [Borneo] Company, and the portion south of that parallel to the Netherlands. [Emphasis added]

Article V

The exact positions of the boundary-line, as described in the four preceding Articles, shall be determined hereafter by mutual agreement at such times as the Netherlands and the British Governments may think fit.

Article VI

The navigation of all rivers flowing into the sea between Batoe-Tinagat and the River Siboekoe shall be free, except for the transport of war material; and no transport duties shall be levied on other goods passing up those rivers.

Article VII

The population of Boelongan shall be allowed to collect jungle produce in the territory between the Simengaris and the Tawao Rivers for 15 years from the date of the signature of the present Convention, free from any tax or duty.

Article VIII

The present Convention shall be ratified, and it shall come into force three months after the exchange of the ratifications, which shall take place at London one month, or sooner, if possible, after the said Convention shall have received the approval of the Netherland States-General.

Done at London, in duplicate, this 20th day of June 1891

Signed: Salisbury (for UK)   C. De Bylandt (for The Netherlands)

Present author's note

[The reader may be interested in viewing the International Court of Justice's *Case Concerning Sovereignty over Pulau Ligitan and Pulau Sipadan* (Indonesia/Malaysia) 17 December 2002 General List 102, available online. Sections 36–58 offers arguments by the Parties to the Dispute and their interpretations of the provision of Article IV of the Treaty.]

SABAH DOCUMENT of 28 September 1915

AGREEMENT BETWEEN GREAT BRITAIN AND THE NETHERLANDS RELATING TO THE BOUNDARY BETWEEN THE STATE OF NORTH BORNEO AND THE NETHERLAND POSSESSIONS IN BORNEO

Signed at London, 28 September 1915

Preamble

[The Parties] having agreed in a spirit of mutual goodwill to confirm the joint report with the accompanying map prepared ... in accordance with Article V of the Convention signed at London on 20 June 1891, for the limitation of the boundary line ... in the island; the undersigned, duly authorised to that effect, hereby confirm the aforesaid joint Report and map, as signed by their Commissioners at Tawao on 17 February 1913.

Followed by the Names of signatories

We have travelled in the neighbourhood of the frontier from 8 June 1912 to 30 January 1913, during which period the Netherland Commission has made the necessary astronomical observations and topographical surveys, the results of which we declare to be correct and sufficient for the determination of the boundary.

Where physical features did not present natural boundaries conformable with the provisions of the Boundary Treaty of 20 June 1891, we have erected the following pillars:

(a) Two pillars on the opposite banks of the Pentjiangan River, both marked "G.P. I."
(b) One pillar on the right bank of the Agisan River, marked "G.P. 3."
(c) One pillar on the left bank of the Seboeda River marked "G.P. 2."
        All being on the parallel 4° 20′ North latitude.

We have determined the boundary between the Netherland territory and the State of British North Borneo, as described in the Boundary Treaty supplemented by the interpretation of Article II of the Treaty mutually accepted by the Netherland and British Governments in 1905 as taking the following course:-

(1) Traversing the island of Sibetik, the frontier line follows the parallel of 4° 10′ north Latitude, as already fixed by Article IV of the Boundary Treaty and marked on the east and west coasts by boundary pillars.

(2) Starting from the boundary pillar on the west coast of the Island of Sebetik, the boundary follows the parallel of 4° 10′ north latitude westward until it reaches the middle of the channel, thence keeping a mid-channel course until it reaches the middle of the mouth of the Troesan Tamboe.

(3) From the mouth of Troesan Tamboe the boundary line is continued up the middle of this Troesoan until it is intersected by a similar line running through the middle of Troesan Sikapal as far as the point where the latter meets the watershed between the Simengaris and Seroedong Rivers (Sikapal Hill) and is connected with this watershed by a line taken perpendicular to the centre of Troesan Sikapal.

(4) From the point where this watershed (Sikapal Hill) meets Troesan Sikapal the boundary line follows the watershed until the latter joins Mount Bemboeding.

NOTE—There is thus included in the Netherland territory all the country that is drained by the Simengaris River and its tributaries, while all the country that is drained by the Seroedong River and its tributaries is included in the territory of British North Borneo.

(5) Leaving the junction point of the Simengaris-Seroedong watershed with Mount Bemboeding, the boundary line follows successively:-

    (a) Mount Bemboeding in a northerly direction.

    (b) Mount Pemantoengan Bagas and Mount Meliat in a westerly direction.

    (c) Mount Keblajoeng in a south-easterly direction.

    (d) The watershed between the Karawangan and Apat Rivers in a south-westerly direction.

    (e) The Inoeloeh Ketek Hill in a northerly direction.

    (f) The watershed between the Loeloewejen and Siangan streams in a westerly direction.

    (g) The most western spur of this watershed, intersected by a straight line running due east from boundary pillar "G.P.2.".

    (h) Along this straight line as far as pillar "G.P.2."

NOTE—There is thus included in the Netherland territory all the country that is drained by any of the following rivers and their tributaries: Seboeloeh, Mesaloei, Tempilan, Apat, and Toelit, together with the Seboeda south of Lat. 4° 20′ N., and its eastern tributaries debouching south of that parallel; and in the territory of British North Borneo all the country drained by the Seroedong and its tributaries, and by the Seboeda north of Lat. 4° 20′ N., and by the tributaries of the Seboeda debouching above that parallel.

(6) From the pillar "G.P.2." the boundary line follows successively:

(a) A straight line running due west as far as the most eastern spur of the watershed between the Linemoejoe and Labau streams, intersected by this straight line.

(b) The above-mentioned spur.

(c) The watershed between the Linemoejoe and Labau streams.

(d) The watershed between the Labau and Baling streams.

(e) The watershed (the Sinogo Ridge) between the Agisan and Seboeda Rivers.

(f) The watershed between the Lakoetan and Makalap streams

(g) The western spur of the latter watershed, intersected by a straight line running due east from boundary pillar "G.P.3".

(h) This straight line itself as far as pillar "G.P.3".

NOTE—There is thus included in the Netherland territory all the country drained by the western tributaries of the Seboeda and the eastern tributaries of the Agisan debouching below Lat. 4° 20′ N., and in the territory of British North Borneo the country drained by the corresponding debouching above that parallel.

(7) From the pillar "G.P.3" the boundary line follows successively-

(a) A straight line running due west as far as the most eastern spur of the watershed between the Klawsan and Mesaloei streams intersected by this line.

(b) The above-mentioned spur

(c) The watershed between the Klawasan and Mesaloei streams.

(d) The watershed (Potoetan Ridge) between the Sesoegon and Agisan streams.

(e) The main watershed between the Sembakoeng and Seboekoe Rivers in a south-westerly direction.

(f) The watershed (Mount Boedjoek Bah) between the Samentebel, with its tributaries, and the Semantaloen, with its tributaries.

(g) The watershed (Mount Boedjoek Bah) between the Samentebel, with its tributaries, and the Seliman and Semangawat, with their tributaries.

(h) The watershed (Poegisiai Hill) between the Semandapi, with its tributaries, and the Sementebel, with its tributaries, as far as the Toenangan Hill.

(i) The watershed between the Semandapi, with its tributaries, and the eastern tributaries of the Pentiangan River debouching below Lat. 4° 20′ N., as far as the intersection of this watershed with a straight line running due east from the pillar "G.P.1" on the left bank of the Pentjiangan.

(j) This straight line itself.

NOTE—There is thus included in the Netherland Territory all the country drained by the Agisan River South of Lat. 4° 20′ N., and by the western tributaries of the Agisan debouching below that parallel, and by the eastern tributaries of the Pentjiangan debouching below the same parallel; and in the territory of British North Borneo all the country drained by the Agisan River north of Lat. 4° 20′ N., and by the western tributaries of the Agisan and eastern tributaries of the Pentjiangan debouching above Lat. 4° 20′ N.

(8) From the Pillar "G.P.1" on the left bank of the Pentjiangan River the boundary follows successively:-

(a) A line running due west to the pillar "G.P.1" on the right bank.

(b) The first hill spur south of the Lombai stream as far as its junction with the main watershed between the tributaries of the Pentjiangan debouching north of Lat. 4° 20′ N., and the tributaries that debouch south of that latitude.

(c) The last-named watershed as far as the Seselatan hill.

(d) The watershed, or series of watersheds, dividing the northern tributaries of the Sedalir that debouch above Lat. 4° 20′ N., from those that debouch below that parallel.

(e) The most western spur of this watershed, or series of watersheds, intersected by parallel Lat 4° 20′ N.

(f) The parallel 4° 20′ N, crossing the Sedalir River until it meets the most eastern spur of the watershed, or series of watersheds, between the southern tributaries of the Sedalir that debouch above parallel 4° 20′ N and those that debouch below that parallel, in conformity with Article 2 of the Treaty.

(g) The last-named watershed, or series of watersheds (and, if necessary, the watershed between the Sedalir and the Sesajap Rivers), until they meet the main watershed described in Article 3 of the Treaty.

NOTE—There is thus included in the Netherland Territory all the country drained by the Pentjiangan below 4° 20′ N, by the Sedalir below that parallel, but the tributaries of both these rivers debouching below 4° 20′ N., and by the Seasajap River; and in the territory of British North Borneo any country that drained by the Pentjiangan north of 4° 20′ N, by the Sedalir north of that parallel, and by the tributaries of both of these rivers debouching north of lat. 4° 20′ N.

To the above we have all agreed and appended our signature at Tawao, British North Borneo, this 17th day of February 1913.

Signatures of: J.H.G. Schepers; E.A. Vreede; H.W.L. Bunbury; G. St. V. Keddell

In witness whereof the undersigned have signed the present Agreement and have affixed thereto their seals.

Done at London, the 28th day of September 1915.

Signatures and seals of E. Grey R. de Marees Van Swinderen

SABAH DOCUMENT of 2 January 1930

Boundary Agreement between Great Britain and the United States

(Abbreviated form)

Convention delimiting the boundary between the Philippines Archipelago and the State of North Borneo

Preamble

Article I

It is hereby agreed and declared that the line separating the islands belonging to the Philippine Archipelago on the one hand and the islands belonging to the State of North Borneo which is under British protection on the other hand shall be and is hereby established as follows:

> From the point of Lat. 4° 45' N. and Lon. 120° 00 E being a point of the boundary defined by the Treaty between the US and Spain, signed at Paris, 10 December 1898 a line due south along Lon. 120° 0' E. to its point of intersection with Lat. 4° 23' N.;
>
> thence due west along Lat. 4° 23' to its intersection with Lon. 119° 0'E;
>
> thence due N. along Lon. 119° 0' to its intersection with Lat. 4° 42' N.;
>
> thence in a straight line approximately 45 ° 54' true to the intersection of Lat. 5° 16' N. Lon. 119° 35' E.;
>
> thence in a straight line approximately 314° 19' True to the intersection of Lat. 6° 0'N., Lon. 118° 50' E.;
>
> thence due west to the intersection of Lat. 6° 0' N., Lon. 118° 20' E.;
>
> thence in a straight line approximately 307° 40' True passing between Little Bakkungaan Island and Great Bakkungaan Island to the intersection of Lat. 6° 17' N., Lon. 117° 58' E.;
>
> thence north along Lon. 117° 58' E to its intersection of Lat. 6° 52' N.,
>
> thence in a straight line approximately 315° 16' True to the intersection of Lat. 7° 24' 45'' N., Lon. 117° 25' 30'' E.;
>
> thence in a straight line approximately 300° 36' true through the Mangsee Channel between Mangsee Great Reef and Mangsee Islands to the intersection of Lat. 7° 40' N., Lon. 117° 0' E., the latter point being on the boundary defined by the Treaty between USA and Spain signed at Paris, on 10 December 1898.

Article II

The line described above has been indicated on Charts No. 4707 and 4720 ... and are attached to the Treaty. It is agreed that if more accurate surveying and mapping of North Borneo, the Philippine Islands, and intervening islands shall in the future show that the line described above does not pass between Little and Great Bakkungaan Islands as indicated on the Chart No 4720, the boundary line shall be understood to be defined in that area as a line passing between the above-named islands as indicated on the Chart, said portion of the line being a straight line approximately 307° 40' True drawn from a point on Lat. 6° 0' N. to a point on Lon. 117° 58' E.

It is likewise agreed that if more accurate surveying and mapping show that the line described above does not pass between Mangsee Islands and Mangsee Great Reef ... the boundary shall be understood to be defined in that area as a straight line drawn from the intersection of Lat 7° 24' 45'' N., Lon. 117° 25' 30'' E., passing through Mangsee Channel as indicated on the Chart to a point on the parallel of Lat. 7° 40' N.

Article III

All islands to the north and east of the said line and all islands and rocks traversed by the said line, should there be any such, shall belong to the Philippine Archipelago and all islands to the south and west of the said line shall belong to the State of North Borneo.

Article IV

[Refers to provisions of Article 19 of the Treaty between USA, British Empire, France, Italy, and Japan limiting naval armament … all islands in the Turtle and Mangsee Groups, etc.]

Article V

[Ratification of Treaty and entry into force]

Signatures of: Henry L. Stimson, Esme Howard

SABAH DOCUMENTS of 1946-48

Exchange of Notes between UK and the Philippine Republic regarding the transfer of the Administration of the Turtle and Mangsee Islands to the Philippine Republic.

Dated: Manila, 19 September 1946

Sir,

By the Convention concluded …2 January 1930, and in an Exchange of … the Government of UK acknowledged the following groups of islands as comprised within the Philippine Archipelago:

(1) Sibaung, Boaan, Lihiman, Langgan, Great Bakkungaan, Taganak, and Baguan in the group of islands known as Turtle Islands;
(2) The Mangsee Islands.

… The Transfer of the administration shall be effected within one year after such notice is given on a day and in a manner to be mutually arranged.

Accept, and c.,

Elpido Quirino

ANNEXATION OF THE CONTINENTAL SHELF

Following World War II and as it became apparent that technological advances would make feasible the exploration and exploitation of the seabed and substratum adjacent to the coastline some maritime states issued unilateral declarations of sovereignty over contiguous submarine territory and even over adjacent seas beyond the traditionally accepted three nautical mile limit. In 1953 a report by the International Law Commission to the United Nations recommended that maritime states limit their claims to the continental shelf only and to a line at the 200-m depth. Moodie (1956) offered a discussion on the problems of maritime boundaries. The issues raised were apparent. Some maps from China, for example, indicate a claim to the continental shelf under a vast area of the South China sea off Borneo. By this stage, no firm set of principles had been accepted that the "present position concerning the legal status of the continental shelf and its superjacent sea was hardly more than one of 'claim-staking'. The post-war 'claim-staking' seems to have derived by the so-called Truman Declaration, a proclamation by the US President on 28 September 1945.

In 1954 Britain extended the boundaries of her colony of North Borneo to include the continental shelf around its shores by the *North Borneo Alteration of Boundaries Order in Council*, 1954. (*Statutory Instruments*, 1954, No 838 and *British and Foreign State Papers*, CLXI, 24-25).

Article 2 of the Order stated:

The Boundaries of the Colony of North Borneo are hereby extended to include the area of the continental shelf being the seabed and its subsoil which lies beneath the high seas contiguous to the territorial waters of North Borneo.

SARAWAK

The original grant, in 1841, of Sarawak by the Sultan of Brunei to James Brooke comprised the territory between Cape Datu and the entrance of the Samarahan River, an area of about half of the present (1960) First Division (province) of Sarawak. In 1843, Brooke sought and received permanent cession of Sarawak to himself and his heirs. In his Last Will, Raja Charles Brooke, on 16 December 1913, gave devise and bequeathed sovereignty of Sarawak to his sons and their heirs.

Deeds of Cession over various portions of land were signed over a period of time. Many Deeds of Cession stated the time of day and date of the signing and often comments, such as, This Agreement can never be changed—The truth is evident— would appear in the document.

The annexation of the coastal strips of land from the Samarahan River northward to Kidurong Point was a direct result of the campaign to suppress the Malay-Iban piracy and inter-tribal warfare. The price for cession varied, for example in the Sarawak Treaty of 24 August 1853, in ceding the country of Sarawak with its dependencies from Tanjong Datu to the entrance of the Samarahan River it made provision for "whoever succeeds to the government must pay four thousand large dollars to the government of the Sultan of Bruni".

At the end of World War II, Raja Vyner Brooke, in seeking aid to re-habilitate Sarawak after the occupation by the Japanese forces, sought assistance from Britain. A provisional Government was established but without much success and in February 1946, the Raja then offered to cede the country to Britain and the new Labour Government of Britain accepted.

The document that authorised cession of Sarawak—Order Number C-24 (Cession of Sarawak) 1946—to His Majesty the King of Great Britain was signed on 18

May 1946 at Kuching. The Instrument of cession was signed on 21 May 1946. It inferred that the interests of the inhabitants of Sarawak and the territory of the State of Sarawak, and the full sovereignty and dominion over the state, should be ceded to Britain.

SARAWAK DOCUMENT of 26 March 1928

Convention between the UK and the Netherlands respecting the Delimitation of the Frontier between the States in Borneo under British Protection and the Netherlands Territory in that Island

Preamble

Being desirous of further delimiting part of the frontier established in Article 3 of the Convention signed at London on 20 June 1891 … have agreed as follows:

Article 1

The boundary as defined in Article 3 of …, is further delimited between the summits of Gunong Api and the Gunong Raja as described in the following Article and as shown on the map attached to this Convention.

Article 2

From the Gunong Api triangulation station, where a pillar has been erected, the boundary line bears approximately WSW for a distance of 450 m to a point on Gunong Api, where a pillar has been erected; thence it follows the watershed downwards in an approximately SW direction for a distance of 650 m, where a pillar has been erected; thence in an approximately W by N direction for a distance of 700 m to a point on the footpath from Gumbang astronomical station to Siding, where a pillar has been erected; thence following the footpath generally in a NE direction for a distance of 500 m to point Batu Aum, where a pillar has been erected; thence in a straight line bearing approximately NW by N to the first stream, tributary of the Odong River, a distance of 600 m; thence following the tributary on its right bank in a generally W. direction to the confluence with the Odong River, where a pillar has been erected; thence a straight line bearing approximately NW to a point on the left bank of the Tring stream, where a pillar has been erected; thence in a generally NW direction to a point on the left bank of the Toepijem River where a pillar has been erected; thence in a generally NNW direction to appoint on the right bank of the Pon stream, where a pillar has been erected; thence following the crest in a generally W by N direction to a point on the left bank of the Meroemo River, where a pillar has been erected; thence across the Meroemo River to a point on its right bank, where a pillar has been erected; thence following the watershed in a generally NW direction to the Gunong Brunei triangulation station, where a pillar has been erected.

From the Gunong Brunei triangulation station the boundary line bears approximately NW in a straight line to the Gunong Jagoi (Poko Payong triangulation station, where a pillar has been erected at the point where this straight line cuts the footpath between Billeh and Jagoi Babang.

From the Gunong Jagoi triangulation station the boundary line follows the right bank of the Boewan River (which has its source within 50 m of the above-mentioned pillar) in a generally N direction to its confluence with the Berenas River, at which point a pillar has been erected. Pillars have also been placed on the right bank of the Boewan River at the point where it is crossed by the footpath between Setaas and Siloewas and where it is crossed by the footpath between Setaas and Gunong Raja. From the point of confluence of the Boewan and Berenas Rivers the boundary follows the Berenas River on its right bank to its confluence with the Separan River, where a pillar has been erected on the right bank of the Separan River opposite to that of the right bank of the river. From the latter pillar the boundary line bears approximately NW in a straight line to the Gunong Raya triangulation station, where a pillar has been erected. From this triangulation station the boundary line bears approximately ENE along the crest of the Gunong Raya to its top, where on account of the nature of the soil no pillar has been erected.

Article 3

The present convention shall be ratified and shall come into force 3 months after the exchange of the acts of ratification, which shall take place at The Hague as soon as possible.

In witness whereof the respective plenipotentiaries have signed the present convention and have affixed thereto their seals. Done in duplicate at The Hague, 26 March 1928.

Signatures of: Granville, Beelaerts van Blokland, and Koningsberger

Annexation of the Continental Shelf

As in the case of North Borneo, Britain in 1954, provided for the incorporation of the continental shelf off the coast within the boundaries of Sarawak (The Sarawak (Alteration of Boundaries) Order in Council, 1954). (*Statutory Instruments*, 1954, No 838 and *British and Foreign State Papers*, CLXI, 24-25; 25-26).

Article 2 of the Order stated:

The Boundaries of the Colony of Sarawak are hereby extended to include the area of the continental shelf being the seabed and its subsoil which lies beneath the high seas contiguous to the territorial waters of Sarawak.

MALAYSIA AND THAILAND

TERRESTRIAL POLITICAL BOUNDARY AGREEMENT, 29 NOVEMBER 1899

The Government of Her Britannic Majesty on the one part, acting in the names and on behalf of the Sultans of Perak and Pahang, and the Government of His Siamese Majesty on the other part, considering that it is desirable to settle all frontier disputes in the Malay Peninsula, and to define the boundaries between the abovenamed States of Perak and Pahang, on the one side, and the Siamese province of Raman and the Siamese dependencies of Kedah, Kelantan and Tringanu on the other, Her Britannic Majesty's Minister Resident and His Siamese Majesty's Minister of Foreign Affairs, duly authorized to that effect, have agreed as follows:

I. The boundary between Perak and Kedah is as follows:

> From the point on the Krian River near Bukit Toongal along the Krian River to its
> Source in Bintang as shown in the map annexed to this Agreement, and marked (A to B).

II. The boundary between Perak and Raman, as shown on the map annexed to this Agreement, and marked (B, C, D, E, F) is as follows:
(1) A straight line from Bintang to Kenderung, from (B to C).
(2) A straight line from Kenderung to a point on the River Rui, about four miles above its mouth, from (C to D).
(3) From the point marked (D) to a straight line to the end of the spur on the Perak River near Jeram Pala, marked (E), which marks the northern drainage of the River Sengo.
(4) The line of the northern drainage of the River Sengo to the main watershed, from (E to F).
III. The boundary between Perak and Pahang on the one side, and Kelantan on the other, is the main watershed.
IV. The boundary between Pahang and Tringanu is -
(1) The main watershed.
(2) Then the southern drainage of the Kemanan River until it meets the watershed of the Chendar River.
(3) Then the northern drainage of the Chendar River to Tanjong Glugor on the south coast.

In witness whereof the undersigned have signed the same in duplicate and have affixed thereto their seals at Bangkok on the 29th day of November in the year 1899 of the Christian era, corresponding to the 118th year of Ratanakosindr.

George Granville and Devawongse Varoparkar

References

Allen, J. de V., Stockwell, A. J. and Wright, L. R. (1981) *A collection of treaties and other documents affecting the states of Malaysia* 1761–1963, *two volumes*. Oceana Publications, Inc.

Braddell, R. St. John (1982) *The law of the straits settlements* (p. 278). Oxford University Press.

Lamb, A. (1968) *Asian frontiers studies in a continuing problem* (p. 246). Pall Mall Press.

Moodie, A. E. (1956). Maritime boundaries. In W. Gordon East, & A. E. Moodie (Eds.), *The changing world* Chapter XL. Harrap.

Moorhead, F. J. (1957). *A history of Malaya and her neighbours* (Vol. 1 and 2) Longman, Green and Co. Ltd.

Prescott, J. R. V. (1975). *Map of mainland Asia by treaty* (518 pp). Melbourne University Press.

Prescott, J. R. V., Collier, H. J., & Prescott, D. F. (1975). *Frontiers of Asia and Southeast Asia* (p. 106). Melbourne University Press.

List of Treaties

1. Agreement with the King of Quedah (Kedah), for the Cession of Prince of Wales' Island, 1786 (contains six Articles)
2. Treaty with the King of Quedah, 1791—Signed sealed and executed, in Fort Cornwallis, on Prince of Wales' Island, 1 May 1791 (contains nine Articles)
3. Treaty with the King of Quedah, 1800—Approved and confirmed by the Governor- General in Council, November 1802 (contains 14 Articles)
4. Preliminary Agreement with the Dato Temengong of Johore, 1819—Sealed and signed 30 January 1819 (contains five Articles)
5. Johore 1819 Treaty of Friendship and alliance concluded between the Hon. Sir Thomas Stamford Raffles …for the Hon. East India Company on the one part, and their Highnesses Sultan Hussain Mahummed Shah, Sultan of Johore, and Dato Tumumgong Sri Maharajah Abdul Raham, Chief of Singapore, and its Dependencies, on the other part. Signed and concluded 6 February 1819 (contains nine Articles).
6. Johore 1819 Arrangements made for the Government of Singapore, in June 1819.
   Placement of seals and signatures made on 26 June 1819 (contains seven Articles)
7. Johore 1823 Memorandum by Sir Stamford Raffles. Concluded June 1823 (seven points stated here).
8. A treaty of Friendship and Alliance between the Hon. EIC … and their High-nesses the Sultan and Tumungong of Johore … concluded on Second Day of August (1824) … on behalf of the said Hon. EIC. Ratified on 19 November 1824 (contains 14 Articles)
9. Treaty between Great Britain and the Netherlands, respecting Territory, and Commerce in the East Indies, signed at London, 17 March 1824. (Contains 17 Articles)
10. Treaty with Perak, signed on 18 October 1826
11. Engagement entered into by the Chiefs of Perak at Pulo Pangkor, 20 January 1874 (16 sections and carrying the official seals of nine dignitaries).
12. 12 Treaty of friendship and Commerce between Her Majesty and the Sultan of Borneo. Signed, in the English and Malay languages, at Bruni, 27 May 1847. (Contains 13 Articles.)

ACTS

1. *The Admiralty Offences (Colonial) Act,* 1849; 12& 13 Vic. C.96
   An Act to provide for the Prosecution and trial in Her Majesty's Colonies of offences committed within the Jurisdiction of the Admiralty, 1 August 1849
2. *The Colonial laws Validity Act,* 1865; 28 & 29 Vic. C. 63 29 June 1865
   An Act to Remove Doubts as to the Validity of Colonial Laws, 19 June 1865
3. *The Government of the Straits Settlements Act,* 1866 29 & 30 Vic. C. 115
   An Act to provide for the Government of the "Straits Settlements" 10 August 1866
4. *The Courts (Colonial) Jurisdiction Act,* 1874 37 & 38 Vic. c.27
   An Act to Regulate the Sentences imposed by Colonial Courts where jurisdiction to try is conferred by Imperial Acts 30 June 1874
5. *The Straits Settlements Offences Act,* 1874 37 & 38 Vic. c. 38
   An Act to extend Jurisdiction … to certain Crimes and Offences committed out of the Colony 30 July 1874
6. *The Territorial Waters Jurisdiction Act,* 1878 41 & 42 Vic. c. 73 16 August 1878
   An act to regulate the Law relating to the Trial of Offences on the Sea within a certain distance of the Coasts of Her Majesty's Dominions 16 August 1878.
7. *The Colonial Courts of Admiralty Act,* 1890; 53 &54 Vic. c. 27
   An Act to amend the law respecting the exercise of Admiralty Jurisdiction in Her Majesty's Dominions and elsewhere out of the United Kingdom, 25b July 1890

LETTERS PATENT

1. Letters patent passed under the Great Seal of the United Kingdom, constituting the Office of Governor and Commander-in-Chief of the Straits Settlements and their Dependencies [17 February 1911]
2. Instructions passed under the Royal Sign manual and Signet to the Governor and Commander-in-Chief of the Straits Settlements and their Dependencies [17 February 1911]
3. Standing Rules and Orders of the Legislative Council

FEDERAL TREATIES

1. Federal Treaty of July 1895
2. Federal Treaty of January 1948
3. Malayan Treaty of 12 October 1957
4. Malayan (Malaysian) Treaty of 9 July 1963
5. Malayan (Malaysian) Treaty of 11 September 1963
6. Dutch Treaty of 17 March 1824
7. Dutch Treaty of 2 November 1871
8. Siamese Treaty of 31 July 1825
9. Siamese treaty of 20 June 1826
10. Siamese Treaty of 10 March 1909

Annex IV
Malaysia as Party to Conventions and Treaties

Treaties and Conventions to which Malaysia is a Member	
Short Title	Long Title
–	ASEAN Agreement on the Conservation of Nature and Natural Resources
APFIC	Asia Pacific Fishery Commission
World Heritage Convention	Convention concerning the Protection of the World Cultural and Natural Heritage
CBD	Convention on Biological Diversity
Living Res. of the High Seas Convention	Convention on Fishing and Conservation of the Living Resources of the High Seas
CITES	Convention on International Trade in Endangered Species of Wild Fauna and Flora
Continental Shelf Convention	Convention on the Continental Shelf
Basel Convention	Convention on the Control of Transboundary Movements of Hazardous Wastes and their Disposal
High Seas Convention	Convention on the High Seas
IMO Convention	Convention on the International Maritime Organization
Territorial Sea Convention	Convention on the Territorial Sea and the Contiguous Zone
Ramsar Convention	Convention on Wetlands of International Importance especially as Waterfowl Habitat
IOFC	Indian Ocean Fishery Commission
IPFC	Indo-Pacific Fishery Commission
IPPC	International Plant Protection Convention
MP	Montreal Protocol for the Protection of the Ozone Layer

(continued)

© The Editor(s) (if applicable) and The Author(s), under exclusive license to Springer Nature Switzerland AG 2025
V. L. Forbes, *Malaysia's Maritime Jurisdictional Limits*,
https://doi.org/10.1007/978-3-031-78783-6

(continued)

Treaties and Conventions to which Malaysia is a Member	
Short Title	Long Title
NACA	Organization for the Network of Aquaculture Centres in Asia and the Pacific
SEAFDEC	Southeast Asian Fisheries Development Centre
—	Treaty Banning Nuclear Weapon Tests in the Atmosphere, in Outer Space and under Water
—	Treaty on the Prohibition of the Emplacement of Nuclear Weapons and Other Weapons of Mass Destruction on the Sea-Bed and the Ocean Floor and in the Subsoil Thereof
UNCLOS	United Nations Convention on the Law of the Sea
UNFCCC	United Nations Framework Convention on Climate Change

Government agency for marine fisheries: Ministry of Agriculture; Fisheries management entity: Department of Fisheries; Government agency for protection of the marine environment: Environment protection Department Major law/legislation: Environment Protection Enactment, 2002

Annex V
Vessel Traffic Information System in the Straits of Malacca and Singapore

Introduction

The International Maritime Organisation (IMO) adopted the joint proposal by Indonesia, Malaysia, and Singapore to introduce a mandatory ship reporting system in the Straits of Malacca and Singapore that will contribute towards navigational safety, efficiency of navigation and the protection of the marine environment in the Straits.

From 1 December 1998, ships plying the southern sector of the Straits of Malacca and Singapore have to participate in STRAITREP by reporting to the shore-based authorities of the coastal states. This will enable shore-based authorities to advise transiting ships on the traffic situation in the Straits, as well as contribute positively towards search-and-rescue (SAR) operations and responses to marine incidents.

The concept of compulsory reporting of the ship's position, course, and speed at a nominated time each day was introduced for ships operating within Australia's Search and Rescue Area (AUSSAR) during 1986 when legislation was enacted in the Australian Parliament. The Australian Ship Reporting System (AUSREP) was established in accordance with provisions contained in the *International Convention for Safety of Life at Sea* (SOLAS).

The geographical coverage is identical to the International Civil Aviation Organisation Search and Rescue regions around Australia. AUSREP has been effective since its inception. Its objectives are threefold: (1) to limit the time between the loss of a vessel and the initiation of search and rescue action, in cases where no distress signal is sent out; (2) to limit the search area for a rescue action; and (3) to provide up-to-date information on shipping resources available in the area in the event of search and rescue incident.

© The Editor(s) (if applicable) and The Author(s), under exclusive license to Springer Nature Switzerland AG 2025
V. L. Forbes, *Malaysia's Maritime Jurisdictional Limits*,
https://doi.org/10.1007/978-3-031-78783-6

The format of reporting is based on IMO Resolution A.153 (13) of 17 November 1983. The Australian *Navigation Act* (Commonwealth) provides penalties involving fines of up to $10,000 for infringement of the reporting provisions.

On 1 December 1998, at 0800 h (8.00am) Western Standard Time (UTC + 8 h), a Vessel Traffic Information System (VTIS) was introduced in the Straits of Malacca and Singapore. The VTIS is a ship reporting system that will operate in the southern half of the Malacca Strait and throughout the length of the Straits of Singapore.

The International Maritime Organisation (IMO) adopted the system, as proposed by Indonesia, Malaysia, and Singapore, in its *Resolution* MSC.73 (69) on 19 May 1998. The mandatory system will be known as STRAITREP.

Sectors

The operational area of STRAITREP covers the Straits between Longitudes 100° 40′ and 104° 23′ East. The area includes the ship routeing system in the Straits. The area is divided into nine sectors each has an assigned VHF (Very High Frequency) channel. The limits of the sectors are illustrated on Map 1.

The VTIS authorities for the STRAITREP are as follows:

Sectors 1–5   Klang VTIS

Sector 6   Johor VTIS

Sectors 7–9   Singapore VTIS

Singapore's VTIS

In 1996, a VTIS was established to monitor vessel traffic in the Traffic Separation Scheme (TSS) within the Straits of Singapore.

The stated objectives of Singapore's VTIS are:

To improve safety of navigation in the Straits of Singapore;
To facilitate safety of vessel traffic flowing in and out of Singapore; and
To provide navigational information to vessels using the Straits of Singapore.

All vessels of 300 gross tonnage and over when transiting the Straits of Singapore are encouraged to participate in the VTIS reporting procedures.

Built at a cost of $ 63 million, the VTIS project was implemented in two phases. Phase 1 (VTIS 1), which was completed in 1990, uses 5 radars to monitor and regulate shipping traffic in the Singapore Strait. This coverage was extended to include Singapore's port waters in 1995, with the addition of four radars under Phase 2 (VTIS 2). The radars, located on the offshore islands and along the southern coast of Singapore, are controlled by the Maritime and Port Authority's (MPA) Port Operations Control Centre.

VTIS operational area is delineated along Longitude 103 degrees 51.2 min East into three sectors, namely, VTIS EAST, VTIS CENTRAL and VTIS WEST. The West operates on VHF Channel 73, and East and Central on VHF Channel 10. Depending on the sector in which the vessel is located, every report must be made to VTIS EAST or VTIS WEST accordingly when the vessel passes the appropriate reporting points. The locations of the VTIS reporting points are contained in *Port Marine Circular* No. 2 of 1996.

Details to be included in the report include: name of the vessel; its call-sign; and its geographical coordinates (present position) at the instant of the transmission of the report.

STRAITREP

Categories of ships required to participate in the system include:

all vessels of 300 gross tonnes and above;
vessels of 50 m or more in length;
vessels engaged in towing or pushing with a combined GT of 300 and above, or with a combined length of 50 m or more;
vessels of any tonnage carrying hazardous cargo;
all passenger vessels that are fitted with VHF, regardless of length or GT; and
any category of vessels less than 50 m in length or less than 300 GT which are fitted with VHF and in an emergency, uses the appropriate traffic lane or separation zone, in order to avoid immediate danger.

The stated objectives of the STRAITREP are:

to enhance the safety of navigation;
to protect the marine environment;
to facilitate the movement of vessels; and
to support Search and Rescue and oil-pollution response operations.

Reporting Format

Every Master (of vessels) in providing information to, or receiving from STRAITREP is not relieved from any of his/her duties and responsibilities in regard to the good practice of seamanship and navigation.

The format is a prescribed form in which is stated the ship's name and call sign; the geographical coordinates and bearing and distance from a clearly identifiable point (land mark) of the ship's position at the time of making the report; and the ship's true course and speed. In addition, the report should indicate whether hazardous cargo is being carried on board; brief detail of defects, deficiencies or other limitations; and description of pollution or dangerous goods lost overboard.

STRAITREP in turn will provide information to ships about specific and critical situations that could cause conflicting traffic movements and other information concerning safety of navigation. Information of general interest to ships' Masters will be broadcast on VHF Channel 16.

The language used for communication will be in English, using the IMO Standard Marine Communication Phrases where necessary. Reports will be based on VHF voice radio-communication and will be inter-active. Information of commercial confidentiality may be transmitted by non-verbal means.

Reporting Procedures

To facilitate reporting, every report shall be made to VTIS EAST or VTIS WEST accordingly when a vessel enters the VTIS operational area and passes any one of the following reporting points:

(1) Reporting via VTIS East (VHF CH 10)

   i.   Approaching from the East (South China Sea) when Horsburgh Lighthouse (01° 19.817′ N 104° 24.444′ E) is abeam (is at right angles to the ship's course);

   ii. Approaching from the South via Selat Riau when Karang Galang Light (01° 09.580′ N 104° 11.470′ E) is abeam; or

   iii. Approaching from the East Johor Strait when Eastern Buoy (01°17.868′ N 104°05.999 ' E) is abeam

(2) Reporting via VTIS West (VHF CH 73)

   Approaching from the South via Selat Durian when Pulau Jangkat Beacon (00° 57.898′ N 103°42.720′ E) is abeam

   Approaching from the West (Malacca Strait) when Pulau Iyu Kechil (01°11.483′ N 103°21.239′ E) is abeam

(3) Vessels approaching from another direction other than that specified above shall on approaching the VTIS operational area call the appropriate VTIS operator (East or West) and provide the vessel's position by bearing and distance from one of the following reference points:

   i.   Pu Iyu Kechil (01°11.470′ N 103°21.240′ E)
   ii. Sultan Shoal Light (01°14.381′ N 103°38.985′ E)
   iii. Raffles Light (01°09.609′N 103°44.552′ E)
   iv. Sakijang Light Beacon (01°13.308′ N 103°51.378′ E)
   v.   Bedok Light (01°18.548′ N 103°56.068′ E)
   vi. Tg Setapa Light (01°20.578′ N 104°08.240′ E)
   vii. Horsburgh Light (01°19.817′ N 104°24.444′ E)

Rules and Regulations

The rules and regulations in force in the area of the system include the *International Regulations for Preventing Collisions at Sea*, 1972, especially Rule 10 and the *Rules for Vessels Navigating Through the Straits of Malacca and Singapore*, as approved by IMO. Ships entering the operational area must report when crossing the lines at points defined by geographical coordinates and or features and as delineated on the map or when leaving port or anchorages in the area or before joining the traffic lane of the TSS.

For ships approaching from any direction shall on reaching Sectors 7, 8 or 9 as appropriate report by giving the vessel's position in terms of bearing and distance from one of the seven nominated points.

Changes to the Existing System

Four major changes to the existing system are highlighted:

A continuous north-west and south-east traffic lanes stretch half the length of the Malacca Strait, about 200 nautical miles (400 kms), thus connecting the TSS at One-fathom Bank to the existing TSS in the Straits of Singapore.

An Inshore Traffic Zone (ITZ) has been introduced in the Malacca Strait off the coast of Malaysia. The ITZ is not continuous but interrupted by two precautionary areas, namely, off Port Dickson and Malacca (Tanjong Kling). The purpose of an ITZ is to permit local maritime traffic to operate along the littoral.

Seven 'Precautionary Areas' have been identified in the new scheme. A precautionary area is a zone in a TSS where ships must navigate with particular caution and within the direction of traffic flow as recommended. The primary objective is to alert masters of ships that crossing traffic including regional ferries may be encountered in the areas. This would render the ships to be in a maximum state of manoeuvring readiness in the defined areas.

An extension to the Deep-Water Route (DWR) in the Straits of Singapore especially off Tanjong Medang. The DWR will enhance safety of deep draught vessels transiting this area. The DWR is approximately 44 nautical miles. Vessels plying the Medang DWR are allowed to overtake and navigate at a safe speed where the rule of maximum speed of 12 knots does not apply.

Speed Limit

The speed limit of 12 knots for VLCCs (Very Large Crude Carriers) and Deep Draught Vessels is confined to the One-fathom Bank TSS; Deep-water routes in the Phillip Channel and the Straits of Singapore; and the west-bound lanes in the TSS south of Singapore. There is no stipulated speed limit for these vessels outside these areas. Overtaking is restricted only in the DWR in the Straits of Singapore and the Phillip Channel area and is permitted in the DWR off Tanjong Medang.

The MPA

The Port of Singapore is run by the Maritime and Port Authority of Singapore (MPA), which acts as the sole regulatory body for the Republic's port and maritime affairs.

Singapore's strategic position as a meeting point of major sea routes linking Asia and Australasia with Europe and ports further west means that it is naturally dependent on international shipping and trade for its economic development. The Maritime and Port Authority of Singapore (MPA) is aware of its responsibilities and is committed to creating a blueprint for Singapore's physical development as a hub port and as an international maritime centre. Singapore has been a member of the IMO Council since 1993.

The Port of Singapore has played a critical role in the Republic of Singapore's transformation into a global trading power. Its strategic geographical location is one of the crucial factors that have made Singapore as a hub for shipping that it is today. Other contributory factors include a strong and stable government, good infrastructure, a transparent legal system, and state-of-the-art telecommunications and more importantly, the quick turn-around for ships bringing in and taking out cargo from the region.

The port's location at the crossroads of the main shipping routes has facilitated the Republic's development into a principal centre for shipping activities in Southeast Asia. It is a focal point for about 400 shipping lines linking Singapore to more than 750 ports in 120 countries worldwide. There are more than 900 ships in the port at any one time.

These ships can look forward to a variety of services, including cargo handling, ware-housing, distribution, bunkering and ship supplies. Where necessary, the port also provides pilots and tugs to ships that may not be familiar with navigation within the Straits of Singapore. The Port also provides round the clock security, environmental control, and fire-fighting services at its six terminals at Tanjong Pagar, Keppel, Brani, Pasir Panjang, Sembawang and Jurong. These terminals can accommodate all types of vessels—container ships, bulk carriers, cargo freighters, coasters, lighters, and passenger liners.

Depending on their cargo, these vessels will either call at the oil terminals run by the petroleum companies, or the terminals run by the PSA Corporation Limited and the Jurong Town Corporation (JTC). PSA Corporation operates the terminals at Brani, Keppel, Pasir Panjang, Sembawang and Tanjong Pagar, which deal in container and conventional cargo. Jurong Port, which handles conventional and bulk cargo, comes under the purview of the JTC.

As a free port with minimal tariff protection, Singapore is well poised to take advantage of the ASEAN Free Trade Area (AFTA) which is expected to come into effect by the year 2003.

Generally, a ship enters or leaves the Straits of Singapore every two minutes. Intra-regional trade, especially fast passenger-carrying ferries add to the volume of maritime traffic criss-crossing this busy congested waterway. Two major incidents during 1997 highlight the important role MPA has in protecting the marine environment. In January 1997, the *Song San*, an oil tanker, discharged oil into the sea. The owners, agents, shipmaster, and Chief Officer of the ship were fined a total of more than $1 million and the Ship's Master and Chief Officer were jailed.

On 15 October 1997, a collision between two oil tankers, the *Orapin Global* and the *Evoikos*, in the Straits of Singapore, caused the worst oil spill in Singapore's history.

A total of 28,463 ton of marine oil (380cst) was discharged following the collision. The MPA with assistance from other agencies successfully contained the damage.

The Port Operations Control Centre (POCC) of the MPA facilitates vessel traffic in the Straits of Singapore and the port waters. A "real-time" image on the monitors displays the traffic movement in the busy waterways 24 h a day. The images are captured by the POCC's multi-radar VTIS, which tracks vessels, and the traffic situations are displayed on high-resolution graphic computer terminals. The VTIS relies on a sophisticated ship database system for information on vessels' particulars.

MPA's Initiative on Research and Development

The MPA and the Nanyang Technological University (NTU) have a Memorandum of Understanding (MOU) on Research and Development (R&D) which permits extensive work on R&D projects pertinent to Singapore's port and maritime sector.

Since 1997, the R&D collaboration between the two organisations has produced significant results. Some of the collaborative projects that have been launched since the signing of the first MOU include the following:

Sea-Space Capacity Model

The Sea-Space Capacity Model is an analytical model developed to determine the capacities of the berthing facilities, anchorages, and shipping lanes within a generalised sea-space network. The model would allow port planners and operators to assess the capabilities of the existing facilities and shipping channels and to study various future scenarios in order to predict how well the port would function under these scenarios. The model is not only useful in the investigation of the impacts of various planning and operational concepts before actual implementation, but it also aids to identify potential problems, formulate appropriate solutions, and optimise sea space planning and operation thus, enhancing port performance.

An Integration of the Numerical Modelling and Geographic Information System

This is a coastal zone resource management system that will be operated as a transparent shell with inter-linking facilities encompassing both a Geographic Information System (GIS) and numerical modelling module. The GIS will serve as a platform to collect and collate information, which forms the core for a dynamic information system. The shell would facilitate the operation of the numerical model and the movement and spread of oil slicks can be simulated. Coastal resource maps and information which are useful in pollutant mitigation and resource management for the coastal zone can also be produced. Furthermore, the system would be able to map out areas and activities affected, to formulate response strategies and assess the damages in the case of an oil spill incident.

Computerised Anchorage Management System

The Computerised Anchorage Management System is designed with the aim of optimising the use of anchorage space in the Port of Singapore. This system would be able to forecast the utilisation of anchorage space and the knowledge attained could be used to assign new arrivals to anchorage locations. This system not only leads to more efficient use of space, but also improves safety in the port by reducing unnecessary ship movements within the port and between anchorage areas.

Maritime Conflict Prediction and Scheduling System (CPSS)

The Maritime Conflict Prediction and Scheduling System (CPSS) is a computer-based system proposed for the prediction of potential maritime conflicts and conflict resolution for sensitive vessels such as passenger ships, oil or chemical carrying tankers etc. in the Port. When integrated with the current Vessel Traffic Information System, real time conflict prediction and scheduling can be achieved and thus aid in the avoidance of potential maritime accidents.

Electronic Enforcement System for High-Speed Craft (EES)

Currently, high-speed regional ferries plying between Singapore and its neigh-bouring countries have to comply with certain international codes or conventions when navigating within port waters and any infringements are monitored manually upon receiving information from the Vessel Traffic Information System. The EES, however, will automate the monitoring of infringements and generate infringement reports using inputs from the ferry's radio transponders.

Conclusion

The implementation of STRAITREP and the aforementioned initiatives will enhance the navigational safety of passing ships and allow the maritime authorities of Indonesia, Malaysia, and Singapore to respond effectively to search and rescue and oil pollution incidents in the busy waterways of the Straits of Malacca and Singa-pore. The cooperation of shipowners, agents, masters, and personnel of the ships is required at all times.

The littoral states of the Straits have demonstrated their cooperative initiative in proposing the system to IMO. The effectiveness of the system will be dependent on the ship-owners and the mariners that ply these waters.

AMENDED RULES FOR VESSELS NAVIGATING THROUGH THE STRAITS OF MALACCA AND SINGAPORE

I Definitions

For the purpose of these Rules the following definitions shall apply:

1. A vessel having a draught of 15 m or more shall be deemed to be a deep draught vessel.
2. A tanker of 150,000 dwt and above shall be deemed to be a very large crude carrier (VLCC).

Note: The above definitions do not prejudice the definition of "vessel constrained by her draught" described in Rule 3(h) of the International Regulations for Preventing Collisions at Sea, 1972.

II. General Provisions

1. Deep draught vessels and VLCCs shall allow for an under-keel clearance of at least 3.5 m at all times during the entire passage through the Straits of Malacca and Singapore and shall also take all necessary safety precautions, when navigating through the Traffic Separation Schemes.
2. Masters of deep draught vessels and VLCCs shall have particular regard to navigational constraints when planning their passage through the Straits.
3. All deep draught vessels and VLCCs navigating within the TSS are recommended to use the pilotage service of the respective countries when they become available.
4. Vessels shall take into account the precautionary areas where crossing the traffic may be encountered and be in a maximum state of manoeuvring readiness in these areas.

III. Rules

Rule 1

Eastbound deep draught vessels shall use the designated deep-water routes.

Rule 2

Eastbound deep draught vessels navigating in the deep-water routes in Phillip Channel and Singapore Strait and shall as far as practicable, avoid overtaking.

Rule 3

All vessels navigating within the TSS shall proceed in the appropriate traffic lane in the general direction of traffic flow for that lane and maintain as steady a course as possible, consistent with safe navigation.

Rule 4

All vessels having defects affecting operational safety shall take appropriate measures to overcome these defects before entering the Straits of Malacca Singapore.

Rule 5

In the event of an emergency or breakdown of a vessel in the traffic lane, the vessel shall, as far as practicable and safe, leave the lane by pulling out to the starboard side.

Rule 6

(a) Vessels proceeding in the westbound lane of the TSS "In the Singapore Strait" when approaching Raffles Lighthouse shall proceed with caution, taking note of the local warning system, and, compliance with Rule 18(d) of the International Regulations for Preventing Collisions at Sea, 1972, avoid impeding the safe passage of a vessel constrained by her draught which is exhibiting the signals required by Rule 28 and which is obliged to cross the westbound lane of the scheme in order to approach the single point mooring facility (in approximate position 01°11′.42N, 103° 47′.50E, from Phillip Channel).

(b) Vessels proceeding in the TSS when approaching any of the precautionary areas shall proceed with caution, taking note of the local warning system, and, in compliance with Rule 18(d) of the International Regulations for Preventing Collisions at Sea, 1972, avoid impeding the safe passage of a vessel constrained by her draught which is exhibiting the signals required by Rule 28 and which is obliged to cross that precautionary area.

(c) Information relating to the movement of ships constrained by their draught as referred to in paragraphs (a) and (b) above will be given by radio broadcasts. The particulars of such broadcasts are promulgated by Notices to Mariners. All vessels navigating in the area of the TSS should monitor these radio broadcasts and take account of the information received.

Rule 7

VLCCs and deep draught vessels navigating in the Straits of Malacca and Singapore shall, as far as it is safe and practicable, proceed at a speed of not more than 12 knots over the ground in the following areas:

(a) At One Fathom Bank TSS.

(b) deep-water routes in the Phillip Channel and in Singapore Straits; and

(c) Westbound lanes between positions 01° 12′.51 N, 103° 52′.25 E and 01° 11′.59 N, 103° 50′.31 E and between position 01° 11′.13 N 103° 49′.18 E and 01° 08′.65 N, 103° 44′.40 E.

Rule 8

All vessels navigating in the routeing system of the Straits of Malacca and Singapore shall maintain at all times a safe speed consistent with safe navigation, shall proceed with caution, and shall be in a maximum state of manoeuvring readiness.

Rule 9

(a) Vessels which are fitted with VHF radio communication are to participate in the ship reporting system adopted by the Organization.

(b) VLCCs and deep draught vessels navigating in the Straits of Malacca and Singapore are advised to broadcast, eight hours before entering the TSS, navigational information giving name, deadweight tonnage, draught, speed, and times of passing One Fathom Bank Lighthouse, Raffles Lighthouse and Horsburgh Lighthouse. Difficult and unwieldy tows are also advised to broadcast similar information.

Rule 10

All vessels navigating in the Straits of Malacca and Singapore are requested to report by radio to the nearest shore authority any damage to or malfunction of the aids to navigation in the Straits, or any aids out of position in the Straits.

Rule 11

Flag States, owners and operators should ensure that their vessels are adequately equipped in accordance with the appropriate international conventions and recommendations.

IV Warning

Mariners are warned that local traffic could be unaware of the internationally agreed regulations and practices of seafarers and may be encountered in or near the TSS and should take any precautions which may be required by the ordinary practice of seamen or by the special circumstances of the case.

Definition: "Laden tanker" means any tanker other than a tanker in ballast having in its cargo tanks residual cargo only.

Marine Electronic Highway (MEH)

Implementation of the Marine Electronic Highway (MEH) Demonstration Project in the Straits of Malacca and Singapore (operational in 2007 following the signing of a US$6.86 million grant agreement (on 19 June 2006) between the Global Environment Facility (GEF) of the World Bank and the International Maritime Organization (IMO).

The four-year regional demonstration project aims to link shore-based marine information and communication infrastructure with the corresponding navigational and communication facilities aboard transiting ships, while being also capable of incorporating marine environmental management systems.

The overall objectives are to enhance maritime services, improve navigational safety and security and promote marine environment protection and the sustainable development and use of the coastal and marine resources of the Straits' littoral States, Indonesia, Malaysia, and Singapore. In addition to the US$6.86 million assigned to IMO for the regional MEH demonstration project, the GEF/World Bank has also agreed to grant US$1.44 million to Indonesia for the procurement of equipment for a differential global positioning system (DGPS) station and automatic ship identification (AIS) stations, as well as tidal instruments and an ocean data buoy.

The MEH is being built upon a network of electronic navigational charts using electronic chart display and information systems (ECDIS) and environmental management tools, all combining in an integrated platform covering the region that allows the maximum of information to be made available both to ships and shipmasters as well as to shore-based users, such as vessel traffic services. The overall system—which would also include positioning systems, real-time navigational information like tidal and current data, as well as providing meteorological and oceanographic information—is designed to assist in the overall traffic management of the Straits and provide the basis for sound marine environmental protection and management.

The implementation of the demonstration project follows a preparatory phase, from 2001 to 2005, involving IMO, the littoral States, and other partners, which was also funded by the GEF/World Bank (amounting to US$473,000). Start-up activities of the regional component of the MEH demonstration project will commence in July with the recruitment of a Project Manager and consultants to establish the Project Management Office in Batam, Indonesia; preparation of the first Project Steering Committee Meeting, to be held within the year; preparation of bidding documents for various goods and services required by the project; as well as assistance to Indonesia in the procurement of maritime safety facilities. The project's experts will also prepare the bidding document for a hydrographic survey, scheduled to take place in 2007, of the Traffic Separation Scheme of the Malacca Strait Routing System from One Fathom Bank to Pulau Iyu Kecil, using multi-beam technology, with the aim of producing electronic navigation charts of the Straits. The financial go-ahead for the project from the GEF/World Bank followed the signing, in Jakarta, Indonesia, in September 2005, of agreements to co-operate and collaborate to implement the MEH Project. Signatories to those agreements included the three littoral States, IMO, the International Hydrographic Organization (IHO), the International Association of Independent Tanker Owners (INTERTANKO) and the International Chamber of Shipping (ICS).

ASSIGNED VHF CHANNELS AND VTS AUTHORITIES IN STRAITREP

SECTOR	VHF CHANNELS	VTS AUTHORITIES
Sector 1	VHF 66	KLANG VTS
Sector 2	VHF 88	KLANG VTS
Sector 3	VHF 84	KLANG VTS
Sector 4	VHF 61	KLANG VTS
Sector 5	VHF 88	KLANG VTS
Sector 6	VHF 88	JOHOR VTS
Sector 7	VHF 73	SINGAPORE VTS
Sector 8	VHF 14	SINGAPORE VTS
Sector 9	VHF 10	SINGAPORE VTS

REPORTING CONTENT IN STRAITREP

DESIGNATOR	FUNCTION	INFORMATION REQUIRED
A	Ship	Name and call sign
C	Position	A 4-digit group giving latitudes in degrees and minutes suffixed with N (north) or S (south) and a 5-digit group giving longitudes in degrees and minutes suffixed with E (east) or W (west); or
D	Position	True bearing (first 3 digits) and distance given in nautical miles from a clearly identifiable point (state landmark)
E	True Course	A 3-digit group

(continued)

(continued)

DESIGNATOR	FUNCTION	INFORMATION REQUIRED
F	Speed in knots and tenths of knots	A 3-digit group
P	Hazardous cargo on board	Indicate "Yes" or "No" to whether vessel is carrying hazardous cargo. If "Yes" the class if applicable
Q	Defects/ damage/ deficiencies/ other limitations	Brief details of defects, deficiencies, or other limitations
R	Description of pollution or dangerous goods lost overboard	Brief details of type of pollution (oil, chemicals, (etc.) or dangerous goods lost overboard; position expressed as in (C) or (D)

Malaysia to Impose Stricter Enforcement of Traffic Lanes

3 May 2002

The Director-General of the Malaysian Marine Department, Captain Raja Malik announced that his department will enforce stricter adherence to the traffic lanes in the Malacca Straits to ensure safer passages through the congested sea route. He said it was noted that larger vessels were encroaching the Malaysian inter-coastal zone and ignoring the traffic separation scheme and his department would be seeking to prosecute those offenders. Enforcement of laws for vessels crossing the separation scheme in non-designated zones will also be "beefed up".

Another concern was that of high-speed vessels such as ferries, fishing vessels and some smaller craft plying between Malaysia, Indonesia, and Singapore. They posed the biggest threat to commercial shipping in the straits and a regulation calling for those vessels to show passage plans indicating the use of the precautionary crossing zones if they wanted to berth at Malaysian terminals, will soon be a requirement.

Citing an incident "in 1992 a small Taiwanese trawler *Teh Fu* 51 collided with the cruise liner *Royal Pacific* and nine of its 355 passengers were killed and the ship sank in just 45 min is an indication that we should not underestimate the damage a small vessel can do," said Captain Malik.

Captain Malik iterated that, the introduction and integration of an electronic navigation system through the Marine Electronic Highway (MEH), a system being conceived by Malaysia, Singapore, Indonesia and the International Maritime Organisation (IMO). The MEH would reduce the risk factor by providing the information to mariners and watchkeepers and coupled with a differential global positioning system presently on trial in Lumut would allow shore-based traffic controllers to fix a vessel's position within two meters.

The major set-back the MEH system faces are funding. The cost is at least US$30 million and IMO, at a recent meeting in Singapore failed to secure the full funding for a US$10 million demonstration project.

ROUTEING SYSTEM

The Routeing System

The International Maritime Organisation (IMO) is recognised as the only international body responsible for establishing and recommending measures on an international level concerning ship's routeing.

The purpose of ship's routeing is to improve the safety of navigation in converging areas and in areas where the density of traffic is great or where freedom of movement of shipping is inhibited by restricted sea-room, the existence of obstructions to navigation, limited depths, and unfavourable meteorological conditions.

Traffic Separation Scheme—A routeing measure aimed at the separation of opposing streams of traffic by appropriate means and by establishment of traffic lanes.

Separation zone and lines—A zone or line separating the traffic lanes in which ships are proceeding in opposite or nearly opposite directions; or separating a traffic lane from the adjacent sea area; or separating traffic lanes designated for particular classes of ship proceeding in the same direction.

Traffic lane—An area within defined limits in which one-way traffic is established. Natural obstacles, including those forming separation zones, may constitute a boundary.

Inshore traffic zone—A routeing measure comprising a designated area between the landward boundary of a traffic separation scheme and the adjacent coast, to be used in accordance with the provision of Rule 10(d), as amended, of the International Regulations for Preventing Collision at Sea, 1972 (Collision Regulations).

Deep-water route—A route within defined limits which has been accurately surveyed for clearance of sea bottom and submerged obstacles as indicated on the charts.

Precautionary areas—A routeing measure comprising an area within defined limits where ships must navigate with particular caution and within which the direction of traffic flow may be recommended.

At One Fathom Bank

Off Port Klang

Port Klang to Port Dickson

Off Port Dickson

Port Dickson to Tanjung Keling

Off Malacca/Dumai

Malacca to Iyu Kecil

Off Sultan Shoal Lighthouse

In The Singapore Strait (Main Strait)

Singapore Strait (Off Pulau Sebarok/Pulau Belakang Padang)

Singapore Strait (Off St. John's Island)

Singapore Strait (Off St. John's Island/Pulau Sambu)

Singapore Strait (Off Changi/Pulau Batam)

Off Tanjung Stapa/Pulau Batan

At Horsburgh Lighthouse Area

AT ONE FATHOM BANK

(Reference chart: British Admiralty 3946 Ed. 1996 Note: This chart is based on Revised Kertau Datum)

Description of the traffic separation scheme

(a) A separation zone is bounded by a line connecting the following geographical positions:

(1) 03° 00.7'N 100° 47.4'E	(5) 02° 43.4'N 101° 10.0'E
(2) 02° 53.7'N 100° 55.8'E	(6) 02° 49.0'N 100° 59.5'E
(3) 02° 49.5'N 100° 59.5'E	(7) 02° 53.4'N 100° 55.4'E
(4) 02° 43.9'N 101° 10.3'E	(8) 03° 00.3'N 100° 47.1'E

(b) A traffic lane for north-west bound traffic is established between the separation zone and a line connecting the following geographical positions:

(9) 03° 02.7'N 100° 48.8'E	(11) 02° 46.3'N 101° 11.5'E
(10) 02° 52.5'N 101° 00.0'E	

(c) A traffic lane for south-east bound traffic is established between the separation zone and a line connecting the following geographical positions:

(12) 02° 54.7'N 100° 43.1'E	(13) 02° 41.2'N 101° 08.8'E

OFF PORT KLANG

Description of the Precautionary areas

(a) A precautionary area is established by the line connecting the following geographical positions:

(14) 02° 46.3′N 101° 11.5′ E   (16) 02° 39.4′N 101° 15.0′ E

(15) 02° 44.3′ N 100° 15.0′ E   (17) 02° 41.2′ N 101° 08.8′E

PORT KLANG TO PORT DICKSON

(Reference chart: British Admiralty 3946 Ed. 1996 Note: This chart is based on Revised Kertau Datum)

Description of the traffic separation scheme

(a) A separation zone is bounded by a line connecting the following geographical positions:

(18) 02° 42.0′ N 101° 13.8′ E   (21) 02° 26.5′ N 101° 36.8′ E

(19) 02° 35.0′ N 101° 27.1′ E   (22) 02° 35.2′ N 101° 25.8′ E

(20) 02° 27.1′ N 101° 37.3′ E   (23) 02° 41.6′ N 101° 13.6′ E

(b) A traffic lane for north-west bound traffic is established between the separation zone and a line connecting the following geographical positions:

(24) 02° 44.3′ N 101° 15.0′ E   (26) 02° 29.0′ N 101° 38.8′ E

(25) 02° 37.4′ N 101° 28.0′ E

(c) A traffic lane for south-east bound traffic is established between the separation zone and a line connecting the following geographical positions:

(27) 02° 39.4′ N 101° 12.4′ E   (29) 02° 24.6′ N 101° 35.3′ E

(28) 02° 34.0′ N 101° 23.3′ E

INSHORE TRAFFIC ZONE

The areas between the landward boundary of the traffic separation scheme and the Malaysian coast between a line drawn from position (24) 02° 44.3′ N, 101° 15.0′ E in a direction of 027 to meet the coast and a line drawn from position (26) 02° 29.0′ N, 101° 38.8′ E in a direction of 034 to meet the Malaysian coast.

OFF PORT DICKSON

(Reference chart: British Admiralty 3946 Ed. 1996, 3947 Ed. 1997 Note: These charts are based on Revised Kertau Datum)

Description of the Precautionary areas

(a) A precautionary area is established by the line connecting the following geographical positions:

(30) 02° 29.0′ N 101° 38.8′ E   (32) 02° 21.4′ N 101° 39.0′ E
(31) 02° 25.8′ N 101° 42.9′ E   (33) 02° 24.6′ N 101° 35.3′ E

PORT DICKSON TO TANJUNG KELING

Description of the traffic separation scheme

(a) A separation zone is bounded by a line connecting the following geographical positions:

(34) 02° 23.9′ N 101° 41.4′ E   (36) 02° 09.0′ N 101° 59.0′ E
(35) 02° 09.7′ N 101° 59.6′ E   (37) 02° 23.2′ N 101° 40.9′ E

(b) A traffic lane for north-west bound traffic is established between the separation zone and a line connecting the following geographical positions:

(38) 02° 25.8′ N 101° 42.9′ E   (39) 02° 11.6′ N 102° 01.0′ E

(c) A traffic lane for south-east bound traffic is established between the separation zone and a line connecting the following geographical positions:

(40) 02° 21.4′ N 101° 39.4′ E   (41) 02° 7.1′ N 101° 57.5′ E

(d) A deep water route for south-east bound traffic is established by connecting the following geographical positions:

(42) 02° 21.4′ N 101° 39.4′ E   (46) 02° 12.3′ N 101° 36.8′ E
(43) 02° 13.8′ N 101° 39.3′ E   (47) 02° 22.2′ N 101° 36.8′ E
(44) 02° 05.1′ N 101° 55.9′ E   (48) 02° 24.0′ N 101° 36.1′ E
(45) 02° 03.0′ N 101° 54.2′ E

Inshore Traffic Zone

The areas between the landward boundary of the traffic separation scheme and the Malaysian coast between a line drawn from position (38) 02° 25.8′ N, 101° 42.9′ E in a direction of 059 to meet the Malaysian coast and a line drawn from position (39) 02° 11.6′ N, 102° 01.0′ E in a direction of 034 to meet the Malaysian coast.

OFF MALACCA/DUMAI

(Reference chart: British Admiralty 3947 Ed. 1997, 3833 Ed. 1988, 2403 Ed. 1983
Note: These charts are based on Revised Kertau Datum)

Description of the Precautionary areas

(a) A precautionary area is established by the line connecting the following geographical positions:-

(49) 02° 11.6′ N 102° 01.0′ E   (51) 02° 00.0′ N 101° 59.8′ E

(50) 02° 07.2′ N 102° 06.2′ E   (52) 02° 03.0′ N 101° 54.2′ E

MALACCA TO IYU KECIL

Description of the traffic separation scheme

(a) A separation zone is bounded by a line connecting the following geographical positions:

(53) 02° 05.4′ N 102° 04.6′ E   (59) 01° 10.5′ N 103° 27.5′ E

(54) 01° 55.7′ N 102° 15.4′ E   (60) 01° 13.2′ N 103° 23.4′ E

(55) 01° 40.0′ N 102° 48.3′ E   (61) 01° 23.2′ N 103° 12.4′ E

(56) 01° 23.0′ N 103° 12.4′ E   (62) 01° 39.1′ N 102° 48.0′ E

(57) 01° 13.8′ N 103° 24.0′ E   (63) 01° 54.8′ N 102° 14.8′ E

(58) 01° 12.2′ N 103° 28.5′ E   (64) 02° 04.6′ N 102° 03.8′ E

(b) A traffic lane for north-west bound traffic is established between the separation zone and a line connecting the following geographical positions:

(65) 02° 07.2′ N 102° 06.2′ E   (68) 01° 25.5′ N 103° 15.0′ E

(66) 01° 57.0′ N 102° 16.6′ E   (69) 01° 15.2′ N 103° 25.3′ E

(67) 01° 38.4′ N 103° 00.0′ E   (70) 01° 14.3′ N 103° 29.7′ E

(c) A traffic lane for south-east bound traffic is established between the separation zone and a line connecting the following geographical positions:

(71) 02° 02.8′ N 102° 02.2′ E   (74) 01° 22.0′ N 103° 11.1′ E

(72) 01° 52.6′ N 102° 13.3′ E   (75) 01° 11.6′ N 103° 22.8′ E

(73) 01° 36.8′ N 102° 46.9′ E   (76) 01° 09.2′ N 103° 26.8′ E

(d) A deep-water route for south-east bound traffic is established by connecting the following geographical positions:

(77) 02° 01.9′ N 102° 01.5′ E

(78) 01° 59.7′ N 102° 05.6′ E

(79) 01° 52.0′ N 102° 13.3′ E

(80) 02° 00.0′ N 101° 59.8′ E

INSHORE TRAFFIC ZONE

The area between the landward boundary of the traffic separation scheme and the Malaysian coast between a line drawn from position (65) 02° 07.2′ N, 102° 06.2′ E, to Pulau Undan Lighthouse (Lat. 02° 02.9′N, Long. 102° 20.1′ E) then in a direction of 040° to meet the Malaysian coast and a line drawn from position (70) 01° 14.3′ N, 103° 29.7′ E in a direction of 038° to meet the Malaysian coast.

OFF SULTAN SHOAL LIGHTHOUSE

(Reference charts: British Admiralty 2598 Ed. 1990, 2556 Ed. 1994, 3833 Ed. 1988, 2403 Ed. 1983 Note: These charts are based on Revised Kertau Datum)

Description of the Precautionary areas

(a) A precautionary area is established by the line connecting the following geographical positions:-

(81) 01° 14.28′ N 103° 29.73′ E

(82) 02° 12.62′ N 103° 36.24′ E

(83) 01° 05.94′ N 103° 32.30′ E

(84) 01° 09.23′ N 103° 26.76′ E

Annex VI
Oil Spills: Contingency Plans and Events off the Malaysian Coast

Oil spills at sea pollute the marine environment to a varying degree during large oil tanker accidents, especially when they occur close to the coast. On the other hand, oil release from ships, when cleaning tanks, may prove to be a much worse pollution. When taking into account how often such spills occur during regular ship operation. The effects of the pollution can be minimized if the spills can be detected early. Remote sensing technology has been employed to help detect oil spills.

The ability to predict oil spills/slicks movement will assist in the assessment of damage. Prediction is hampered by the present lack of capability to quantify the combined effects of winds, waves, and currents on the movement of oil spills/slicks near the coast.

It is important that predictive analyses based on mathematical models are made to evaluate the fate of possible oil spills for adequate risk assessments. This can be done on a real time basis to assist the containment, control, and recovery of the oil. Models are used for estimating the environmental impact due to an actual spill or potential spills and used to find probabilities of oil contaminating certain areas. Furthermore, it is vital that contingence plans are in hand at regional, national, and local levels, for example at anchorages and ports.

An oil spill contingency plan (OSCP) is no longer a simple document with instructions on what to do if an oil spill occurs. An OSCP now is a representation of the corporate responsibility of a company and for that matter Government Agencies and Authorities and the respect it shows for its neighbouring community. Corporations increasingly are required by law to take responsibility for the products they use and distribute. This perspective provides the base upon which this module is built. The mandatory contents of an OSCP are presented with an explanation of the need of these components. Since an OSCP must be supported from the highest echelon in an agency or corporation, a policy statement must be up front and there must be a clear

V. L. Forbes, *Malaysia's Maritime Jurisdictional Limits*, https://doi.org/10.1007/978-3-031-78783-6

statement of the purpose and scope of the plan. The module then explains the components and issues involved in pre-emergency planning. These include community and legislative issues. This is followed by outlining the traditional and still necessary measures that must be taken during emergency response. In order for a plan to be and to remain effective, it must be supported by an active and sustained training program punctuated with practice drills and exercises. The importance of scheduled plan evaluation and updating is of utmost importance.

Regional ASIAN-OSPAR Contingency Plan

The ASEAN-OSPAR Project stands for "Project on Oil Spill Preparedness and Response in the ASEAN Seas Area". Formerly known as OSPAR Project, it was initiated in 1993 by the Ministry of Transport of Japan (currently the Ministry of Land, Infrastructure and Transport of Japan) and financially assisted mainly by The Nippon Foundation. Its main objective is to improve the capability of ASEAN countries to deal with large-scale oil and Hazardous and Noxious Substance (HNS) spill incidents in the ASEAN region, based on the ASEAN Oil Spill Response Action Plan (OSRAP).

Bilateral/Multilateral

Straits of Malacca Contingency Plan—Malaysia/Indonesia/Singapore

Sulu Sea Contingency Plan—Malaysia/Indonesia/Philippines

Brunei Bay Contingency Plan—Malaysia/Brunei

National

National Contingency Plan for the Control of Oil Spill—Malaysia Water (incl. EEZ)

Straits of Malacca Contingency Plan—Straits of Malacca

Area of Sabah Contingency Plan—Sabah

Area of Sarawak Contingency Plan—Sarawak

Area of Johor Contingency Plan—Johor

South China Sea Contingency Plan—South China Sea

Oil Spill Response Equipment

Region	Type of Equipment	Number of Equipment
Central Region Headquarters	Kepner Boom System Kepner Skimmer System ASI Boom System Surface Dispersant Spraying System Water Pressure Washer Aerial Dispersant Spray Bucket Temporary Storage Tank Equipment Storage Container and Spare Parts Beach Cleaning PowerVac	Two sets Three sets One sets One sets One sets Two sets One sets One unit One unit
Northern Region Headquarters	Kepner Boom System Kepner Skimmer System Water Pressure Washer Temporary Storage Tank Equipment Storage Container and Spare Parts	Two sets Three sets One sets One sets One unit
Southern Region Headquarters	Kepner Boom System Kepner Skimmer System ASI Boom System Surface Dispersant Spraying System Water Pressure Washer Temporary Storage Tank Equipment Storage Container and Spare Parts	Two sets Three sets One sets One sets One sets One sets One unit

Oil Spill Response Vessel

Name of the vessel	Type of the vessel
LANG TIRAM	Work Boat
LANG SIPUT	Work Boat
LANG KANGOK	Tug Boat
LANG HINDEK	Tug Boat
LANG SEWAH	Tug Boat
LANG MERAH	Tug Boat
LANG RAJAWALI	Patrol Boat
DOE I	Barge
DOE II	Barge
DOE III	Barge
DOE IV	Barge
DOE V	Barge
DOE VI	Barge

Oil Spill Incidents in Malaysian Water Year 1975–1997

Year	Name of ship	Location	Cause	Type and quantity of oil spill
1975	*SHOWA MARU*	Straits of Singapore	Grounding	Crude oil 4000 ton
1975	*TOLA SEA*	Straits of Singapore	Collision	Fuel oil 60 ton
1976	*DIEGO SILANG*	Malacca Strait	Collision	Crude oil 5500 ton
1976	*MYSELLA*	Straits of Singapore	Grounding	Crude oil 2000 ton
1976	*CITTA DI SAVONNA*	Straits of Singapore	Collision	Crude oil 1000 ton
1977	*ASIAN*	Malacca Straits	Collision	Fuel oil 60 ton
1978	*ESSO MERSIA*	South China Sea	Collision	Fuel oil 505 ton
1979	*FORTUNE*	South China Sea	Collision	Crude oil 10,000 ton
1980	*LIMA*	Straits of Singapore	Collision	Crude oil 700 ton
1981	*MT OCEAN TREASURE*	Malacca Straits	Human Error	Fuel oil 1050 ton
1984	*BAYAN PLATFORM*	South China Sea	Human Error	Crude oil 700 ton
1986	*BRIGHT DUKE* and MV *PANTAS*	Malacca Straits	Collision	–
1987	MV *STOLT ADV*	Straits of Singapore	Grounding	Crude oil 2000 ton
1987	*ELHANI PLATFORM*	Straits of Singapore	Grounding	Crude oil 2329 ton
1988	*GOLAR LIE*	Straits of Singapore	Grounding	–
1992	*NAGASAKI SPIRIT* and *OCEAN BLESSING*	Malacca Strait	Collision	Crude oil 13,000 ton
1997	*EVOIKOS/ORADIN GLOBAL*	Straits of Singapore	Collision	Fuel oil 25,000 ton
1997	*AN TAI*	Malacca Straits	Material Fatigue	Fuel oil 237 ton

2 September 2000—MALAYSIA—A major clean-up was underway following an oil spill from a sunken Chinese cargo ship at Tanjung Po anchorage point at the Sarawak River mouth. The ill-fated 5000 ton Kingston registered vessel *Double Brave* was loaded with about 116 ton of diesel oil when it sank after a collision with a barge being towed by a tugboat. About 60 workers from the Marine Department, Department of Environment, and the Kuching Port Authority assisted in the clean-up operation.

2000 October 4—INDONESIA—An estimated 7.000 ton of oil has been spilled in Indonesian waters off the Batu Berhanti Beacon, after a Panama registered tanker, *Natuna Sea*, ran aground damaging four of its cargo tanks. An oil boom of 300 m length was deployed around the tanker, but this could not prevent the oil from escaping in a South-westerly direction. Some patches of the oil reached Singapore, and clean-up actions, amongst others, commenced on the beaches of Sentosa, Singapore's famous tourist resort.

2001 May 28—MALAYSIA—An oil tanker with some 67 ton of fuel, including diesel and 1500 ton of bitumen, sunk after it was crashed from behind by a super tanker about 7.5 nautical miles off Pulau Undan, near Malacca. Officials said the crash caused MT *Singapura Timur* to take in water and remained half-submerged in the sea floating southwards. Diesel and bitumen have started to spill into the sea and is spreading to about one nautical mile from the collision spot.

2001 June 13—MALAYSIA—An Indonesian tanker laden with a toxic chemical has capsized off Malaysia's southern Johor state, just across from Singapore. The 533-ton MV *Endah Lestari* was on its way to East Kalimantan in Indonesia with some 600 ton of the poisonous industrial chemical phenol, and 18 ton of diesel. Newspaper reports said the toxic spill had killed thousands of fish and cockles reared in 85 offshore cages, and Singapore authorities have also warned its citizens to stay away from nearby waters. Officials said it would be tough to mop up the phenol, as it is soluble in water.

2002 June 12—SINGAPORE—A collision between Thailand-registered freighter MV *Hermion* and Singapore-registered bunker tanker *Neptank VII* has caused about 450 ton of marine fuel oil to spill into the south-eastern waters of Singapore. Clean-up operations by the Maritime and Port Authority of Singapore (MPA) have largely contained the oil spill, but broken patches of oil remained visible in the Singapore Straits, and oil booms were placed off the waters of Marina Bay and Sentosa.

2002 December 5—SINGAPORE—A potentially disastrous crude oil spill in Singapore waters was contained to just 350 ton when a small general cargo vessel collided with a heavily laden single-hulled tanker in the middle of the Singapore Straits. Two oil slicks measuring 2.5 km × 300 m and 2.5 km × 500m were spotted in Indonesian waters off the island of Bintan.

2003 June 12—SINGAPORE—The MV APL Emerald, a 40.077-ton container ship, spilled about 150 ton of fuel oil when it ran aground near Horsburgh Lighthouse, in the eastern approaches of the Singapore Straits. Six anti-pollution craft were involved in the clean-up, and oil booms have been laid around the vessel to contain the spill, which has since been contained. The authorities said there was no chance of any fuel reaching Singapore's shores, about 46 km away.

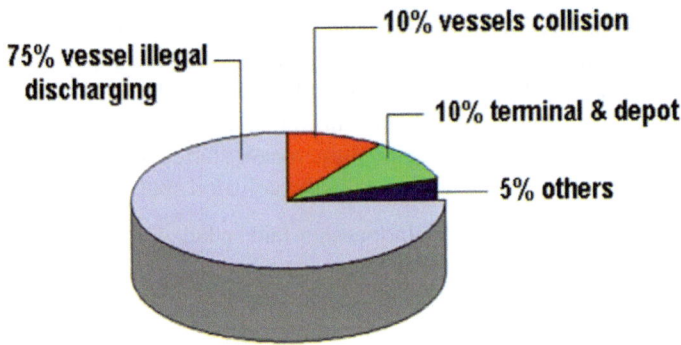

SOURCES OF OIL SPILL IN SEAS OFF MALAYSIA

Oil Discharge Malacca Strait

In excess of 4 million barrels/day in transported 65,649 ships passed through the Strait in 2006.

Intra-Strait maritime trade not included.

Offshore exploration and production

Terminal and Depot

Other land-based sources

Annex VII
Glossary of Malay Words in the Maritime Context

Glossary of Malay language words found on Nautical Charts and Sailing Directions

Bahasa Malay	English (approximate meaning)
Ayer	Stream
Alangan	Bar (sand)
Anak ayer	Small stream
Api Api	Type of mangrove
Arus	Current (ocean)
Ayer masin	Salt water
Bagan	Landing place
Bahru	New
Bandar	Port
Barat	West
Batang	River
Batu	Rock, stone
Bender	Port, trading town
Besar	Large, great
Beting	Shoal, bank, sand
Bukit	Hill, height
Burong	Bird
Busong	Islet, sandbank
Changkat	Hill, shoal
Chetak	Shallow
Dalam	Deep, depth, inside
Danau	Lake

(continued)

V. L. Forbes, *Malaysia's Maritime Jurisdictional Limits*, https://doi.org/10.1007/978-3-031-78783-6

(continued)

Bahasa Malay	English (approximate meaning)
Darat	Land, the interior
Gelong	Channel over a bar
Gili	Islet, rock
Gunong	Mountain
Gusong	Shoal, reef
Hijau	Green
Hitam	Black, dark
Hutan	Jungle, forest
Jalan	Road (as in street, avenue). Perhaps not in the nautical context
Jambatan	Mole, jetty, bridge, gangway
Jeram	Rapids (in a river)
Kaler	North
Kali	River
Kampong	Village
Karang	Coral, reef
Kechil (Kecil)	Small
Keramat	Shrine
Kering	Dry
Kidul	South
Kota	Fort, fortified town
Kuala	Mouth of River
Kubang	Waterhole
Kulon	West
Kuning	Yellow
Labuhan	Anchorage, harbour, roadstead
Lama	Old
Larangan	Prohibited
Laut	Sea
Lebar	Wide, broad
Lor	North
Lumpur	Mud
Malang	Rock, reef, shoal
Mas	Golden
Mendara	Minaret, watchtower
Merah	Red
Muara	Mouth of River
Napu	Reef

(continued)

(continued)

Bahasa Malay	English (approximate meaning)
Negeri	Town, state
Nusa	Island
Padang	Open space
Panchang	Pole, stake, pile
Pangkalan	Landing place
Panjang	Long
Pantai	Coast, seaboard, beach
Parit	Ditch, stream
Pasanggrahan	Government rest-house
Pasir	Sand, sandbank, beach
Paya	Marsh, swamp
Pekan	Town, market
Pelabuhan	Anchorage, port
Perhentian	Railway station
Perigi	Well, spring
Pisang	Banana
Pohon, pokok	Tree
Ponchak	Summit, peak
Praja	Town
Prau	Boat, ship
Puteh	White
Ras	Head (headland, promontory)
Rawa	Marshy ground
Rawang	Swamp
Rendah	Low
Riam	Waterfall
Rimba	Virgin jungle
Rumah	House
Sawang	Narrows, strait
Selat	Strait, channel, narrows
Selantan	South
Sumber	Spring of water
Sumur	Well
Sungai	River
Taka, takat	Shoal, reef, rock
Tanah	Land, country
Tanjong	Cape, headland, promontory

(continued)

(continued)

Bahasa Malay	English (approximate meaning)
Tasek, tasik	Lake
Telok	Bay, bight, bend of river
Tenang	Calm, smooth
Tepi laut	Coast, seaboard
Terumbu	Dangerous, hidden shoal, reef which dries
Terusan	Connecting channel
Timor	East
Tinggi	High, height, lofty
Tinjau	Look-out
Titiyan	Mole, jetty, footbridge
Tohor	Shallow
Tokong	Rocky treeless islet, large rock
Tua	Old
Tukun	Sunken rock
Utara	North
Wai, waj	Stream, creek
Wetan	East

Numerals

Satu	One	Dua	Two
Tiga	Three	Empat	Four
Lima	Five	Enam	Six
Tujoh	Seven	Delapan	Eight
Sembilan	Nine	Sa-puloh	Ten
Sa-belas	Eleven	Dua-belas	Twelve
Dua-puloh	Twenty	Dua-puloh-satu	Twenty-one
Sa-ratus	100	Dua-ratus	200
Sa-ribu	1,000	Dua-ribu	2000

Annex VIII
Laws of Malaysia Pertaining to Maritime Jurisdiction-Related Issues

Ordinance No. 7 of August 1969   Enabling legislation for straight baselines.

Act 57	*Continental Shelf Act* 1966	
Act 83	*Continental Shelf Act of* 1972	In force
	Continental Shelf (Amendment) Act 2009	
Act 95	*Petroleum Mining Act* 1966	In force
Act 129	*Geological Survey Act* 1974	In force
Act 311	*Exclusive Economic Zone Act* 1984 (As Amended)	1 January 2013
	Exclusive Economic Zone (Amendment) Act 2007	
Act 317	*Fisheries Act* 1985	1 November 2012
Act 488	*Port Authorities Act* 1963	In force
Act 527	*Carriage of Goods by Sea Act* 1950	In force
Act 633	*Malaysian Maritime Enforcement Agency Act* 2004	1 January 2006
Act 660	*Baselines of Maritime Zones Act* 2006	
Act 670	*Anti-Trafficking in Persons...Migrants Act* 2007	In force
Act 750	*Territorial Sea Act* 2012	In force
Act 799	*Malaysian Border Security Agency Act* 2017	In force
Act A1316	*Merchant Shipping (Amendment & Extension) Act* 2007	In force
MSAA2017	*Merchant Shipping (Amendment) Act* 2017	

The Case Concerning Land Reclamation by Singapore in and around the Straits of Johor (Malaysia v. Singapore). ITLOS Press 84 of 8 October 2003

Joint Submission to the Commission on the Limits of the Continental Shelf pursuant to Article 76, paragraph 8 of the United Nations Convention of the Law of the Sea 1982 in respect to the Southern part of the South China Sea of May 2009 made by the Governments of Malaysia and the Socialist Republic of Vietnam.

© The Editor(s) (if applicable) and The Author(s), under exclusive license
to Springer Nature Switzerland AG 2025
V. L. Forbes, *Malaysia's Maritime Jurisdictional Limits*,
https://doi.org/10.1007/978-3-031-78783-6

Annex IX
Landfall Lighthouses of Malaysia

NAME	Lat. N	Lon. E	Characteristic	Elevation (m)	Range (M)
Malaysia west coast-Malacca Strait					
Pulau Dangli	6° 26′ 59″	99° 46′ 11″	Fl W 10s	86	15
Pulau Enggang	6° 14′39″	99° 52′ 39″	Fl W 8s	60	15
Pulau Sigal	6° 19′34″	99° 55′ 40″	Fl W 5s	28	15
Pulau Kedera	6° 15′26″	99° 46′ 20″	Fl W 5s	79	15
Muka Head	5° 28′ 24″	100° 10″ 54″	Fl W 10s 2	42	25
Fort Cornwallis	5° 25′ 18″	100° 20′ 42″	Fl R 2s	27	16
Penang Br. Chan N	5° 21′17″	100° 20′ 52″	Oc WRG 10s	37	W10, R8, G8
Penang Br. Chan S	5° 21′16″	100° 20′ 52″	Oc WRG 10s	37	W10, R8, G8
Pulau Rimau	5° 14′ 48″	100° 16′ 36″	Fl(2) W 10s	39	22
Pulau Talang	4° 25′ 20″	100° 34′ 41″	Fl W 7s	21	8
Batuan Putih	4° 00′ 24″	100° 30′ 30″	Fl(2) W 6s	21	15
Kuala Selangor	3° 20′ 29″	101° 14′ 47″	Fl(2) W 15s	73	18
Angas Bank	3° 20′ 01″	100° 59′ 59″	Q W	21	15
Pulau Angsa	3° 11′ 12″	101° 13′ 36″	Fl W R 10s	36	W 22, R 15
Tanjung Tentaram	3° 00′ 14″	101° 12′ 55″	Fl W 3s	9	21
Bukit Jugra	2° 50′ 06″	101° 25′ 06″	Fl W 10s	146	24
One Fathom Bank	2° 53′ 18″	100° 59′48″	Fl(4) W 20s	34	23
Port Dickson	2° 34′ 59″	101° 42′ 53″	Fl WRG	19	W15R10G10
Tanjung Tuan	2° 24′ 24″	101° 51′ 12″	Fl(3) W 7s	118	23
Pulau Undan	2° 02′ 54″	102° 20′ 06″	Fl(2) W 15s	53	18
Tanjung Tohor	1° 50″ 48″	102° 42′ 12″	Fl W 5s	16	18
Bukit Segenting	1° 47′ 30″	102° 53′ 24″	Fl(4) W 30s	86	23
Pulau Pisang	1° 28′ 12″	103° 15′ 24″	Fl W 10s	150	22

(continued)

© The Editor(s) (if applicable) and The Author(s), under exclusive license
to Springer Nature Switzerland AG 2025
V. L. Forbes, *Malaysia's Maritime Jurisdictional Limits*,
https://doi.org/10.1007/978-3-031-78783-6

(continued)

NAME	Lat. N	Lon. E	Characteristic	Elevation (m)	Range (M)
Tanjung Piai	1° 15' 37"	103° 30' 33"	Fl W 3s	15	10

Landfall lighthouses within the Straits of Singapore are not included in the following list.

MALAYSIA EAST COAST

Pulau Tioman	2° 43' 19"	104° 13' 02"	Fl(3) W 15s	17	10
Batu Boya	2° 45' 23"	103° 36' 41"	Fl W 5s	8	8
Kuala Rompin	2° 48' 34"	103° 28' 50"	Fl W 5s	8	9
Kuala Pandan	3° 04' 48"	103° 26" 18"	Fl W 8s	10	6
Pulau Berhala	3° 14' 42"	103° 39" 36"	Fl(2) W 5s	31	10
Kuala Pahang	3° 32' 06"	103° 27' 54"	Fl W 7s	12	10
Tanjung Tembelling	3° 48' 16"	103°22' 27"	Fl(2) W 10s	69	15
Tanjung Gelang	3° 57' 49"	103° 26' 19"	Fl(3) W 10s	85	25
Bukit Pejajat	4° 14' 18"	103° 27' 24"	Fl W 12s	105	15
Tanjung Labuhan	4° 31' 18"	103° 28' 18"	Fl(2) W 4s	76	15
Tanjung Dungun	4° 47' 00"	103° 26' 24"	Fl W 10s	69	20
Bukit Puteri	5° 20' 12"	103° 08' 18"	Fl W 3s	35	15
Tok Bali	5° 53' 24"	102° 29' 18"	Fl R 5s	7	8
Bachok	6° 03' 42"	102° 24' 12"	Fl W 5s	12	10
Tumpat	6° 13' 13"	102° 09' 18"	Fl W 7s	11	12

SARAWAK

Tanjung Datu	2° 04' 48"	109° 38' 18"	Fl W 15s	171	18
Tanjung Po	1° 43' 31"	110° 31' 13"	Fl(2) W 10s	131	24
Tanjung Jerijeh	2° 09' 23"	111° 11' 57"	Fl(3) W 15s	38	23
Loba Ketan	2° 04' 29"	111° 11' 57"	Fl(2) W 6s	24	18
Tanjung Sedi	2° 26' 01"	111° 17' 48"	Fl W 5s	21	9
Tanjung Sirik	2° 46' 48"	111° 19' 06"	Fl W 10s	38	28
Tanjung Kidurong	3° 16' 30"	113° 03' 49"	Fl(2) W 15s	90	23
Bintulu Port	3° 15' 22"	113° 04' 08"	Iso W 12s	30	15
Tanjung Lobang	4° 21' 44"	113° 57' 43"	Fl W 10s	47	16
Tanjung Baram	4° 35' 45"	113° 58' 26"	Fl(3) W 20s	39	25

SABAH

Tanjung Kubong	5° 23′ 36″	115° 14′ 54″	Fl(2) W 15s	46	20
Pulau Tiga	5° 44′00″	115° 38′ 09″	Fl(4) W 20s	74	23
Pulau Gaya	6° 02′ 29″	116° 00′ 35″	Fl W 5s	168	35
Pulau Usukan	6° 23′48″	116° 19′ 38″	Fl(2) W 15s	58	17
Pulau Mantanani	6° 42′ 48″	116° 18′ 20″	Fl W 10s	47	15
Pulau Kalampunian	7° 03′ 11″	116° 44′ 45′	Fl W 20s	26	15
Pulau Silingan	6° 10′ 24″	118° 03′ 42″	Fl W 10s	26	15
Berhala (Sandakan)	5° 52′ 12″	118° 08′ 54″	Fl W 6s	194	28
Telukan Kinabatangan	5° 39′ 00″	118° 36′ 24″	Fl W 10s	24	15
Tanjung Terang	5° 25′ 10″	119° 12′ 45″	Fl W 6s	26	25
Tanjung Labian	5° 08′ 37″	119° 13′ 00″	Fl W 10s	26	15
Tanjung Tungku	4° 59′ 42″	118° 51′ 42″	Fl(3) W 20s	24	15
Bagahak	4° 56′ 35″	118° 38′ 27″	Fl W 5s	70	15
P. Katung Kalungan	4° 55′ 24″	118° 15′ 54″	Fl W 10s	17	17
Pulau Mataking	4° 34′ 41″	118° 56′ 48″	Fl W 20s	27	15
Si Amil	4° 18′ 53″	118° 52′ 36″	Fl(2) W 15s	107	15
Ligitan Reef	4° 09′ 48″	118° 53′ 04″	Fl W 5s	10	11
Pulau Sipadan	4° 06′ 36″	118° 37′ 54″	Fl(3) W 20s	24	14
Batu Tinagat	4° 13′ 35″	117° 58′ 47″	Fl WR 7.5s	99	W25 R16

SOUTH CHINA SEA

P. Layang Layang (Swallow Reef)	7° 22′ 38″		113° 50′ 58″	Fl W 5s	8	10
T. Semarang B B	6° 56′ 30″ (Royal Charlotte Reef)		113° 35′ 38″	Fl(2) W 10s	8	10
T. Semarang Barat	6° 20′ 00″		113° 14′ 24″	Fl W 10s	8	10

WITHIN THE STRAITS OF SINGAPORE

Pulau Iyu Kecil	1° 11′ 30″	103° 21′ 30″	Fl(3) W 15s	41	18
Sultan Shoal	1° 14′ 23″	103° 38′ 59″	Fl(2) W 15s	20	22
Pulau Nipa	1° 10′16″	103° 39′ 52″	Q W	12	12
Raffles, P. Satuma	1° 09′ 12″	103° 44′ 33″	Fl(3) W 20s	32	22
Takong Kecil	1° 06′ 18″	103° 43′ 12″	Fl(2) WR	48	18
Jangkat	0° 57′ 24″	103° 42′ 24″	Fl W 5s	37	17
Batu Berhanti Rf	1° 11′ 06″	103° 52′ 54″	Fl W 10s	10	13

(continued)

(continued)

Nongsa Island	1° 12′ 18″	104° 04′ 36″	Fl W 8s	40	20
Lazarus Is	1° 13′ 19″	103° 51′ 19″	Fl W 2.5s	59	15
Tg Berakit P Bintan	1° 13′ 12″	104° 34′ 30″	Occ W 3s	68	20
Horsburgh Lt.	1° 19′ 49″	104° 24′ 26″	Fl W 10s	31	22
Pulau Mungging	1° 21′ 42″	104° 17′ 54″	Fl W 3s	24	15
Pulau Damar	2° 44′ 30″	105° 22′ 54″	Fl W 3s	88	22

Annex X

Sempadan Pelantar Benua Malaysia Ditakrifkan Dengan Kordinat-Kordinat Geografis Yang Berpandukan Daripada Carta-Carta Admiralty NO. 771, 793, 1353, 1358, 2414, 2660A DAN 2660B

CONTINENTAL SHELF BOUNDARIES OF MALAYSIA

DEFINED BY THE FOLLOWING GEOGRAPHICAL COORDINATES

BASED ON ADMIRALTY CHARTS Nos. 771, 793, 1353, 1358, 2414, 2660A & 2660B

POINT	LATITUDE North	LONGITUDE East
TITIK	*LINTANG Utara*	*BUJAU Timur*
1	6° 18.4′	99° 27.5′
2	6° 16.3′	99° 19.3′
3	6° 18.0′	99° 06.7′
4	5° 57.0′	98° 01.5′
5	5° 27.0′	98° 17.5′
6	4° 55.7′	98° 41.5′
7	3° 59.6′	99° 43.6′
8	3° 47.4′	99° 55.0′
9	2° 51.6′	101° 00.2′
10	2° 41.5′	101° 12.1′
11	2° 15.4′	101° 46.5′
12	1° 55.2′	102° 13.4′
13	1° 41.2′	102° 35.0′
14	1° 19.5′	103° 03.9′
15	1° 15.0′	103° 22.8′
16	1° 13.45′	103° 26.8′
17	1° 08.45′	103° 32.05′
18	1° 11.0′	103° 34.2′
19	1° 15.15′	103° 34.95′

(continued)

© The Editor(s) (if applicable) and The Author(s), under exclusive license to Springer Nature Switzerland AG 2025
V. L. Forbes, *Malaysia's Maritime Jurisdictional Limits*,
https://doi.org/10.1007/978-3-031-78783-6

(continued)

POINT	LATITUDE North	LONGITUDE East
20	1° 16.37′	103° 37.38′
21	1° 15.85′	103° 36.1′
22	1° 17.63′	104° 07.5′
23	1° 17.42′	104° 02.9′
24	1° 17.3′	104° 04.6′
25	1° 16.2′	104° 07.1′
26	1° 15.65′	104° 09.47′
27	1° 13.65′	104° 12.67′
28	1° 16.2′	104° 16.15′
29	1° 16.5′	104° 19.8′
30	1° 15.55′	104° 28.45′
31	1° 16.95′	104° 29.33′
32	1° 23.9′	104° 29.5′
33	1° 38.0′	104° 53.0′
34	1° 54.4′	105° 05.2′
35	2° 22.5′	105° 01.2′
36	2° 55.2′	104° 51.5′
37	3° 50.1′	104° 46.5′
38	4° 03.0′	104° 51.9′
39	5° 04.7′	105° 28.8′
40	5° 40.6′	105° 47.1′
41	6° 05.8′	105° 49.2′
42	6° 48.25′	104° 30.0′
43	7° 49.0′	103° 02.5′
44	7° 10.25′	102° 29.0′
45	6° 50.0′	102° 21.2′
46	6° 27.8′	102° 09.6′
47	6° 27.5′	102° 10.0′
48	2° 05.0′	109° 38.8′
49	3° 00.0′	109° 54.5′
50	4° 40.0′	110° 02.0′
51	5° 31.2′	109° 59.0′
52	6° 18.2′	109° 38.6′
53	7° 07.75′	111° 34.0′
54	8° 23.75′	112° 30.75′
55	8° 44.42′	113° 16.25′
56	8° 33.92′	113° 39.0′

(continued)

(continued)

POINT	LATITUDE North	LONGITUDE East
57	8° 24.42′	113° 47.75′
58	8° 24.43′	113° 52.42′
59	8° 23.75′	114° 19.83′
60	8° 30.25′	114° 29.17′
61	8° 28.17′	114° 50.12′
62	8° 55.0′	115° 10.58′
63	8° 49.08′	115° 38.75′
64	8° 19.92′	115° 54.08′
65	8° 01.5′	116° 03.5′
66	7° 40.0′	116° 00.0′
67	7° 40.0′	117° 00.0′
68	7° 24.75′	117° 25.5′
69	6° 52.0′	117° 58.0′
70	6° 17.0′	117° 58.0′
71	6° 00.0′	118° 20.0′
72	6° 00.0′	118° 50.0′
73	5° 16.0′	119° 35.0′
74	4° 42.0′	119° 00.0′
75	4° 23.0′	119° 00.0′
76	4° 23.0′	120° 00.0′
77	3° 02.75′	120° 15.75′
78	3° 01.5′	119° 53.0′
79	3° 06.0′	118° 57.5′
80	3° 08.67′	118° 46.17′
81	3° 39.0′	118° 22.0′
82	4° 03.65′	118° 01.1′
83	4° 08.0′	117° 56.95′
84	4° 10.0′	117° 53.97′

AN ANALYSIS OF THE CONTINENTAL SHELF TURNING POINTS

AS DEFINED IN THE 1979 MAP

Points 1–21 inclusive are located in the Malacca Strait and western approaches to the Straits of Singapore. Points 22–32 are in the vicinity of the eastern approaches to the Straits of Singapore. Indeed, Points 32–41 inclusive are coincident with Turning Points 11–20 of the Indonesia-Malaysia Continental Shelf boundary in the south-western sector of the South China Sea. Turning Points 41–47 encompass the polygons that comprise Agreed Common Area negotiated between Malaysia and Vietnam and the Joint Development Area between Malaysia and Thailand. These polygons were defined long after the publication of the 1979 Map.

Points 48–52 inclusive are coincident with the geographical coordinates of the Indonesia-Malaysia continental shelf boundary which commences at Tanjong Datu which is the western terminal point of the terrestrial boundary between the two countries on Borneo Island. Points 52–66 delineate the unilateral continental shelf limit of Malaysia. Turning Point 53 is located in an area of overlapping claim by Brunei.

Thirteen Turning Points, namely TP 54–66, were employed to define the edge of its continental shelf of the Sabah coast bordering the South China Sea. The justification thereof is based on the equidistance principle measured from the following features, all part of the Spratly Islands group. They are:

TP 54   Amboyna Cay (Pulau Kecil Amboyna) and Spratly Island
TP 55   Barque Canada Reef and Cuareteron or Pearson Reef
TP 56   Barque Canada Reef and Pearson or Alison Reef
TP 57   Barque Canada Reef and Cornwallis South Reef
TP 58   Terembu Mantanani (Mariveles Reef) and Cornwallis South Reef
TP 59   Terumbu Siput (Erica Reef) and Cornwallis Reef
TP 60   Terumbu Peninjau (Investigator Reef) and Tennent Reef
TP 61   Terumbu Peninjau and Tennent Reef
TP 62   Terumbu Laksamana (Commodore Reef) and Alicia Annie Reef
TP 63   Commodore Reef and First Thomas Shoal
TP 64   Commodore Reef and (unclear)
TP 65   Commodore Reef and Western Shoals (west of Balabac Is)
TP 66   A point agreed upon in the Treaty of Paris of 1898

Turning Points 66–76, inclusive follow the general alignment of a boundary which is generally perceived as an international limit of Philippine sovereignty as outlined in the Treaty of Paris, 1898 signed by Spain and the United States of America.

The above analysis and list of geographical coordinates should be read in relation to the map in the atlas which depicts the actual and potential maritime boundaries of Malaysia.

Annex XI
Malaysia's Territorial Sea Base Points List of Geographical Coordinates of Points

No.	Base Point	Area (Name)	Geo Coord Base Point WGS84	Distance	Azimuth
			Latitude (N) Longitude (E)	Km/NM	Degree
1.	SM 01	Batu Puteh	06° 25′ 22.0″ 100° 05′ 31.8″	19.8/**10.69**	263
2.	SM 02	P. Tg. Dendang	06° 24′ 11.0″ 99° 54′ 51.0″	3.88/**2.09**	009
3.	SM 03	P. Tg. Dendang (i)	06° 26′ 15.0″ 99° 55′ 10.8″	0.49/**0.26**	325
4.	SM 04	PTg. Dendang (ii)	06° 26′ 16.3″ 99° 55′ 09.9″	3.82/**2.06**	300
5.	SM 05/1	Tg. Langun	06° 27′ 18.8″ 99° 53′ 22.5″	7.53/**4.07**	289
6.	SM 06	Tg. Kemarong	06° 28′ 39.5″ 99° 49′ 30.9″	20.5/**11.07**	258
7.	SM 07	Tg. Chinchin	06° 26′ 16.8″ 99° 38′ 41.0″	0.27/**0.15**	246
8.	SM 08/1	Tg. Chinchin (i)	06° 26′ 13.1″ 99° 38′ 32.8″	0.28/**0.15**	222
9.	SM 09	Tg. Chinchin (ii)	06° 26′ 06.3″ 99° 38′ 26.6″	114/**61.6**	223
10.	SM 10/1	P. Perak	05° 41′ 08.1″ 98° 56′ 12.3″	111/**59.9**	090
11.	SM 11/1	P. Perak (i)	05° 41′ 03.4″ 98° 56′ 09.8″	229/**124**	146
12.	SM 12/2	P. Jarak	03° 58′ 32.2″ 100° 05′ 43.8″	169/**91.3**	133

(continued)

(continued)

No.	Base Point	Area (Name)	Geo Coord Base Point WGS84	Distance	Azimuth
			Latitude (N) Longitude (E)	Km/NM	Degree
13.	SM 13	Beting Kepah	02° 55′ 16.2″ 101° 11′ 25.8″	41.4/**22.4**	129
14.	SM 14	Tg. Gabang	02° 41′ 03.3″ 101° 28′ 43.4″	51.9/**28.0**	127
15.	SM 15	TTuan(Rachado)	02° 24′ 14.7″ 101° 51′ 08.7″	66.5/**35.9**	127
16.	SM 16	P. Undan	02° 02′ 49.4″ 102° 19′ 57.3″	146/**78.8**	124
17.	SM 17/1	P. Kukup	01° 18′ 19.5″ 103° 24′ 49.8″	9.09/**4.91**	120
18.	SM 18	Tg. Piai	01° 15′ 51.9″ 103° 29′ 04.7″	0.72/**0.39**	110
19.	SM 19	Tg. Piai (i)	01° 15′ 43.8″ 103° 29′ 26.6″	1.77/**0.96**	101
20.	SM 20	Tg. Piai (ii)	01° 15′ 32.2″ 103° 30′ 22.8″	0.37/**0.20**	088
21.	SM 21	Tg. Piai (iii)	01° 15′ 32.6″ 103° 30′ 34.6″	12.5/**6.75**	063
22.	SM 22	P. Merambong	01° 18′ 45.9″ 103° 36′ 38.3″	0.11/**0.06**	054
23.	SM 22/1	P. Merambong(i)	01° 18′ 48.0″ 103° 36′ 41.3″	8.90/**4.81**	025
24.	SM 23	Tg. Bunga	01° 23′ 08.7″ 103° 38′ 44.1″	3.13/**1.69**	027
25.	SM 24	Bahan	01° 24′ 38.8″ 103° 39′ 30.5″	1.05/**0.57**	043
26.	SM 25	Tg. Tuan	01° 25′ 03.5″ 103° 39′ 53.7″	0.34/**0.18**	036
27.	SM 25/1	Tg. Tuan (i)	01° 25′ 12.3″ 103° 40′ 00.2″	0.26/**0.14**	031
28.	SM 26	Tg.TebingRuntuh	01° 25′ 19.5″ 103° 40′ 04.5″	1.17/**0.63**	018
29.	SM 27	Tg. Kijang	01° 25′ 55.3″ 103° 40′ 16.8″	2.31/**1.24**	059
30.	SM 28	Tg. Tajam	01° 26′ 33.9″ 103° 41′ 21.0″	4.90/**2.65**	105
31.	SM 29	Tg. Danga	01° 27′ 50.3″ 103° 42′ 55.0″	0.23/**0.12**	097
32.	SM 30	Istana Besar	01° 27′ 10.6″ 103° 45′ 28.5″	0.10/**0.05**	072

(continued)

(continued)

No.	Base Point	Area (Name)	Geo Coord Base Point WGS84	Distance	Azimuth
			Latitude (N) Longitude (E)	Km/NM	Degree
33.	SM 31	Menara Johor	01° 27′ 09.6″ 103° 45′ 36.0″	0.86/**0.46**	072
34.	SM 31/1	Menara Johor (i)	01° 27′ 10.6″ 103° 45′ 39.1″	0.93/**0.5**	057
35.	SM 32	Tambak Barat	01° 27′ 19.4″ 103° 46′ 05.5″	4.02/**2.17**	051
36.	SM 33	Jeti Johor	01° 27′ 35.8″ 103° 46′ 30.8″	2.47/**1.33**	094
37.	SM 34	Kuala Tebrau	01° 28′ 57.5″ 103° 48′ 12.0″	5.04/**2.72**	116
38.	SM 35	Tg. Lunchoo	01° 28′ 52.0″ 103° 49′ 31.7″	0.07/**0.03**	151
39.	SM 37/1	Jeti IDEMITSU	01° 27′ 41.6″ 103° 51′ 58.9″	0.12/**0.06**	184
40.	SM 37/2	Jeti IDEM. (i)	01° 27′ 39.6″ 103° 52′ 00.0″	0.44/**0.24**	138
41.	SM 37/3	Jeti IDEM. (ii)	01° 27′ 36.9″ 103° 52′ 02.8″	0.50/**0.27**	140
42.	SM 37/4	Jeti MSE	01° 27′ 26.3″ 103° 52′ 12.4″	0.99/**0.53**	143
43.	SM 37	Jeti MSE (i)	01° 27′ 13.9″ 103° 52′ 22.7″	0.21/**0.11**	139
44.	SM 38	Jeti TEN	01° 26′ 48.0″ 103° 52′ 42.1″	0.65/**0.35**	141
45.	SM 39	Jeti TEN (i)	01° 26′ 42.9″ 103° 52′ 46.6″	0.17/**0.09**	136
46.	SM 41/2	Merbahaya jeti	01° 26′ 26.3″ 103° 52′ 59.7″	0.08/**0.04**	122
47.	SM 41/4	Merbahaya (i)	01° 26′ 22.4″ 103° 53′ 03.5″	0.20/**0.10**	118
48.	SM 41/3	Merbahaya (ii)	01° 26′ 21.0″ 103° 53′ 05.7″	0.76/**0.41**	114
49.	SM 41/1	Merbahaya (iii)	01° 26′ 17.9″ 103° 53′ 11.5″	0.27/**0.15**	105
50.	SM 41	Merbahaya (iv)	01° 26′ 07.7″ 103° 53′ 33.9″	0.60/**0.32**	107
51.	SM 42	Merbahaya (v)	01° 26′ 05.5″ 103° 53′ 42.2″	0.04/**0.02**	102
52.	SM 43/1	Jeti Minyak	01° 25′ 59.8″ 103° 54′ 00.9″	0.02/**0.01**	097

(continued)

(continued)

No.	Base Point	Area (Name)	Geo Coord Base Point WGS84	Distance	Azimuth
			Latitude (N) Longitude (E)	Km/NM	Degree
53.	SM43/4	Jeti Minyak (i)	01° 25′ 59.5″ 103° 54′ 02.1″	0.34/**0.18**	098
54.	SM43/2	Jeti Minyak (ii)	01° 25′ 58.1″ 103° 54′ 13.1″	0.28/**0.14**	074
55.	SM 44 J	Pasir Gudang (i)	01° 26′ 00.6″ 103° 54′ 21.8″	0.30/**0.16**	099
56.	SM 45	Pasir Gudang (ii)	01° 25′ 59.1″ 103° 54′ 31.4″	2.09/**1.13**	091
57.	SM 45/1	Pasir Gudang (iii)	01° 25′ 52.3″ 103° 55′ 14.1″	1.38/**0.75**	106
58.	SM 46	Tg. Gemuk	01° 25′ 50.6″ 103° 56′ 21.6″	0.99/**0.53**	096
59.	SM 47	Batuan Dawes	01° 25′ 37.8″ 103° 57′ 04.5″	1.87/**1.00**	091
60.	SM 48	Kg. Kabong	01° 25′ 34.2″ 103° 57′ 36.5″	0.52/**0.28**	078
61.	SM 49	P. Nanas	01° 25′ 33.1″ 103° 58′ 37.2″	2.01/**1.1**	066
62.	SM 50	P. Nanas (i)	01° 25′ 36.7″ 103° 58′ 53.5″	2.01/**1.1**	067
63.	SM 51	Tg. Kopok	01° 26′ 02.5″ 103° 59′ 53.2″	6.96/**3.76**	076
64.	SM 52	Tg. Belungkor	01° 26′ 57.0″ 104° 03′ 31.8″	0.22/**0.12**	200
65.	SM 52/1	Tg. Belungkor (1)	01° 26′ 50.4″ 104° 03′ 29.4″	0.14/**0.07**	165
66.	SM 53	Tg. Belungkor (2)	01° 26′ 46.0″ 104° 03′ 30.6″	0.09/**0.04**	090
67.	SM 54	Tg. Belungkor(3)	01° 26′ 42.1″ 104° 03′ 33.7″	0.41/**0.22**	122
68.	SM 55	Tg. Belungkor (4)	01° 26′ 35.0″ 104° 03′ 44.9″	1.28/**0.69**	101
69.	SM 56	Baoh Kechil	01° 26′ 27.4″ 104° 04′ 25.6″	2.02/**1.09**	148
70.	SM 57	Karang Si Ajar	01° 25′ 31.9″ 104° 05′ 00.4″	2.73/**1.47**	134
71.	SM 58	Pasir Bunga	01° 24′ 30.2″ 104° 06′ 03.9″	3.51/**1.89**	200
72.	SM 58/3	KD S. Ismail	01° 22′ 43.6″ 104° 05′ 24.1″	0.22/**0.11**	205

(continued)

(continued)

No.	Base Point	Area (Name)	Geo Coord Base Point WGS84	Distance	Azimuth
			Latitude (N) Longitude (E)	Km/NM	Degree
73.	SM 58/2	KD S Ismail (i)	01° 22′ 37.3″ 104° 05′ 21.1″	1.11/**0.59**	189
74.	SM 58/1	KD S Ismail (ii)	01° 22′ 34.8″ 104° 05′ 20.4″	0.97/**0.52**	170
75.	SM 59	Tg. Pengelih	01° 21′ 59.4″ 104° 05′ 15.0″	6.19/**3.34**	117
76.	SM 61	M. Berendam	01° 21′ 28.4″ 104° 05′ 20.2″	4.02/**2.17**	180
77.	SM 62	Tg. Setapa	01° 19′ 56.2″ 104° 08′ 18.3″	0.64/**0.34**	108
78.	SM 63	Tg. Kapal	01° 19′ 49.7″ 104° 08′ 37.9″	3.24/**1.75**	036
79.	SM 64	Tubir Selatan	01° 17′ 50.8″ 104° 23′ 35.7″	3.23/**1.75**	034
80.	SM 65	Batuan Tengah	01° 19′ 15.6″ 104° 24′ 36.6″	125.2/**62.6**	007
81.	SM 67	P. Pinang	02° 26′ 15.3″ 104° 33′ 04.7″	0.15/**0.08**	340
82.	SM 67/1	P. Pinang (i)	02° 26′ 20.0″ 104° 33′ 03.0″	1.97/**1.06**	327
83.	SM 68	P. Aur	02° 27′ 13.9″ 104° 32′ 28.8″	63.1/**34.07**	320
84.	SM 69	Tg. Letak Hitam (Pulau Tioman)	02° 53′ 31.0″ 104° 10′ 47.2″	70.2/**37.90**	305
85.	SM 70/1	P. Berhala	03° 15′ 04.9″ 103° 39′ 35.4″	170/**91.79**	001
86.	SM 71	Batu T Daik	04° 46′ 56.1″ 103° 41′ 21.2″	3.41/**1.84**	003
87.	SM 72	Tg. Pak Kiok (P. Tenggol)	04° 48′ 37.7″ 103° 41′ 26.8″	5.65/**3.05**	339
88.	SM 73	B. T Kemudi	04° 51′ 28.8″ 103° 40′ 22.2″	104/**56.2**	327
89.	SM 74/1	P. Yu Besar	05° 38′ 33.2″ 103° 09′ 14.0″	18.2/**9.83**	325
90.	SM 75	P. Limau	05° 46′ 32.4″ 103° 03′ 39.6″	12.72/**6.36**	277
91.	SM 76	Tg. Gua Kawah (P. Redang)	05° 47′ 22.7″ 103° 02′ 37.8″	4.88/**2.63**	310
92.	SM 77/1	Tg. Batu Tokong	05° 49′ 05.6″ 103° 00′ 37.3″	86.1/**46.5**	298

(continued)

(continued)

No.	Base Point	Area (Name)	Geo Coord Base Point WGS84	Distance	Azimuth
			Latitude (N) Longitude (E)	Km/NM	Degree
93.	SM 78	Kg. Pantai Dasar	06° 10′ 49.1″ 102° 19′ 20.6″	10.8/**5.83**	295
94.	SM 79/1	K. Besar	06° 13′ 18.0″ 102° 14′ 02.9″	16.1/**8.69**	279
95.	SM 80	Pengkalan Kubor	06° 14′ 37.1″ 102° 05′ 25.1″		
Sarawak offshore					
96.	SWK 01	Tg. Datu	02° 04′ 53.7″ 109° 38′ 41.9″	0.55/**0.3**	035
97.	SWK 01A	Tg. Datu (i)	02° 05′ 08.1″ 109° 38′ 52.1″	111/**59.94**	111
98.	SWK 02	Tg. Patong	01° 45′ 16.9″ 110° 29′ 57.4″	148/**79.9**	036
99.	SWK 03	Tg. Sirik	02° 49′ 21.1″ 111° 17′ 27.1″	5.39/**2.91**	050
100.	SWK 04	Tg. Sirik (i)	02° 51′ 13.7″ 111° 19′ 40.9″	209/**113**	074
101.	SWK 05	Batuan Likau	03° 22′ 09.5″ 113° 08′ 03.5″	164/**88.6**	033
102.	SWK 06/1	K. Baram	04° 36′ 22.3″ 113° 56′ 44.2″	1.85/**1.0**	047
103.	SWK 06A	K. Baram (i)	04° 37′ 03.7″ 113° 57′ 28.01″	1.86/**1.0**	046
104.	SWK 07	Sg. Tujuh	04° 35′ 21.0″ 114° 04′ 22.0″	13.1/**7.07**	104
105.	SWK 10A	Tg. Angareteng	04° 52′ 11.7″ 115° 03′ 58.4″	114/**61.6**	074
106.	SWK 10	Tg.Angareteng (i)	04° 51′ 37.1″ 115° 03′ 34.0″	1.31/**0.71**	215
107.	SWK 11	Tg. Tubu-Tubu	04° 51′ 01.1″ 115° 02′ 54.0″	1.66/**0.9**	228
108.	SWK 12	M S Pandaruan	04° 49′ 51.1″ 115° 02′ 01.9″	16.7/**9.02**	052
109.	SWK 13	K. Bangau	04° 55′ 16.4″ 115° 09′ 14.3″	5.76/**3.11**	088
110.	SWK 14	Beting Sunda	04° 58′ 20.9″ 115° 09′ 41.8″	2.91/**1.57**	031
111.	SWK 15	Beting Sunda (i)	04° 59′ 41.5″ 115° 10′ 30.5″		

(continued)

(continued)

No.	Base Point	Area (Name)	Geo Coord Base Point WGS84	Distance	Azimuth
			Latitude (N) Longitude (E)	Km/NM	Degree
Sabah offshore					
112.	SBH 01	K. Sg. P Damit	05° 14′ 01.6″ 115° 23′ 39.9″	29.33/ **15.84**	257
113.	SBH 02	B R Besar	05° 10′ 29.9″ 115° 08′ 08.5″	3.28/**1.77**	333
114.	SBH 03	B Keraman	05° 12′ 04.5″ 115° 07′ 19.9″	107/**57.8**	180
115.	SBH 04	P. Keraman	05° 13′ 54.0″ 115° 07′ 10.3″	119/**64.3**	025
116.	SBH 05	P. Mangalum	06° 12′ 17.6″ 115° 34′ 48.0″	1.80/**0.97**	046
117.	SBH 06	P. Mangalum(i)	06° 12′ 58.2″ 115° 35′ 30.2″	186/**100.4**	050
118.	SBH 07	Tg. Timohing	07° 17′ 17.2″ 116° 53′ 04.2″	15.4/**8.32**	056
119.	SBH 08	Beting Siagut	07° 21′ 58.6″ 116° 59′ 58.2″	28.4/**15.3**	059
120.	SBH 09	T Mangsee B	07° 29′ 46.4″ 117° 13′ 15.9″	2.62/**1.41**	063
121.	SBH 10	T Mangsee B(i)	07° 30′ 24.5″ 117° 14′ 32.0″	1.15/**0.62**	078
122.	SBH 11	T Mangsee B(ii)	07° 30′ 32.2″ 117° 15′ 08.6″	0.58/**0.31**	096
123.	SBH 11A	Ter Mangsee	07° 30′ 30.3″ 117° 15′ 27.4″	1.13/**0.61**	105
124.	SBH 12	Ter Mangsee(i)	07° 30′ 20.6″ 117° 16′ 02.9″	0.91/**0.49**	134
125.	SBH 13	Ter Mangsee(ii)	07° 29′ 59.9″ 117° 16′ 24.0″	14.5/**7.83**	140
126.	SBH 14	Banggi 1 Rf	07° 23′ 48.5″ 117° 21′ 25.0″	1.24/**0.67**	154
127.	SBH 15	Bang. ONERf(i)	07° 23′ 12.4″ 117° 21′ 42.8″	57.0/**30.7**	145
128.	SBH 16	Ter Minna	06° 58′ 03.7″ 117° 39′ 35.6″	58.5/**31.59**	149
129.	SBH 17	P. Langkayan	06° 30′ 43.4″ 117° 55′ 31.6″	44.4/**23.97**	170
130.	SBH 18	P. Bonting	06° 07′ 10.2″ 117° 59′ 49.2″	9.46/**5.11**	047

(continued)

(continued)

No.	Base Point	Area (Name)	Geo Coord Base Point WGS84	Distance	Azimuth
			Latitude (N) Longitude (E)	Km/NM	Degree
131.	SBH 19	P. Silingan	06° 10′ 40.8″ 118° 03′ 32.9″	16.7/**9.01**	007
132.	SBH 20	P. Silingan (i)	06° 10′ 41.9″ 118° 03′ 38.7″	5.56/**3.0**	100
133.	SBH 21	P. Bakungan K	06° 10′ 10.4″ 118° 06′ 36.9″	0.21/**0.11**	115
134.	SBH 22	P. Bakungan K(i)	06° 10′ 07.6″ 118° 06′ 43.0″	137/**74.0**	128
135.	SBH 23	P. Tambisan	05° 24′ 21.4″ 119° 04′ 58.4″	6.56/**3.54**	013
136.	SBH 24	P. Tambisan (i)	05° 27′ 48.1″ 119° 05′ 48.1″	1.97/**1.06**	071
137.	SBH 25	P. Tambisan (ii)	05° 28′ 08.4″ 119° 06′ 49.0″	1.94/**1.04**	094
138.	SBH 26	P. Tambisan (iii)	05° 28′ 04.1″ 119° 07′ 51.9″	1.19/**0.64**	103
139.	SBH 27	P. Tambisan (iv)	05° 27′ 55.2″ 119° 08′ 29.4″	7.01/**3.79**	118
140.	SBH 27A	Tg. Unsang	05° 26′ 06.3″ 119° 11′ 49.4″	2.03/**1.09**	084
141.	SBH 28	Tg. Trang	05° 25′ 12.8″ 119° 12′ 55.1″	3.32/**1.79**	180
142.	SBH 29	Tg. Trang (i)	05° 23′ 25.1″ 119° 14′ 26.2″	4.19/**2.26**	180
143.	SBH 30	Tg. Trang (ii)	05° 21′ 09.6″ 119° 15′ 30.6″	4.51/**2.44**	180
144.	SBH 31	Kg. Anteamu	05° 18′ 43.5″ 119° 15′ 56.9″	8.83/**4.77**	180
145.	SBH 32	Kg. Sahabat19	05° 13′ 57.5″ 119° 16′ 10.5″	0.22/**0.11**	180
146.	SBH 32A	Kg. Sahabat19	05° 13′ 50.3″ 119° 16′ 09.9″	1.90/**1.03**	180
147.	SBH 33	Kg. Sahabat19/2	05° 12′ 48.8″ 119° 15′ 58.0″	68.3/**36.88**	180
148.	SBH 34	P. Mataking K.	04° 35′ 59.3″ 118° 56′ 56.5″	1.25/**0.68**	163
149.	SBH 35	P. Mataking K(1)	04° 35′ 20.5″ 118° 57′ 08.4″	2.43/**1.31**	169
150.	SBH 36	P. Mataking K(2)	04° 34′ 03.2″ 118° 57′ 23.6″	12.8/**6.91**	184

(continued)

(continued)

No.	Base Point	Area (Name)	Geo Coord Base Point WGS84	Distance	Azimuth
			Latitude (N) Longitude (E)	Km/NM	Degree
151.	SBH 37	P. Boheian	04° 27' 08.6" 118° 56' 54.3"	25.9/**13.98**	188
152.	SBH 38	P. Ligitan	04° 13' 17.6" 118° 54' 57.8"	8.77/**4.74**	195
153.	SBH 39	P. Ligitan (i)	04° 08' 43.5" 118° 53' 42.9"	0.92/**0.49**	197
154.	SBH 39A	P. Ligitan (ii)	04° 08' 13.1" 118° 53' 33.6"	0.62/**0.33**	241
155.	SBH 40	P. Ligitan (iii)	04° 08' 03.2" 118° 53' 16.0"	28.4/**15.3**	263
156.	SBH 41	P. Sipadan	04° 06' 13.2" 118° 38' 00.9"	50.7/**27.3**	274
157.	SBH 42	BatuanTangan	04° 08' 24.4" 118° 10' 40.8"	30.0/**16.2**	275
158.	SBH 43	P. Sebatik	04° 10' 00.5" 117° 54' 31.8"		

Summary

Total number of defined Basepoints:	158
Total number of Straight Baselines:	155
Straight baselines encompassing Pen. Malaysia:	94
Straight baselines enclosing Sarawak coastline:	15
Straight baselines encompassing Sabah's coast:	46
Longest length-points 11–12- P Perak to P Jarak	124M (nautical miles (M)) (1 km = 0.539957M)

Category	Segments
0–0.99	118
10–10.99	2
11–19.99	6
20–20.99	5
30–39.9	6
40–49.9	1
50–59.9	4
60–69.9	4

(continued)

(continued)

Category	Segments
70–79.9	3
80–89.9	1
90–99.9	2
100–109.9	2
110–119.9	0
Over 120	1

Bibliography

Academy of Sciences, Malaysia. (2022). *Position paper on blue economy: Unlocking the value of the oceans* (p. 216). Academy of Sciences.

Allianz. (2021). *Safety and shipping review* 2021. https://www.agcs.allianz.com

Armstrong, P. H. (1986). *Ecology and ecosystems* (61 pp). Bookland Pty. Ltd.

Armstrong, P. H., & Forbes, V. L. (2005). The exploitation of resources in the coastal waters of countries in the Indo-Pacific region: Ecological and politico-legal approaches. In *Perspectives in resource management in developing countries* (Vol. 2).

Asian Development Bank. (2014). *State of coral triangle: Malaysia, Philippines* (127pp). ADB.

Auburn F. M., Ong, D., & Forbes. V. L. (1994). Dispute resolution and the Timor gap treaty. *Occasional Paper, 35*, 74. IOCPS, UWA, Perth.

Awalluddin, A. (2020). *Marine biotechnology research in Malaysia.* MIMA.

Azan, A. H. M. et al. (2017). Satellite-based offshore wind energy resource mapping in Malaysia. *Journal of Marine Science and Application, 9* (online version).

Azam, A. H. M., et al. (2023). Malaysia's blue economy: Position, initiatives, and challenges. *Policy Brief* (4 p). Economic Research Institute for ASEAN and East Asia.

Basiron, M. N., & Dastan, R. (Eds.). (2004). *Building a comprehensive security environment in the Straits of Malacca* (p. 318). K.L.

Beckman, R. C., Grundy-Warr, C., & Forbes, V. L. (1994). Acts of piracy in the Malacca and Singapore Straits. *Maritime Briefing, 1*(4), 37.

BIMCO. (2014). *Safe passage: The Straits of Malacca and Singapore.* www.bimco.org

Bueger, C. (2014). What is maritime security. *Marine Policy, 53*, 156–164. https://doi.org/10.1016/j.marpol.2014.12.005

Butcher, J. G., & Elson, R. E. (2017). *Sovereignty and the sea how Indonesia became an archipelagic state* (p. 527). NUS Press.

Colgan, C. S., Forbes, V. L., & Mwanyoka, I. (2021). Measuring the blue economy (Ch. 10). In Emeritus Professor Donald Sparks (Eds.), *The blue economy of Sub-Saharan Africa working for a sustainable future* (pp. 197–214). Routledge.

Defense Mapping Agency US. (1991). *Sailing directions for Southeast Asia* (271pp). DMA. (This Organization also produces maps and other graphics).

Department of Environment (Malaysia). (2006). Quarterly update on environment, development and sustainability. *IMPAK, 3*, 3.

Forbes, V. L. (1995a). *The maritime boundaries of the Indian Ocean region* (267 p). Singapore University Press.

Forbes, V. L. (1995b). *Indonesia's maritime boundaries* (100 p). Malaysian Institute of Maritime Affairs.

Forbes, V. L. (1995c). *Border disputes in module 7: Cooperation on non-military threats to security, final report* (pp. 43–51). Centre for Marine Policy, The Illawarra Technology Corporation Ltd.

© The Editor(s) (if applicable) and The Author(s), under exclusive license to Springer Nature Switzerland AG 2025
V. L. Forbes, *Malaysia's Maritime Jurisdictional Limits,*
https://doi.org/10.1007/978-3-031-78783-6

Forbes, V. L. (1997). The geopolitics of energy: Natuna oil and gas field. *Geopolitics of Energy,* *19*(10), 6–11.

Forbes, V. L. (1998). Cooperative approaches to the utilisation of marine resources in the ASEAN region. In V. R. Savage et al., (Eds.), *The Naga awakens* (pp. 113–126). Times Academic Press.

Forbes, V. L. (1999). Vessel traffic separation system in the straits of Malacca and Singapore. *Indian Ocean Review, 12*(1), 12–16.

Forbes, V. L. (2001). *Conflict and cooperation in managing maritime space in semi-enclosed seas* (pp. 382 + xviii). Singapore University Press.

Forbes, V. L. (2004). ECDIS and potential legal implications: Proceeding with caution. *The Hydrographical Journal, 111*, 1–11.

Forbes, V. L. (2005a). Geopolitical change: Direction and continuing issues. In L. S. Chia (Ed.), *Southeast Asia transformed, a geography of change* (Ch. 2, pp. 47–94). Institute of South-East Asian Studies.

Forbes, V. L. (2005b). Territorial claims and Southeast Asian Regional cohesiveness. In V. R. Savage & M. Tan-Mullins (Eds.), *The Naga challenged* (pp. 102–140). Marshall Cavendish Academic.

Forbes, V. L. (2005c). Sunda Shelf: Indonesia and Vietnam maritime boundary delimitation. *MIMA Bulletin, 12*(1), 5–9.

Forbes, V. L. (2006). The Malacca Strait in the context of the ISPS code. In M. N. Basiron, & Dastan (Eds.), *Building a comprehensive security environment in the Straits of Malacca* (pp. 290–318). MIMA.

Forbes, V. L. (2007a). The territorial sea datum of Malaysia. *MIMA Bulletin, 14*(4), 3–8.

Forbes, V. L. (2007b). Ensuring security in the Straits of Malacca: Solutions offered and suggested implementations. In N. Khalid (Ed.), *Enhancing security in the Straits of Malacca: Amalgamation of solutions to keep the straits open to all* (pp. 50–65). Maritime Institute of Malaysia.

Forbes, V. L. (2008a). Archipelagic baseline systems: Indonesia and The Philippines. *MIMA Bulletin, 15*(1), 3–12.

Forbes, V. L. (2008b). Regional overview of disputed and divided territories: Islands, islets, reefs and rocks. *MIMA Bulletin, 15*(1), 33–37.

Forbes, V. L. (2008c). The Judgment over Pulau Batu Puteh and associated rocks: Analysis and commentary. *MIMA Bulletin, 15*(2), 23–27.

Forbes, V. L. (2008d). Cartography, hydrography, charts and maps. *MIMA Bulletin, 15*(3), 4–13.

Forbes, V. L. (2008e). The International Court of Justice (ICJ) gives, it takes and it waivers. In J. S. Sidhu & K. S. Balakrishnan (Eds.), *The seas divide: Geopolitics and maritime issues in Southeast Asia*, Monograph Series 5 (pp. 67–82). Institute of Earth and Ocean Sciences (IOES), University of Malaya.

Forbes, V. L. (2009a). Philippines archipelagic baseline system. *MIMA Bulletin, 16*(2), 12–15.

Forbes, V. L. (2009). The Indian Ocean fishery; resources and exploitation within and outside national jurisdiction. In D. Rumley and et al. (Eds.), *IORG and ISEAS* (Ch. 5, pp. 72–103).

Forbes, V. L. (2010). India's look east policy. In D. Rumley, & D. Gopal (Eds.), *Globalisation and regional security* (pp. 107–131). Shipra Publications, Ch. 8.

Forbes, V. L. (2012). Refashioning geography for terrestrial increase and incremental maritime jurisdictional creep. *MIMA Bulletin, 19*(2), 8–19.

Forbes, V. L. (2013a). Malaysia and China: Economic growth overshadows sovereignty dispute. In B. A. Elleman, S. Kotkin, & C. Schofield (Eds.), *Beijing's power and china's borders twenty neighbours in Asia*, Ch. 11 (pp. 155–167). M.E. Sharpe.

Forbes, V. L. (2013b). Geopolitics, energy security and 'Soft-Shoe Diplomacy in the South China Sea. *The Journal of Diplomacy and Foreign Relations, 14*(1), 39–64.

Forbes, V. L. (2014a). *Indonesia's delimited maritime boundaries* (pp. 266 + vii). Springer.

Forbes, V. L. (2014b). An exclusive economic zone boundary in the Sulawesi Sea. *MIMA Bulletin, 21*(1).

Forbes, V. L. (2015a). Artificial Islands in the South China Sea: Rationale for terrestrial increase, incremental maritime jurisdictional creep and military bases. *The Journal of Defence and Security, 6*(1), 30–55.

Forbes, V. L. (2015b). ASEAN, China and Malaysia: Cautious diplomacy, trade and a complex sea. *International Journal of China Studies, 6*(2), 129–148.

Forbes, V. L. (2017a). China-Malaysia relations: Navigating a complex geopolitical sea, part one: Deep economic ties. *Associate paper, future directions international* (Published: Online on 31 August).

Forbes, V. L. (2017b). China-Malaysia relations: Navigating a complex geopolitical sea, part two: The South China Sea. In *Associate paper, future directions international.* (Published online on 5 September).

Forbes, V. L. (2017c). Territorial sea limits in the Singapore Strait'. *Journal of Territorial and Maritime Studies, 4*(2), 119–134 (Summer/Fall).

Forbes, V. L. (2017d). Archival research to enhance territorial and sovereignty claims. In Ch. 3, *South China Sea disputes—Historical, geopolitical and legal studies* (pp. 65–99).

Forbes, V. L. (2018). Straits of Malacca and Singapore: Sharpening the focus. *Journal of Diplomacy and Foreign Relations, 17*(1), 7–32.

Forbes, V. L. (2019a). Malaysia's maritime limits: Concerns and implications for defence policy. In Hamzah, B. A., Leong, A., & Fadzil, M. (Eds.), *Malaysia's defence: Selected strategic issues* (pp. 13–23). CDiSS.

Forbes, V. L. (2019b). Maritime security and Sea Lanes of communication: Geopolitical perspective on the belt and road initiative. In K. Zou, S. Wu, & Q. Ye (Eds.). *The 21st century maritime Silk Road challenges and opportunities for Asia and Europe* (pp. 78–95). Routledge.

Forbes, V. L. (2019c). Celebes and Sulu Seas: Maritime jurisdictional limits, Ch. 2. *Maritime security and the Sulu Zone: Readings on history, peace making and terrorism* (pp. 6–25).

Forbes, V. L. (2019d). Malaysia's maritime limits: Concerns and implications for defence policy'. In B. A. Hamzah, A. Leong, F. Mokhtar (Eds.), Ch. 2, (pp. 13–27).

Forbes, V. L. (2020a). Malaysia's actual and potential maritime boundaries in the South China Sea. In Hamzah, B. A., Leong, A., & V. L. Forbes (Eds.) *Malaysia and South China Sea.* UPNM: CDiSS.

Forbes, V. L. (2020b). *Maritime boundary delimitation in the Timor Sea: Australia's experience during five decades* (28pp). CDiSS Commentary.

Forbes, V. L. (2021a). Tensions brewing at Luconia shoals: Harmful for China-Malaysia Relations. *Commentary, 13*, 5 (Published Online) CDiSS, UPNM.

Forbes, V. L. (2021b). The South China Sea: Geographical overview (Chapter 1) In Z. Keyuan (Ed.), *Routledge handbook of the South China Sea* (pp. 1–26). Routledge

Forbes, V. L. (2021c). Coastal and offshore energy and mineral resources (Chapter 3). In Emeritus Professor Donald Sparks (Eds.), *The blue economy of sub-Saharan Africa. Working for a sustainable future* (pp. 67–82). Routledge.

Forbes, V. L. (2021d). Ports, shipping and transport, Chapter 5, of book: Titled *The blue economy of sub-Saharan Africa working for a sustainable future. Emeritus Professor Donald Sparks* (pp. 97–111). Routledge.

Forbes, V. L. (2022). Marine awareness and ocean governance: Arafura and Timor seas, and Torres Strait. SHIMBA. https://doi.org/10.4108/eai.18-9-2022.2326048

Forbes, V. L. (2004). The Malacca straits in the context of the ISPS code. In Basiron and Darum (Eds.), *Building a comprehensive security environment* (pp. 290–299). MIMA.

Forbes, V. L., & Basiron, N. M. (1998). *Malaysia's maritime space—Analytical atlas of environments and resources* (100 p). Maritime Institute of Malaysia.

Forbes, V. L., & Mohd. Nizam Basiron. (2009). *Malaysia's maritime realm atlas* (pp. 231 + x). Maritime Institute of Malaysia.

Forbes, V. L., & Basiron, N. (2010). Unresolved maritime boundaries and implications for maritime security in Southeast Asia. *MIMA Bulletin, 17*(1), 3–13.

Forbes, V. L., & Permal, S. (2015). Geopolitical trends in the South China Sea: 2013–2015. *The Journal of Diplomacy and Foreign Relations, 15*(1), 5–19.

Forbes, V. L., & Sakhuja, V. (2004). Challenging acts of marine transboundary transgressions in the Indian Ocean Region. *MIMA Issue Paper, 1*, 22.

Forbes, V. L., & Zulkifil, I. Cdr (Retd.). (2011). Hydrographical surveying and chartering of the coastline of Malaysia: An historical perspective to 1958. *MIMA Bulletin, 18*(2), 4–11.

Freemen, C. P. (2016). The fragile global commons in a world of transition. *SAIS Review of International Affairs, 36*(1), 17–28 (Winter-Spring).

Gewirtz, P. (2016). Limits of Law in the South China Sea. In *East Asia Policy Paper, 8*, 21.

Global Maritime Forum. (2023). *Annual progress report on green shipping corridors*, 36pp.

Gunaratna, R. (2018). Global threat forecast. *Counter Terrorist Trends and Analyses, 10*(1), 1–6.

Haller-Trost, R. (1993). Historical legal claims: A study of disputed sovereignty over Pulau Batu Puteh/Pedra Branca. *Maritime Briefing, 1*(1), 36.

Haller-Trost, R. (1995). The territorial dispute between Indonesia and Malaysia over Pulau Sipidan and Pulau Ligitan in the Celebes Sea: A study in international law, IBRU. *Boundary and Territory Briefing, 2*(2), 40.

Haller-Trost, R. (1998). *The contested maritime and territorial boundaries of Malaysia: An international law perspective* (p. 575). Kluwer Law International.

Hallwood, P. (2014). *Economics of the oceans* (p. 298). Routledge.

Hamid, A. G. (n.d.). *Malaysia's commitments under international conventions and the need for a harmonized legal regime regulating marine pollution.* Paper presented at the first national Maritime Convention.

Hamzah, B. A., & Basiron, M. N. (1997). *The straits of Malacca: Some funding proposals* (48 pp). Maritime Institute of Malaysia.

Hamzah, B. A. et al. (Eds.), *Malaysia's selected strategic issues* (pp. 13–27).

Hamzah, B. A., Forbes, V. L., Jalil, J. A., & Basiron, M. N. (2014). The maritime boundaries of Malaysia and Indonesia in the Malacca strait: An appraisal. *Australian Journal of Maritime and Ocean Affairs, 6*(4), 207–226.

Hamzah, B. A., & Forbes, V. L. (Eds.). (2019). *Maritime security and the Sulu zone: Readings on history, peace making and terrorism* (p. 537). Publisher.

Hamzah, B. A., Adam Leong, A., & Fadzil Mokhtar (Eds.). (2019). *Malaysia's defence: Selected strategic issues* (80 p). CDiSS (UPNM).

Hamzah, B. A., Leong, A., & Forbes, V. L. (Eds.) (2020) *Malaysia and South China sea: Policy, strategy and risks* (202 p). Centre for Defence and International Security Studies, UPNM.

Hamzah, B. A., & Forbes, V. L. (2021). *The territorial sea of Malaysia: Preliminary observations* (33pp). CDiSS.

Harun, R., & Ja'afar, S. (2021) *Malaysia a maritime nation* (406 pp). MIMA.

Hew, K. (2013). *Coral triangle initiative ecosystem approach to fisheries management country position paper—Malaysia, USA* (pp. 60). AID and Malaysia's National Coordinating Committee.

Hydrographer of the Navy, UK. (1961). *Eastern archipelago pilot* (Vol. II, 821pp). HMSO.

Hydrographer of the Navy, UK. (1971). *Malacca strait and west coast of Sumatra pilot* (477pp). HMSO.

Hydrographer of the Navy, UK. (1978). *China sea pilot* (Vol. 1, 304pp). HMSO.

Hydrographer of the Navy, UK. (1982). *China sea pilot* (Vol. 11, 218pp). HMSO.

Hydrographic Office, U. S., & Dep, N. (1932). *Sailing directions for Malacca Strait and Sumatra* (p. 612). US Govt. Printing Office.

International Hydrographic Organization. (1953). *Limits of oceans and sea*, S.P. (Vol. 23, 3rd edn).

ICC IMB. (2024). *Piracy and armed robbery against ships report for 1 January to 31 December 2023* (55 p). London.

International Maritime Organisation. (2024). 'IMO Monthly Piracy Report'. A 3-page summary of acts of piracy and armed robbery allegedly committed against ships in the Malacca Strait during the last quarter of 2023.

Jayakumar, S., & Koh, T. (2009). *Pedra Branca the road to the world court* (p. 195). NUS Press in association with MFA Diplomatic Academy.

Jenne, N. (2017). Managing territorial disputes in Southeast Asia: Is there more than the South China Sea? *Journal of Current Southeast Asian Affairs, 3*, 35–61.

Kadir, A. M. (2009). Malaysia's territorial disputes—Two cases at the ICJ, IDFR: Kuala Lumpur, 110 pages (Relating to the Malaysia/Singapore Case over Batu Puteh and two other Marine Features and the Indonesia/Malaysia over the Ligitan and Sipadan Islands).

Kaur, C. R. (2022). *Navigation hazards as an evolving threat to Southeast Asia's maritime security*, S. Rajaratnam School of International Studies, Singapore.

Kepli, M. Y. Z. (2023). *Shipping and logistics in Malaysia* (558pp). MIMA and P&I Club.

Kusmuk, N., & Forbes, V. L. (2016). The scourge of piracy. *MIMA Occasional Paper, 2*(1), 152.

Lee, H. K. (2005). Mapping the law of legalizing maps: The implications of the emerging rule on map evidence in international law. *Pacific Rim Law and Policy Journal Association, 14*(1), 159–188.

Liow, J. C. (2018). Shifting Sands of Terrorism in Southeast Asia. *RSIS Commentary, 025*, 5pp.

Malaysia's Marine Department. (2017). *Spurring national growth* (136pp). Marine Department. (Narrative and graphics contributed by MIMA personnel, including Forbes, V.L.)

Malaysia's Marine Department. (2019). Strategic plan—2019 to 2023, KL, 37pp. (In Bahasa)

Malaysia's Marine Department. (2022). Pelan Strategic Pendigitalan [Strategic Plan] (2021–2025), KL, 54pp.

Malaysia. (2017). *Malaysia partial submission part I: Executive summary*, MYS_ES_DOC-0!

Malaysia. (2017). National activities on standardization of Malaysia's Geographical Names' E/Conf.105/119/CRP.110.

Malaysia Ministry of Economy. (2022). *The Malaysian economy in figures 2022*, Fact Sheet

Malaysia's MIDA. (u.d.). *Maritime transport: Accelerating international trade* (online)

Malaysia's Ministry of Transport. (2017). *Malaysia shipping master plan 2017 to 2021: Revitalizing shipping for a stronger economy*, Kuala Lumpur.

Malaysia's Federal Government Gazette 'Water Services Industry (Prohibited Effluent) *Regulations* 2021. 9pp.

Mansell, J. N. K. (2009). The concept of flag state. In *Flag state responsibility* (Ch. 2, pp. 13–23). Springer. https://doi.org/10.1007/978-3-540-92933-8_2

Marine Department of Malaysia. (2014). *Lights by the Sea* Beacons in the Straits, Edited by Captain Mohamad Halim Bin Ahmed, 56pp.

Maritime Institute of Malaysia. (2021). The impact of COVID-19 pandemic on Malaysia's maritime sectors and way forward. *MIMA Issue Paper*, Kuala Lumpur.

Maritime Institute of Malaysia. (2021). *The effectiveness of the Cabotage policy liberalization in Sabah and Sarawak*. MIMA.

MIMA. (2021a). *Issues and prospect of bunkering industry in Malaysia—A preliminary study*. MIMA.

Li, M., et al. (2024). Potential offshore wind energy on Malaysia: An investigation into wind and bathymetry conditions and site selection', *ENERGIES MDPI, 17*

Mekhilef, S., & Chandrasegaram, D. (2011). *Assessment of off-shore wind farms in Malaysia* (pp. 1351–1355). TENCON 2011. IEEE 978-1-4577-6/11.

Muniandy, S. C. (2021). The case for internationalising the South China Sea. *Sea Power Soundings* (33pp), Royal Australian Navy.

Netherlands Ministry of Foreign Affairs (n.d.) *Port Development in Malaysia*, 26pp.

National Geospatial-Intelligence Agency (2015) *Sailing Directions for Borneo, Jawa…* Pub. 163, 13th Ed. Springfield, Virginia 389pp.

National Geospatial-Intelligence Agency. (2016). *Sailing Directions Strait of Malacca and Sumatera*, Pub. 174, 193pp.

National Geospatial-Intelligence Agency. (2017). *Sailing directions South China Sea and the Gulf of Thailand*, Pub. 161, 246pp.

National Geospatial-Intelligence Agency. (2022). Sailing directions. *Pacific Ocean and South-East Asia Malaysia, 120*, 264.

National Hydrographic Centre. (2019). Chart catalogue of Malaysia. MAL 2. Graphic online.

National Hydrographic Centre. (2024). Annual summary of Malaysian Notices to Mariners, NHC (99 pp).

Nemeth, S. C., Mitchell, S. M., Nyman, E. A., & Hensel, P. R. (2014). UNCLOS and the management of maritime conflicts. In *Paper prepared for presentation at the 2007 Annual Meeting of the International Studies Association*, Chicago, Ill., 28 February to 3 March, 37pp.

O'Connor, A. C. et al. (2020). Economic impacts of submarine fiber optic cables and broadband connectivity in Malaysia. In *Working paper* 0214363.292.9 RTI International (16pp).

Oxman, B. H. (1994). The 1994 agreement and the convention. *American Journal of International Law, 88*(4), 687–686. [Letters of Transmittal and Submittal and Commentary regarding the Law of the Sea Convention]/

Plischke, E. (1973). Treatment of "Diplomacy" in international relations textbooks. *World Affairs, 135*(4), 328–344.

Pratt, M. (2009). The scholar-practitioner interface in boundary studies. In *Proceedings of the Slavic Research Centre at Hokkaido University* (pp. 29–36).

Prescott, V. (1996). *The South China Sea: Limits of national claims* (55pp). Maritime Institute of Malaysia.

Prescott, V. (1998). *The gulf of Thailand: Maritime limits to conflict and cooperation* (104pp). Maritime Institute of Malaysia.

Raharja, D. P., & Karim, M. F. (2022). Re-territorialization and the governance of ocean frontiers in Indonesia. *Territory, Politics, Governance*. Regional Studies Association, Routledge. https://doi.org/10.1080/21622671/2022.2118824

Rahman, A. A. A., et al. (1997). *The maritime economy of Malaysia* (p. 119). Pelanduk Publications.

Raham, N. S. F. A., et al. (2016). A descriptive method for analyzing the Kra Canal decision on maritime business patterns in Malaysia. *Journal, of Shipping and Trade, 17* (online version).

Reed, M. (n.d.). National and international jurisdiction and boundaries, Ch. 1. American Bar Association, online version (42 pp).

ReCAAP. (2024). Annual Report for 2023. Including Weekly and Monthly Reports over many years.

Ridzuan, M. R., et al. (2022). Blue economy in Malaysia: An endeavour of achieving the sustainable development goals (SDGs). *Journal of Academic Research in Economics and Management and Sciences, 11*(3), 289–309.

Sahabat, Alam Malaysia. (n.d.). *Impacts of coastal reclamation* (20pp). SAM.

Saieed, Z. (2023). Malaysian maritime agency detains 4 ships, including two Singapore-registered vessels. *The Star/Asia News Network*, 22 May (online version).

Thayer, C. A. (2017). 'China to Launch 10 Satellites over South China Sea, 2019–21', *Thayer Consultancy Background Brief*, December 22, 2017, 3pp.

The Geographer. (1973). 'Territorial Sea Boundary Indonesia-Malaysia', *Limits in the Sea* Series A, US Department of State.

The Geographer. (2014). 'China's Maritime Claims in the South China Sea', *Limits in the Seas Series*, No. 143, US Department of State.

The Geographer. (2014). Philippines' Maritime Claims and Boundaries', *Limits in the Seas Series*, No. 142

The Geographer. (2022). Indonesia's Maritime Claims and boundaries, *Limits in the Seas Series*, No. 141, US Department of State.

The Geographer. (2022). People's Republic of China: Maritime claims in the South China Sea. *Limits in the Seas Series, 150*.

The Star Newspaper Online available at https://www.thestar.com.my/... A source of reference.

The Straits Times. (2024, January 14). Attempted Acts of Armed-Robbery in Singapore Strait.

Tufte, E. R. (1983). *The visual display of quantitative information* (p. 197). Graphics Press.

UKMTO. (2023). *Weekly report: 28 December 2022 to 03 January 2023'*. This and similar Weekly Reports were constantly consulted prior to and during the compilation of this study.

United Nations. (1982). *Final act of the third UN conference on the law of the sea.* Full text: https://www.un.org/Depts/los/convention_...

United Nations. (2021). *The second world ocean assessment* (Vol. I and II, 570 pp). UN.

UNCTAD. (2020). *Review of maritime transport.* UNCTAD.

UNCTAD. (2021). Maritime profile: Malaysia, *UNCTAD Statistics* (available online).

UNCTAD. (2022). *Review of maritime transport 2021.* UNCTAD.

UN Environment. (2019). *Lessons from a decade of emissions gap assessments* (18pp).

UN Environment. (2021). *'The Heat is On.' A world of climate promises not yet delivered.* Emissions Gap Report 2021, UNEP DTU (112 pp).

Environment, U. N. (2005). *Millenium ecosystem assessment reports.* Island Press.

World Meteorological Organization. (2023). *State of the global climate 2022.* WMO.

Yang Lai Fong and others. (2020). Different countries, different perspectives: A comparative analysis of the South China sea disputes coverage by Malaysian and Chinese newspaper. *China Report, 56*(1), 39–59.

Zaman, A. A. A., et al. (2018). Satellite-based offshore wind energy resource mapping in Malaysia. *Journal of Marine Science and Application.* https://doi.org/10.1007/s11804-019-00066w

Further Readings

The author sourced the following reliable online news media daily, over several decades, in quest for material to support this academic research. They included BBC World News; *China Daily, The New Straits Times, The Star, Bernama* all from Malaysia; *Safety4Sea* and *Google Alert* for articles relating to the South China Sea dispute, other geopolitical issues of East and Southeast Asia and Armed-Robbery and Piracy. To list each item sourced would entail a voluminous tome.

The manufacturer's authorised representative in the EU is Springer
Nature Customer Service Centre GmbH, Europaplatz 3, 69115 Heidelberg,
Germany. If you have any concerns regarding our products, please
contact ProductSafety@springernature.com

Printed and bound by CPI Group (UK) Ltd, Croydon, CR0 4YY

28/05/2025
01885638-0001